# From Ultra Rays to Astroparticles

Brigitte Falkenburg · Wolfgang Rhode
Editors

# From Ultra Rays to Astroparticles

## A Historical Introduction to Astroparticle Physics

Springer

*Editors*
Brigitte Falkenburg
Philosophy and Political Science, Faculty 14
Technische Universitaet Dortmund
Dortmund, Germany

Wolfgang Rhode
Physics, Faculty 2
Technische Universitaet Dortmund
Dortmund, Germany

ISBN 978-94-007-5421-8 ISBN 978-94-007-5422-5 (eBook)
DOI 10.1007/978-94-007-5422-5
Springer Dordrecht Heidelberg New York London

Library of Congress Control Number: 2012954697



Printed on acid-free paper

Springer is part of Springer Science+Business Media (www.springer.com)

# Preface

In 1912, Victor Hess discovered cosmic rays. His discovery opened the skies in many regards: for detecting extraterrestrial particles, for making energies beyond the MeV scale of nuclear physics accessible, for interpreting all kind of astrophysical data in terms of cosmic messenger particles, and finally, for giving cosmology an empirical basis. In the 1920s, it turned out that the so-called *Höhenstrahlung* has an extraterrestrial origin and contains charged particles such as the electron. The discovery of the positron in 1932 inaugurated the detection of a plethora of new subatomic particles. With the rise of the big particle accelerators in the early 1960s, cosmic ray studies shifted from particle physics to astrophysics. The cosmic microwave background discovered in 1965 gave support to the *big bang* model of cosmology and made cosmology an empirical science. In the late 1980s, with the experiments that measured the solar neutrino flux, part of the particle physics community moved back to cosmic ray studies and astroparticle physics began. The experiments of astroparticle physics use particle detectors arranged to telescopes. Hence, astroparticle physics is doing particle physics by means of telescopes, and vice versa, doing astrophysics by means of particle detectors. Cosmic rays are messenger particles that carry information about exploding and collapsing stars (in particular, supernovae, their remnants, and black holes), the large-scale structure of the universe, and the microwave afterglow of the *big bang*. Their investigation is one of the most fascinating fields of modern physics.

This book may be read on its own as an introduction to a fascinating multi-faceted field of research, but also used in addition to undergraduate or graduate lectures in astroparticle physics. It covers the historical experiments and lines of thought to which lectures cannot give sufficient attention. The material presented here makes the bridge from the beginnings of radioactivity research, particle physics, astrophysics, and cosmology in the early days of quantum theory and relativity, to the current foundations of physical knowledge, and to the questions and methods of a future physics. It shows that fundamental research is fascinating and of great importance, and that it seems worth tremendous efforts to the physicists.

At the centenary of the pioneering phenomenon found by Victor Hess, we present a historical introduction into astroparticle physics here. We think that the historical approach is a good thread for understanding the many experimental methods, phenomena, and models employed in astroparticle physics, the ways in which they are linked to each other, as well as their relations to their neighboring disciplines of particle physics, astrophysics, and cosmology. Each of these fields on its own is highly complex, and to learn a mixture of them before getting to the bottom of any may be confusing for beginners. We hope that this complex body of knowledge is made more transparent by a historical account of the different research traditions which come together in current astroparticle physics.

Astroparticle physics has emerged from several distinct fields of research. Indeed, these fields have not completely grown together as long as physics does not dispose of a unified "theory of everything". Nevertheless the models and experiments of astroparticle physics are much more than provisional or piece-meal physics. They are no less and no more than surveys and maps of our knowledge of the universe at a small scale and at a large scale. On the way in *terra incognita*, careful cartography of the details is indispensable. Indeed astroparticle physics aims at establishing as many experimental details as possible about cosmic rays, their particle nature, their spectrum, their astrophysical sources, and the mechanisms of their acceleration. But in distinction to other scientific disciplines, this gathering of details does not give rise to increasing specialization. Quite to the contrary, the history of the different branches of physics grown together in astroparticle physics shows the merging of very distinct scientific traditions.

The book is addressed to undergraduate and graduate students of physics and to their teachers. It may serve as background material for lectures. It mays also serve the students and teachers of other faculties, in particular philosophers and historians of science, and everybody interested in a fascinating field of research in physics. To historians and philosophers of science it gives an overview as well as detailed information of a new sub-discipline of physics that has not been studied yet as a whole, but only in partial approaches to the history of particle physics, cosmology, etc., and to their epistemological aspects. Historians of science will read the book as a history written by the physicists, with all the advantages and disadvantages of objective expert knowledge combined with subjective memory. Philosophers of science will find in the book a lot of epistemological material, most of which has been neglected by a philosophy of physics that has traditionally been focusing on the theories rather than the phenomena of physics, even though the latter are most important constraints of the former. The history, the current shape, and the goals of astroparticle physics raise deep epistemological questions about the grounds of a discipline grown together from distinct scientific traditions in search for unified knowledge. But these philosophical questions are kept apart here. They will be discussed in a follower volume on the question of what kind of knowledge astroparticle physics gathers about particles from cosmic sources.

The authors of the book reflect the various approaches to astroparticle physics. All of them substantially contributed to developing the many-faceted methods and to the results of this field of research. We should add that the collection of subjects presented here is far from being complete. We thank to the authors and we apologize for all neglected subjects and all the colleagues and their merits which could not be included in the book.

This volume emerged from a workshop on the history and philosophy of astroparticle physics which took place in Dortmund in October 2009, and which was supported by the German Physical Society. We would like to thank the authors, Kirsten Theunissen from Springer, whose support made this edition possible, and Raphael Bolinger, who prepared Appendices A–D.

Dortmund, Germany

Brigitte Falkenburg
Wolfgang Rhode

# Contents

# Chapter 1
# Introduction

**Wolfgang Rhode**

Taking the discovery of the "Ultra Rays", nowadays called "Cosmic Radiation", by Victor Hess 1912 as a beginning, the research field of Astroparticle Physics celebrates in 2012 its 100th Birthday. It is a unique research field, to which results from a large number of other physical disciplines contribute and which treats fundamental questions of astrophysics, particle physics and cosmology. In this book, the development of Astroparticle Physics since its beginning in the early 20th century is described in contributions of authors who all have left distinctive footsteps in the field they are reporting about. Emphasizing the basic ideas of the development, in this first chapter a tour d' horizon on the connecting path between the following contributions is followed.

## 1.1 Roots and Connections

At the beginning of modern physics, there were the two different approaches of Galileo Galilei and Francis Bacon to the question of how the book of nature should be read. Galilei introduced the analytic–synthetic technique to subdivide every physical problem in its properties, to investigate these properties in detail, to express the experimental results – as good as possible – in the language of mathematics, and finally to synthesize the solution to the original problem from these investigated constituents. Bacon, to the contrary, suggested that the physicist should observe the experiment as a whole, not anatomize the experimental setup but notice every influencing effect and finally formulate the solution like a judge renders the verdict after having come to know all evidence.

The rapid and very successful development of physics is a consequence of an application of Galilei's analytic–synthetic method. By restricting the range of the

W. Rhode (✉)
Physics Department, TU Dortmund University, Otto-Hahn-Str. 4, 44227 Dortmund, Germany
e-mail: wolfgang.rhode@tu-dortmund.de

B. Falkenburg, W. Rhode (eds.), *From Ultra Rays to Astroparticles*,
DOI 10.1007/978-94-007-5422-5_1, © Springer Science+Business Media Dordrecht 2012

analytical interest of the researchers, physical disciplines and branches and later on complete engineering faculties were established. A 'physical discipline' in this way appears as physics within a special branch of interest.

It seems that the analytic–synthetic way of research is perfectly adapted to the needs of physics – so long a further extension of the analysis, a better control of the experimental parameters or an enlargement of the range of measurements in a laboratory is still possible. But how can one proceed, if there is no or only limited control over the experimental parameters, if the improvement of the resolution of the detector does not lead anymore to a better understanding of a subsystem, which again solves the primary question? Then one might feel unwillingly thrown in the position of Bacon's judge. The 'evidence' here means results of measurements, often correlated, in experiments, which can be controlled only partially and in a figurative sense. The 'verdict' relies on a simultaneous solution of the 'evidence'-puzzle with physical, mathematical or even modern statistical methods.

Since new techniques and actual research always are in contradiction with this uncontrolled or uncontrollable part of the experiments, new science is in this sense always at the borderline between Galileo and Bacon. How easy for example questions in particle physics could be answered, if only the physicist had the freedom to only switch on the reaction that he wants to investigate! The subject of this book, the development of Astroparticle Physics as a physical discipline is an example of how physical questions are solved in a much more 'Baconian' than 'Galileian' way.

Also a second approach to the question of the special way of Astroparticle Physics concerns nothing less than the goal of all physical research. The epistemological key question is: Does one intend to draw a picture of the world, which is as true as possible, or should it be as simple and beautiful as possible? Given the small part of the physical world, in which singular measurements are possible, the truth-requirement depicts this world as a patchwork of approximations. The requirement of beauty and simplicity on the other hand leads into that Platonic world in which the nature is described by a small number of forces. The price for this beauty is, however, an inclusion of assumptions and of extrapolations in the mathematical model and finally a loss of provable truth. While Mach's critical positivistic questions helped to identify the right questions in the dusk of classical physics, in the dawn of quantum mechanics and the theory of relativity physicists quickly adopted the goal of finding the unifying world formula.

The heart of most physicists who think about this problem still seems to beat more for the beauty model than for the ugly truth. Therefore it is no surprise that in disciplines like elementary particle physics, the search for a mathematical expression of the unified forces between the particles is one of the driving forces to build step by step larger and faster and more precise and better controlled experiments. Given the success of the standard model of elementary particle physics in describing the results of these high precision experiments, is there still room for Astroparticle Physics on the beauty side of that discussion, or is this discipline condemned to live with its passive measurements in a positivistic patchwork world?

The answer to this question has two sides. First, one has to be aware that a search for more beauty in describing the theory alone is not a sufficient reason to justify those huge investments necessary to build new particle accelerators and experiments. There have to be real problems with the original model. In the case of the standard model of particle physics these problems occur if the model is applied to astrophysical and cosmological processes. The flatness of space, the non-observation of magnetic monopoles, the matter–anti-matter relation, the nature of dark matter and energy indicate such problems. Those measurements, pinning down the corresponding problems of the standard model of particle physics are, however, measurements from the field of Astroparticle Physics. These problems in return direct the laboratory research to experiments revealing hints for the next improvement of the theory. Experiments in Astroparticle Physics are therefore important to ask relevant questions to nature. But can they also answer questions relevant to more than one part on the sky?

This is the second side of the answer: Those processes, which are observed by the experiments of Astroparticle Physics, only in rare cases can be replaced by a laboratory experiment on Earth. So poor the control over the fusion process within a star or its supernova explosion ever may be, in many cases an astrophysical or cosmological setup for an experiment is the single source of information that we have. For example, energy and space surveyed in experiments of Astroparticle Physics exceed the frame of laboratory experiments by many orders of magnitude. Accordingly, the single extension of the standard model of particle physics found up to now, the fact that neutrinos are not massless, was inferred by analysis of the neutrino flux from the sun. We conclude that there are precise answers in Astroparticle Physics, though the puzzles to be solved are in general more complicated than in laboratory experiments. Quite a lot of these experiments are discussed in this book.

During the century covered by this book, the rôle and position of both theory and experiment have changed substantially. On the side of the theory, the general theory of relativity and quantum field theory were developed and elaborated, both describing their subject of interest with impressive precision – also, however, showing for both that they are not trivially to be unified. This structural break between the most fundamental and powerful theories is a problem for the requirement of a description of nature by a unified and beautiful theory. If even the columns on which the monument of the physical description of the world is based are not commensurable, then how can one require that for all special cases? One might ask, however, why such a unification should not be invented in future or why the fact that the physical picture is composed by two theories should mean that it is in truth composed by a giant number of approximations. One way to a solution of these questions leads through measurements at the highest available energies, produced in sources of the largest energy densities in the universe, probing as much of the space as possible, in other words: through experiments of Astroparticle Physics.

These experiments have started in small setups and laboratories, even if the location of those laboratories was always slightly extraordinary: the discipline saw its 'first light' in balloons, glacier adits, vessels fixed under rowing boats and particle counters on the top of high mountains. Within the first half of the 20th century, the investigation of the constituents of cosmic radiation could be told as the history of particle physics to a time where no man-made accelerators could provide the investigated energies above the MeV scale of nuclear physics. The development of particle accelerators then enabled the physicist to build even larger and more precise and better controlled experiments. The complexity of the experiments occasionally exceeded the size, up to which an experiment could be completely understood and executed by a single person or even a few scientists. Finally, the combined knowledge of hundreds and thousands of physicists became necessary to experiment in particle physics. 'Multi-person' is the label that Pedro Waloschek invented for this new super-individual scientist.

Only in niches between particle physics and astronomy, and still driven by a small number of scientists, experiments using the technique of high-energy physics were done without accelerators with growing success until the beginning of the 1990s. With the transition between the LEP and LHC area and also connected to the end of accelerator experiments in high-energy physics laboratories at DESY, SLAC, and Fermilab, the possibility to fill in white spots on the landscape of fundamental science became of increasing interest. Especially collaborations working on experiments with the techniques of charged particle detection, gamma and neutrino astronomy, but also accelerator-less neutrino experiments, grew rapidly. By introducing the science-policy structures of particle physics also in these fields, huge experiments could be planned and realized. The individual astroparticle physicists was adapted to do science as part of a 'multi-person'.

The various techniques necessary to detect different particles from astrophysical and cosmological sources lead to the construction of different experiments in satellites, balloons, with the Cherenkov or fluorescence light of the atmosphere, with particle counters and calorimeters, below the surface – in tunnels through mountains, mine shafts and deep below the water or ice surface. Each of them was built with a dedicated physical program. The full physical picture, however, can only be obtained if all physical information carried by the messenger particles from their astrophysical or cosmological source to the detector on Earth is combined. An end point of this development would be the combined observation of the full sky with all particles at all detectable energies and at all possible times. Given the size and number of experiments reached until now, the last step to a 'world detector' seems viable.

This 'world detector' would, as required by Bacon, observe 'everything' at the same time. The 'multi-person' physicist treats the observational puzzle containing many pieces, not to be further investigated in a controlled experiment on Earth. This, however, also means that more information is barely possible. The world formula, the unified theory of everything that physicists are aiming at, will not have more experimental results from the sky than provided by this detector. The scope of that world formula then would be clear: to explain everything measured by / or relevant to that final detector.

## 1.2 The Path

From Ultra Rays to Astroparticle Physics is the route to be followed in this book. This is a way starting at the birth of the modern physics, leading through one century of development and discoveries, and finally ending in a conceptual view on perspectives on the future.

Astroparticle Physics is – as discussed above – a somewhat different physical discipline, whose origin and presence is embossed by its roots in nuclear physics, particle physics, astronomy, cosmology and plasma physics. As different as the contributing research fields are, so also are the physical questions different whose pure treatment has brought many physicists only later to the insight that through their work they can be identified as 'astroparticle' physicists.

Apart from the historical and personal access to the field, this book also handles the question of the change of the research field as a whole. What were and are the driving questions? With which scientific and science-policy means were and are they answered? Wherein consists the merit of the answer to the scientific questions? Is one constructing the answer to D. Adams' 'ultimate Question of Life, the Universe and Everything' from approximations and models? Or is one still working on a more and more truth approaching and unified transformation, of the world of physical experiments and measurements into the letters of the mathematical book of nature, as suggested by Galilei, Planck, Einstein and others?

The flow from the past to the future of Astroparticle Physics is not linear. Arising from multiple sources, it contains bifurcations and confluences. We depict this by looking through the eyes of the authors from the hills of the cape of science to its past. From different spots of experimental physics, astrophysics and cosmology, this past will appear in a different light. The complexity of the subject implies that a book like this has to deliver a historical introduction rather than a complete history of the field, which we leave to professional science historians. The reader may feel free to choose the order of his own approach to the subject of this book.

In the following, the connections between the chapters are discussed.

### *1.2.1 From the Discovery of Radioactivity to the First Accelerator Experiments*

The experimental investigation of all forms of radiation and their theoretical understanding was one of the main subjects of physics of the 19th century. Since this research yields one of the primary reasons to abandon the classical mathematical picture of nature in favor of the new and uncertain world of quantum statistics, it still belongs to the most discussed subjects of physics lectures of today. In parallel, starting with analyses of the colorful patterns of the gas-discharge tubes (Geissler tubes), first experiments in the physics with the 'electron' elementary particle were done. In cathode-ray tubes, the properties of this radiation type were investigated systematically, always looking at the interaction between these 'rays' and 'matter' as a special key aspect. It were the latter investigations which opened the window to modern physics. At low energies (1 eV), the interaction between light and matter provided (after the investigation of the blackbody radiation) the second step to quantum mechanics in the form of the discovery and theoretical understanding of the photo effect. At moderate energies in the keV region, through Röntgen's discovery of X-rays, new ways of structural information about matter became accessible. By Becquerel's discovery of (MeV) radioactivity the stability of the matter was questioned.

For our purpose, the understanding of the roots of Astroparticle Physics, one more technical side step of these discoveries became important. Beneath the transient fluorescence, the blackening of photographic plates was the first way to construct a detector with spatial resolution. An investigation of the absolute intensity of charged radiation and its temporal change became possible by the analysis of the discharge velocity of electroscopes, essentially consisting e.g. of two flexible gold leaves rejecting and deflecting each other so long they are equally charged. The discharge rate of the electroscopes depends both on the construction of the instrument and on the flux of charged particles through this detector. It was finally the attempt to calibrate these instruments far from the charged products of the radioactive decay of elements in the crust of the Earth which led to the balloon flights of Victor Hess. Unexpectedly, the electroscope discharge rate grew with rising altitude, giving rise to the investigation of this now discovered 'Ultra Radiation'.

From Michael Walter we learn in this chapter how systematic physical research of independent physicists with very different experiments unveiled the nature of this radiation step by step.

We will further notice the dispute between the German and Anglo–American interpretations of the firstly measured signals. Whilst in Germany one tended to be convinced about the wave nature of this radiation, in the Anglo–American region one had no major problems assuming its particle nature. It might be interesting to mention that this dispute was to some extent also a consequence of the philosophy of science discussion in the two regions. In Germany, the 19th century was the time of the so-called 'German Idealism'. One might characterize this as a time in which philosophers tried to construct – similarly like the physicists in the classical mechanics – huge wonderful, closed and more or less logical systems, explaining the complete world. These efforts had positive consequences also for our physical understanding of the world. The concept of energy conservation, e.g. was invented by Robert Mayer from such a philosophical inspiration. However, Idealism contained more than enough elements that were in complete contradiction with the attempt to explain the world from its physical observation and description. No wonder that at the end of this epoch, the opposite thinking of Ernst Mach gained huge influence under German physicists. Mach wanted to establish physics only based on clear observations and without any metaphysical elements. Though his criticism of the concept of the existence of elementary particles (i.e. atoms) partially was overdrawn, in the end it helped to pave the way to a clean formulation of quantum mechanics. His aversion against metaphysically claiming the existence of 'particles' as reasons of observed effects was a living heritage among many German physicists at a time in which the final formulation of quantum mechanics was not yet to be found. Outside of Germany, one had less problems accepting newly invented 'metaphysically as real defined' particles as reason for observations. Therefore the Anglo–American community had much less of a problem to accept the particle nature of the new 'Ultra Radiation' than the German.

After a travel through the beautiful and systematic investigation of the cosmic radiation, leading to the development of new experimental techniques and new insights in the nature of elementary particles, their interaction with electromagnetic radiation, and their theoretical description, this chapter leaves us at the dawn of accelerator physics.

### *1.2.2 Development of Cosmology: From a Static Universe to Accelerated Expansion*

In this second chapter, we direct our view to Astroparticle Physics from the point of view of cosmology. Since the beginning of human thinking, cosmology was a subject of religion; since the beginning of science, a subject of philosophy, and only in parallel to the investigation of cosmic rays it became subject to experimental investigations. Other than in cosmic ray research, this field, however, was primarily driven by theoretical considerations. The birthday of modern cosmology was, as Matthias Bartelmann explains, the day in 1915 on which Einstein's general theory of relativity was finished. The transition from cosmology as a subject of philosophy to a subject of physics was driven by two epistemic discussions. Ernst Mach's positivistic requirement to base physics on measured facts led Einstein on his way from the Newtonian absolute understanding of space and time to special and general relativity. That the theory of relativity, however, was intended to be much more than a construct to explain facts known before its construction follows already from the suggestion of experiments to test the theory by Einstein (1916). These 'classical tests of general relativity' were the questions whether the perihelion precession of Mercury's orbit could be calculated correctly, whether a deflection of light by the Sun could be observed and whether light undergoes a gravitational redshift. This understanding of the theory as a suggested model carrying a deeper truth than its progenitors, which can be tested by experiments is paradigmatic for the epistemic understanding of science by Karl Popper's critical rationalism.

The second important discussion concerns the question if those physical facts measured today on Earth are valid also at other times and other places of the Universe. Do we perform our measurements at a special location? Already Nicolaus Copernicus answered this question in his Copernican Principle with: no. For purposes of cosmology, this principle was complemented by assumptions about the equal distribution of matter in the Universe and its geometry in the Cosmological Principle (i.e. that we do not live at a special location). The Perfect Cosmological Principle even required, additionally, that we do not live at a special time: the Universe as a whole should not undergo any temporal change. The belief of such a perfect homogeneous and static universe was widely spread under physicists at the beginning of the 20th century.

Matthias Bartelmann follows the interplay between theory and experiments through the 20th century, showing the way in which the mentioned primarily metaphysical principles were investigated by experiments. The Perfect Cosmological

Principle was rejected due to the Hubble expansion of the Universe. The validity of the Cosmological Principle was questioned by investigations of the homogeneity of the Universe backwards from the present distribution of matter, stars and galaxies, through the time of emission of the microwave background and to the interactions in the very early Universe. These investigations unveiled deep information about the forms and properties of the matter in the Universe. The geometrical flatness of the space could be shown and epochs with a different rates of expansion could be identified. The essential mechanism of the formation of structure could be established. The present end point of this research is a picture of a world dominated by dark matter and dark energy sustaining its expansion with increasing velocity – and still containing a multitude of questions to be answered by further experiments and research in Astroparticle Physics.

### *1.2.3 Evolution of Astrophysics: Stars, Galaxies, Dark Matter, and Particle Acceleration*

To gain knowledge about the Universe, one needs well understood experimental conditions. In the optimal case one would know the particle accelerators delivering particles with well defined energy spectra. As well as the interaction probability of these particles we would also know the composition of matter and radiation, which the particles cross on their way to the detector – and of course also the properties of the detector. Against this background, by defining clear theoretical predictions, hypotheses can be set up and their validity can be tested in dedicated experiments.

In the case of Astroparticle Physics, the accelerators delivering the particles are astrophysical objects like stars in their different states, and all types of galaxy and types of objects maybe still unknown to us today. Investigations of questions in this field relate to the research in astrophysics. If the matter and radiation that is crossed or the geometry of the space is investigated the research contributes to cosmology, and if the properties of the messenger particles or their interactions are investigated, it is a contribution to particle physics.

How the understanding of experimental and theoretical details in these fields is nested and finally used to ask new relevant questions, is explained by Peter Biermann. The chapter with his contribution is based on the cosmology section, integrating the accumulation of astrophysical knowledge of the last century and pointing forward to the actual research questions in Astroparticle Physics. Still, the acceleration mechanisms of the cosmic radiation at the different energy scales are not completely understood. Charged nuclei are observed up to the highest energies, however, neither their sources are yet unanimously identified nor was a signal observed produced by neutrinos from interactions of these nuclei. Gamma rays are observed from a multitude of galactic and extragalactic sources; however, also these observations leave the contribution of hadronic acceleration to the certainly present leptonic accelerations open. Finally, beside other details from the field of particle physics, also the question of dark matter is still open and subject to experimental investigation.

The discussion of the impact of astrophysics to Astroparticle Physics shows which huge successes we had in the last century solving the difficult Baconian puzzle, and which important pieces still have to be placed.

### *1.2.4 Development of Ultra High-Energy Cosmic Ray Research*

The most frequent particles, detected directly or indirectly in experiments above the Earth's surface, are charged nuclei. In the chapter they wrote, we learn from Karl-Heinz Kampert and Allan Watson, how the investigation of this charged component of cosmic radiation on the Earth's surface developed. Temporal overlapping with Chap. 2 and resuming the scientific questions of Chap. 3, the description starts with the experimental situation in the third decade of the last century. Starting from this situation, the key questions are posed still occupying cosmic ray research today: What and how can we learn about the nature of the involved primary and secondary particles? Which chemical composition and which energy spectra do the charged cosmic rays have? What do we learn from the arrival direction of charged cosmic rays?

The crucial physical property, which one had to discover and to understand to answer these questions, is the development of air showers from the first interaction of the cosmic particles in the high atmosphere to the arrival of the cloud of secondary particles on the surface of the Earth. Therefore the history of Cosmic Ray research starts with the discovery of extended air showers. The development of appropriate detectors from Geiger-counters to phototube-based detectors, from small coincidence experiments fitting in one room to the AUGER detector covering an area of 3 000 $km^2$, is described. Depending on the detector's construction, the energy and nature of the particles arriving on the Earth's surface can be identified and studied. Investigations of the air shower development and insights about the primary particles are, however, only possible if the fluorescence (or Cherenkov) light emitted in the atmosphere is recorded in appropriate telescopes.

We observe in this chapter the interplay between the invention and development of detection methods and the growth of physical knowledge. Thresholds in the measurement quality were overrun, when new techniques or new combinations of detection methods were established. With the detector size, also the maximal energy inferred for the primary particles grew until the scale of the Greisen–Kuzmin–Zatsepin cutoff was reached. Particles in this highest ever observed energy regions will barely be observed in accelerator experiments. Thus, the high-energy end of particle physics will be investigated in the same experimental situation as the one in which particle physics had its roots. Furthermore, a small energy window below the GZK cutoff is interesting, in which the charged primaries are only marginally deflected by the intergalactic magnetic fields. Here, astronomy with charged particles is possible – if particles with sufficiently high energies are accelerated in galaxies close enough. The special conditions, however, to which the use of this window to the universe is connected, restrict the use of this window. Therefore a different technique, applicable also for the detection of low energetic primaries or sources on all distance scales, is necessary to solve the quest to the origins of cosmic rays.

### *1.2.5 Very-High-Energy Gamma-Ray Astronomy: A 23-Year Success Story in Astroparticle Physics*

It may seem to be straightforward to only extend the maximal energy up to which photons are detected from optical and X-ray range to GeV and higher energies, and to try to identify the sources of high-energetic charged primaries by detecting high-energetic gamma's from the same location. In fact, one has to face severe problems. The first comes from the fact that the atmosphere is opaque to photons with energies above the optical window. To a certain extent (from X-ray up to TeV range) this can be compensated for, by the use of satellite telescopes. The energy spectrum of the flux of high-energetic photons may, however, for the moment be thought of as limited from above by a steeply falling power law. As a consequence, the sensitive surface of satellite detectors becomes simply too small for the detection of large numbers of high-energetic photons from a certain source. Like in the case of charged cosmic rays, the sensitive area of the detectors has to be increased by indirectly observing gamma induced air showers from the Earth's surface.

In their chapter, Eckart Lorenz and Robert Wagner describe the long way from the first idea of the possibility of this detection to the first successful gamma detection with the Whipple Cherenkov telescope in 1989 and further on to the multitude of present telescopes working on this field and the planned coming generation of Cherenkov telescope arrays.

The key to gamma astronomy, covering the energy range from roughly 10 GeV to 10 TeV, is to record in clear dark nights the Cherenkov light emitted from the cascades of secondary particles induced by one primary cosmic ray particle. Since the overwhelming number of cosmic particles are charged particles, techniques to distinguish between nuclei and gamma's had to be developed. Today we are able to suppress the hadronic background, by analyzing the air shower direction, its spatial and temporal structure and to observe the same air shower with different telescopes, allowing its spatial reconstruction.

The multitude of galactic and extragalactic sources found with this technique by now, allows us to answer numerous astrophysical, cosmological and to a certain extent even particle physics questions, of which the most important are introduced and discussed in the article. The question of the origin of the highest energy charged particles, however, could up to now not be finally answered with the help of gamma-ray astronomy. The reason lies in the shape of the gamma energy spectra, reconstructed with these ground-based telescopes and combined with all information from other wavelengths. Up to GeV energies, the observed spectra can be explained by the assumption that electrons are accelerated in the sources and that the observed gamma spectrum can be explained by electromagnetic energy losses of these sources. There is at least barely need to require proton acceleration to explain this part of the spectrum. In the TeV region, the spectral behavior becomes crucial. Some spectra are shaped so that an additional proton component might be needed to explain the spectrum. Some structures in the temporal changes of the flux also indicate that a purely leptonic explanation fails and hadrons, as accelerated particles, might be or are needed.

In the 10 TeV region, the only region where the energy reached by the acceleration, respectively, lost by gamma emission could decide this question, the spectra from the sources high-energetic and distant enough to accelerate protons to the highest energies are cut off by interactions with the Infrared Background. On the one side, this cut off opens the possibility to measure this background flux, on the other side it lets us hope for an independent method to identify hadron acceleration: the detection of neutrinos produced as secondaries of proton interactions.

### *1.2.6 Search for the Neutrino Mass and Low Energy Neutrino Astronomy*

The investigation of the charged products of the nuclear decay had in the history of Astroparticle Physics already given rise to the development of the first detectors, which led to the discovery of the Ultra Radiation. In parallel to the development of the cosmic ray research, also the physics of the nuclei was further investigated. Kai Zuber follows for us the stepwise understanding of the properties of the nuclei flowing into the puzzle of how to distribute in beta decays the energy and the spin among the nuclear constituents in such a way that the conservation laws were observed. To solve the puzzle, Wolfgang Pauli uttered in his famous letter in 1930 the 'desperate remedy' that in these decays a neutral and barely detectable particle could be created. This hypothesis of a neutral particle, later called neutrino, became an important, and by decay physics, a better and better supported part of the theory. The small interaction probability, however, prevented the discovery of the first neutrinos for more about quarter of a century until their detection by Clyde Lorrain Cowan and Frederic Reines in 1956.

After the first discovery of a certain particle, physicists immediately wish to observe large numbers of them to determine their properties. One such neutrino source could be the fusion processes claimed, with growing understanding of the nuclear properties, as the energy source of the sun. Through their small interaction cross section, the neutrinos would be able to cross sun and Earth unhampered until, deep enough under the Earth's surface to shield unwanted particles from the cosmic radiation, the neutrinos could be detected in a large volume detector. In the late 1960s, in the Homstake experiment, the first single solar neutrinos were detected. On the one hand this was the first step, even though radiochemical and thus non-directional, to neutrino astronomy and a large success for the solar model. On the other hand, only a third of the expected number of neutrinos was observed. This comparably small deviation was the origin of the so-called solar neutrino problem. Forty years and several detector generations later, and after developing Cherenkov proton decay experiments in water, i.e. the technique telling us how to also reconstruct the neutrino direction, this solar neutrino problem led to the insight that the neutrinos oscillate and are not massless – different from the way they had been defined in the meanwhile developed standard model of particle physics.

The time of proton decay experiments was enlightened in 1987 by a second astrophysical neutrino signal: the signal from the Supernova 1987a. The time difference between the neutrinos arriving from this event could be used together with the upcoming measurements of the fluctuations of the Microwave Background and the different beta-, double-beta and neutrino-less double-beta decay experiments to limit the range of the neutrino masses. Until this mass range becomes a measurement of the mass, these investigations will be continued in future.

It should be noticed that the neutrino story told up to here forms a root of Astroparticle Physics that is widely independent of the cosmic ray and gamma-ray research already discussed; however, it is increasingly connected by the studied subjects, applied methods and nested theories. Astrophysics (the fusion process in the sun), particle physics (neutrino oscillations, respectively, masses) and cosmology (contribution of the hot dark matter particle neutrino) are studied with one family of experiments.

### *1.2.7 From Particle Physics to Astroparticle Physics: Proton Decay and the Rise of Non-accelerator physics*

We have left the particle physics in the situation, when accelerator experiments came up. In this book we will not look on the details of how with these accelerators a large number of particles were discovered. We will also not discuss how the standard model was established and confirmed. One property of the standard model, however, had unexpected consequences for Astroparticle Physics. Its symmetry would have required that matter and anti-matter annihilated in the early universe, so that no world made of 'matter' could have formed. In 1968, Andrei Sacharow found under which conditions the matter–anti-matter asymmetry could have formed. That the universe was not in a state of thermal equilibrium was obvious in big bang cosmology, and the required C and CP violation was found and is still investigated today – for example in the LHC experiment LHCb. Proton decay, also required, however, could be only investigated in large none-accelerator experiments. The size of the first generation of such experiments depended on the idea of the unification of the fundamental forces extending the standard model. In the middle of the 1980s, the most simple extension of the standard model called SU(5) implied a proton life-time of about $10^{29}$ years. With detectors consisting of 1 000-tons of matter and hidden from cosmic radiation as deep under the Earth's surface as possible, one expected to detect several proton decays per year. Hinrich Meyer was one of the leading physicists who constructed the French–German Fréjus iron calorimeter for this purpose.

In this section, he reports on the path leading from accelerator laboratories to underground physics.

Unfortunately, at the end of the 1980s, the attempts to detect proton decay failed, and SU(5) – be it supersymmetric or not – had to be abandoned. With this failed search, the experimental background to proton decay events consisting of neutrinos and muons got into the center of interest. In addition to the already discussed solar neutrinos, also atmospheric muon- and electron-neutrinos were detected and their energy spectrum was measured. The overwhelming flux of atmospheric muons was used for cosmic ray studies. Thus the experiments built primarily as detectors of 'particle physics without accelerators' became entirely and successfully cosmic ray experiments.

The open question of the sources of the cosmic radiation at high energies had led to first theories of the acceleration of protons and their energy loss by proton–proton or proton–gamma interactions. Some of these theories predicted such high neutrino fluxes that they should have been visible on top of the atmospheric neutrino flux in existing experiments. By the non-detection of these neutrinos, these theories, as well as theories predicting a measurable very-high-energy neutrino flux, for example from the decay of topological defects, could be rejected. The observed data sample was further used to set limits on the flux of possible point-like sources. In this way, Cygnus X3, the position at which the Kiel air shower array had claimed to see a signal, was investigated (and with the given sensitivity rejected) as one of the first point source positions.

The epistemological purpose to reject theories was even more successfully fulfilled than the involved physicist would have wished at that time. As a consequence, however, first results in the field of neutrino astronomy published and in nuce methods still in use in actual neutrino telescopes were developed. At the moment of its appearance, the sensitivity of the Fréjus apparatus as a neutrino telescope had entered a range in which physical meaningful results could be published, and also meaningful questions for the construction of the next Astroparticle Physics detector generation and the next generation of particle acceleration theories could be asked.

### *1.2.8 Towards High-Energy Neutrino Astronomy*

The road to Astroparticle Physics to be followed in this section starts with the first neutrino detection in the 1950s. Since then, experiments to detect neutrinos from the interactions of the primary interactions of nuclei in the atmosphere were planned and executed. Like all neutrino detectors, appropriate experiments had to be installed as deep under the surface as possible to shield the flux from atmospheric muons. Then muon events from a direction with a shielding of more than 13 000 mwe (i.e. horizontal events), from which atmospheric muons could not reach the detector, were searched. As Christian Spiering explains, the first successful detection of atmospheric muons was in 1965, nearly at the same time as detection by two experiments (KGF and CWI).

An improvement of the experimental understanding led to the already discussed setup of proton decay experiments, of which the Kamiokande detectors and IMB were working with the water Cherenkov technique: the detection of the Cherenkov light cone emitted by a charged particle moving with a velocity faster than light in that medium. Though leaving only a light signal integrated over all particles belonging to an event, this technique has the tremendous advantage that huge detector volumes can be surveilled. If the light detecting photomultipliers are installed in clear water, even detectors exceeding the size of natural or man-made caves under ground can be realized.

In the 1970s, such a detector was discussed by the initiators of the later DUMAND experiment. They developed the idea and its potential for the measurement of the atmospheric neutrino and muon flux. In 1987, the DUMAND collaboration operated a test string successfully from a ship in the open sea. The following discussions of new technologies to be invented and the construction itself were important for the complete development of the field. Thus, a proposal for a cubic kilometer detector was written. DUMAND, however, could not earn the success of the invented technology. Roughly at the time, when the first atmospheric neutrino energy spectra and exclusion limits to extra-terrestrial contributions were published by the Fréjus Collaborations, the DUMAND efforts ended. In parallel, however, former Russian collaborators, who had been forced by the cold war to set up their own experiment in Lake Baikal, had developed an under-water neutrino telescope, of which the first string was successfully deployed in 1993. At about the same time the work to build a cubic kilometer size detector in the clear ice at the geographic South Pole started, leading first to the AMANDA and later to the IceCube detector, which was finished and has been taking data since December 2010. Within the last 20 years, since the first atmospheric energy spectra were measured, the maximal energy of detected atmospheric muon neutrinos grew with the size of the experiments from the TeV to the PeV region by three orders of magnitude. However, though the smallest detected flux is today more than 6 orders of magnitude smaller than 20 years ago, still no indication for extra-terrestrial contributions and thus no direct evidence for proton acceleration was found. IceCube, though, is still at the beginning of its planned operation of 10 years and not all properties of the detection mechanism, i.e. the light scattering in the ice, are understood, so that neutrino detection from astrophysical sources is not yet excluded. From the correlation of the detected signals with results from other experiments, a sharpening of the signal hypothesis and improvement of the results is expected.

Like in other experiments of Astroparticle Physics, the IceCube detector, designed for one purpose, is used as purely observing detector also for other goals. By the search for magnetic monopoles and the particles of cold dark matter, by precision measurements of particle fluxes to be converted to cross sections, it contributes to the program of particle physics.

Independent of the measures which one has to take to find, some day, neutrinos as evidence for proton acceleration, one may conclude already now that all analyzed data gain their true value through the connection with all other experiments – through the connection of all experiments to the 'world detector'.

### *1.2.9 From Waves to Particle Tracks and Quantum Probabilities*

Up to here, we have discussed a multitude of measurements and experimental results embedded in theories and models, improved with the measurement quality. Depending on the situation, either unexpected experimental results enforced a revision of the theory, or theoretical predictions pointed the way to the next experiment. Some of the roots of Astroparticle Physics point back to the time when the first experiments, later leading to the development of quantum mechanics, were done and the wave or particle nature of the 'Ultra Radiation' was discussed. Elementary particles were discovered in cosmic radiation, their properties were unveiled and astrophysical as well as cosmological results were obtained. Obviously, the discipline is very successful.

But what is the object of those measurements? How is the measurement process defined? Which rules does it follow? Many physicists cheerfully and jauntily follow a heuristic realism allowing a successful technical communication, however, not answering these fundamental questions. From the contribution of Brigitte Falkenburg we learn how historical and present questions and results of Astroparticle Physics are connected to the theory of measurement and how the scientific problems in Astroparticle Physics relate to the scientific problems of other involved disciplines.

The original problem consisted of the question which well defined properties (e.g. wave or particle nature) could be attributed to an investigated object. Rules for how to interpret the results of measurements were developed, so that in a kind of logical decision tree, the searched attributes could be assigned to the investigated object. In many cases the high energy of the investigated objects helped to get along with semi-classical models. However, since the results from quantum field theory are indispensable for an understanding of the results, parts of the measurement theory of Astroparticle Physics are in accordance with particle physics.

One new idea is coming up because, due to the complexity of the observed processes, the classical way to assign measurement results to a certain object is no longer accessible, i.e. it is not possible anymore to reconstruct all single secondary particles which in further steps are synthesized to a primary. Such processes are simulated in Monte Carlo procedures. The analysis of these processes by means of computer science shows that the classical way is connected to a loss of information and therefore to a deterioration of the results. New is the assignment of probability clouds to statistical distributed measurements and a direct reconstruction of the searched distribution from knowing the transfer properties of the measurement process. Also this way of thinking is in principle, but not to the same extent, relevant for other measurements, e.g. in particle physics. Thus, besides all other things, Astroparticle Physics is also driving new methods and insights of measurement theory.

## 1.3 Summary

In this introduction, the dependencies between the chapters of this book, illuminating the history, presence and future of Astroparticle Physics, are presented. With this help the reader may decide which track he wants to follow to obtain a basic knowledge of the history of Astroparticle Physics in the past century.

# Chapter 2
# From the Discovery of Radioactivity to the First Accelerator Experiments

**Michael Walter**

## 2.1 Introduction

The article reviews the historical phases of cosmic ray research from the very beginning around 1900 until the 1940s, when the first particle accelerators replaced cosmic particles as a source for elementary particle interactions. Contrary to the discovery of X-rays or the ionising $\alpha$-, $\beta$- and $\gamma$-rays, it was an arduous path to the definite acceptance of the new radiation. The development in the years before the discovery is described in Sect. 2.2. The following section deals with the work of Victor F. Hess, especially with the detection of extraterrestrial radiation in 1912 and the years until the final acceptance by the scientific community. In Sect. 2.4 the study of the properties of cosmic rays is discussed. Innovative detectors and methods like the cloud chamber, the Geiger–Müller counter and the coincidence circuit brought new stimuli. The origin of cosmic rays was and is still an unsolved question. The different hypotheses of the early time are summarised in Sect. 2.5. In the 1930s a scientific success story started which nobody of the first protagonists might have imagined. The discovery of the positron by C.D. Anderson was the birth of elementary particle physics. The 15 years until a new era started in 1947 with first accelerator experiments at the Berkeley synchro-cyclotron are described in Sect. 2.6.

It is obvious that this article can only cover the main steps of the historical development. An excellent description of the research on the "Höhenstrahlung" in the years between 1900 and 1936 was given by Miehlnickel (1938). Two other volumes are also recommended: Brown and Hoddeson (1983) and Sekido and Elliot (1985). Both summarise the personal views of the protagonists or their coworkers of the early time.

M. Walter (✉)
DESY, Platanenallee 6, 15738 Zeuthen, Germany
e-mail: michael.walter@desy.de

B. Falkenburg, W. Rhode (eds.), *From Ultra Rays to Astroparticles*,
DOI 10.1007/978-94-007-5422-5_2, © Springer Science+Business Media Dordrecht 2012

## 2.2 The Way to the Discovery of Cosmic Rays

The technical revolution in the 19th century was related with an explosion-like development, especially in physics and chemistry. On the other hand, many scientists had the impression that all essential discoveries in physics were done. Phillipp von Jolly, a physics professor in Munich, gave Max Planck 1874 the advice not to study physics, since this science had mainly been completed. It would be interesting to speculate what would have happened with physics if Max Planck and all the other young talents would have followed such an advice. Fortunately, they did not. With the discovery of X-rays by Wilhelm Conrad Röntgen , of radioactivity by Henri Becquerel and with the foundation of quantum physics by Planck, the existing building was shaken and it got a completely new fundament with many new floors. Even now, more than 100 years later, the whole building is still under construction.

### *2.2.1 Conductivity of Air*

In general it was assumed that dry air is a good isolator. Then, in 1785, Charles Augustin de Coulomb observed that a very well isolated electrically charged conductor loses its charge with time. Coulomb's hypothesis was that the charge is taken away from the conductor by the contact of dust and other particles contained in the air. But this explanation was not generally accepted and for more than 100 years there was no clear answer to the question why air becomes conductive.

### *2.2.2 Cathode Rays, X-Rays and Radioactivity*

In 1857 the electrical discharge tube was developed by Heinrich Geißler. It consisted of a glass cylinder with two electrodes inside at both ends. Filled with gases like air, neon or argon at low pressure, and operated at a high voltage of several kilovolts, the tube showed a plasma glow. These effects were first used for entertainment demonstrations, but this discharge tube was finally the basis for the development of cathode, X-ray and neon tubes. William Crookes operated in 1869 a tube at lower gas pressure and found that cathode rays are produced at the negative electrode. At the other end of the tube, close to the positive charged anode, they hit the glass wall where fluorescence light was emitted. Like many others, also W.C. Röntgen investigated the properties of cathode rays using a tube provided by Phillip Lenard. In the end of 1895, he observed an energetic radiation penetrating a black cardboard covering accidentally the tube. With the picture of his wife's hand skeleton the discovery of X-rays reached worldwide publicity within a few weeks. The new radiation and the photographic imaging were a breakthrough for new developments in physics and a revolution in medical diagnostics.

Only two months later H. Becquerel discovered also by chance a new penetrating radiation in uranium minerals. Inspired by Röntgen's discovery, he continued in

**Fig. 2.1** C.R.T. Wilson, J. Elster and H. Geitel, A. Gockel and Th. Wulf

February 1896 his studies on phosphorescence and fluorescence minerals, looking for possible X-ray emissions. Since phosphorescence is initiated by sun light, he wrapped photographic plates in black paper and put it with different minerals on top in the sun. Only the uranium mineral showed an image of its contours on the exposed photography. As he wanted to repeat this experiment, there was a cloudy sky in Paris. Becquerel placed the probe in a drawer and decided a few days later to develop the plate. To his surprise also this probe showed the contours of the uranium mineral which ruled out the working hypothesis of sun light-induced X-ray emission. The only explanation was that uranium emits a new invisible penetrating radiation. In contrast to X-rays, this discovery was not recognised immediately.

Marie Curie started her doctoral thesis in Paris at the end of 1897, investigating the properties of the Becquerel radiation. She studied different minerals and salts containing uranium with an electrometer developed by her husband. With the electrometer, the first device to detect intensities, it was possible to measure the conductivity of air caused by the ionising radiation with high accuracy. Marie and Pierre Curie found with thorium, radium and polonium new elements with higher radiation intensities than uranium. In the presentation of these results she introduced the term 'radioactivity' in 1898 for the first time.

Another important discovery was made by Joseph John Thomson in 1897. He showed that cathode rays consist of electrons. Then, investigating the properties of ionising radiation, Ernest Rutherford and others verified that the radiation consists of three different components: $\alpha$-, $\beta$- and $\gamma$-rays. The fundament for the development of atom and nuclear physics was settled and, no surprise, all of these protagonists were under the first Nobel prize winners.

### 2.2.3 *Penetrating Radiation*

A new effect was discovered when gases were irradiated with $\alpha$-, $\beta$- and $\gamma$-rays: ionisation. The radiation is energetic enough to dissociate atoms and molecules into positively and negatively charged ions, as it was assumed before the atomic structure was known. These ions allow the transport of electricity and make gases conductive.

It was the Scottish physicist Charles Thomson Rees Wilson (Fig. 2.1) who found in 1896 that the formation of clouds and fog is connected with the ionisation of air

**Table 2.1** Absorption coefficients of $\gamma$-rays for different substances and the necessary absorber thickness to reduce the ionisation by a factor of two (Eve, 1906) (It should be emphasised that at this time measured or calculated values were given without errors)

| Substance | Density, $g\,cm^{-1}$ | Absorption Coeff. $\lambda$, $cm^{-1}$ | Absorber Thickness d, cm (for 50 % reduction) |
|---|---|---|---|
| Lead | 11.6 | 0.5 | 1.4 |
| Earth | 2.7 | 0.092 | 7.5 |
| Water | 1.0 | 0.034 | 20.4 |
| Air | 0.0013 | 0.000044 | 15 700 |

molecules. But the work on this topic was continued almost 15 years later, when he developed a cloud chamber which visualised $\alpha$- and $\beta$-rays. At first he investigated the ionisation of gases using an electrometer, the standard detector at this time. It consisted of two thin gold leafs mounted on a metal rod enclosed in a metallic vessel. Charging the rod, the gold leafs move away from each other, because of the equal charge. The distance is then a measure of the amount of charge. In a publication in 1900 (Wilson, 1900), Wilson gave an explanation for the conductivity of air in an isolated vessel. The reason that a charged metallic conductor loses its charge in an isolated chamber filled with air is that there are small quantities of radioactive substances. These can be pollutions embedded in the chamber walls and in the surrounding environment. At the same time, Julius Elster and Hans Geitel (Fig. 2.1), two friends and physics teachers at a German school in Wolfenbüttel came to the same conclusion (Geitel, 1900; Elster and Geitel, 1901). In our days forgotten, both were in the time from 1880 to 1920 with about 200 common publications internationally accepted authorities in the fields of electricity in the atmosphere, photo effect and radioactivity (Fricke, 1992). Several times Elster and Geitel were nominated for the Nobel prize. They did not accept an offer of a professorship at the university but preferred independence with school teaching and working in their private laboratory.

The group around Ernest Rutherford at McGill University in Canada went in 1903 a step further. An electroscope was shielded with different materials, like water and lead, to measure the ionisation in dependence on the absorber thickness. A decrease by about a factor of three was observed, but then the ionisation remained constant. There is obviously radiation of high penetration power, which: "... may have its origin in the radioactive matter which is distributed throughout the earth and atmosphere" (Cooke, 1903). Several authors assumed then that the penetrating radiation is $\gamma$-rays coming from radium in the earth crust and from radium emanations in the atmosphere. With their own measurements and results of others, A.S. Eve estimated the absorption coefficients $\lambda$ of $\gamma$-rays for different substances given in Table 2.1 (Eve, 1906). The dependence of the ionisation $I$ on the distance $d$ to the $\gamma$-ray source is described by $I = I_o \cdot e^{-\lambda d}$. Interesting for later discussions is that 99 % of the $\gamma$-rays from radium emanations will be absorbed by 1 000 m atmosphere.

Wilson in Scottland, Elster and Geitel in Germany were probably the first who investigated radioactivity in the environment outside of laboratories. They performed

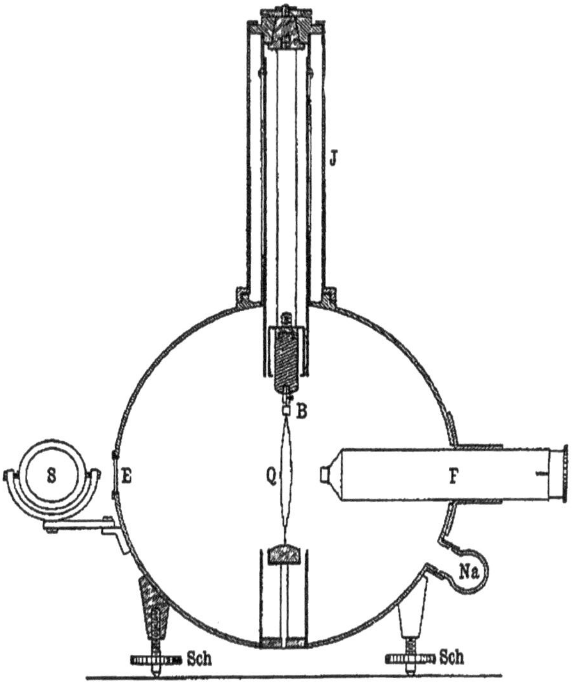


**Fig. 2.2** The two-string electroscope developed by Th. Wulf. The ionisation vessel has a volume of 2.7 liter. $Q$: quartz fibres, $B$: amber for electrical isolation, $J$: container for the metallic rod to charge the fibres, $F$: microscope to measure the fibre distance, $S$: mirror, $E$: windows, $Na$: natrium container to dry the air

measurements in a railway tunnel (Wilson) and in caves, and salt mines (Elster and Geitel). A comparison with the ionisation measured outside in free nature showed different results. Volcanic rock contains in general a higher fraction of radioactive substances than sediment stones. Radioactive pollution is much smaller in rock salt mines and in water.

A new measurement quality was reached by Theodor Wulf (Fig. 2.1), a German Jesuit priest, who studied physics in Innsbruck and Göttingen. As physics lecturer at the Jesuit University Valkenburg in The Netherlands he investigated from 1905 to 1914 the electricity of the atmosphere and radioactivity. Wulf developed a robust, transportable electrometer which became for many years the state of the art instrument. The gold leafs were replaced by two metallised quartz strings. Figure 2.2 shows a schematic view of this two-string electrometer. It was produced by the company Günther & Tegetmeyer in Braunschweig/Germany (Fricke, 2011), as were many of its worldwide distributed succeeding models. In autumn 1908 Wulf performed absorption measurements of $\gamma$-radiation in the area of Valkenburg and concluded (Wulf, 1909a):

> Then particularly observations in balloons and with kite flights could give very valuable information whether the starting point of this radiation is the earth, or the atmosphere, or the stars.

One can only speculate why Wulf himself did not explore the atmosphere with a balloon. But he was focused first of all on detailed investigations inside and outside of buildings and caves, in mines up to 980 m below ground, on lakes and the river

Maas above and below the surface. From these ionisation measurements he came to the following summary (Wulf, 1909b):

> Experiments are presented which demonstrate that the penetrating radiation at the place of observation is caused by primary radioactive substances which are located in the upper Earth's layers, up to 1 m below the surface. If a part of the radiation comes from the atmosphere, then it is so small that it is not detectable with the used methods. The time variations of the $\gamma$-radiation can be explained by the shift of emanation-rich air masses in the Earth in larger or smaller depths due to variations of the air pressure.

To prove this hypothesis he did at least a small step into the atmosphere. Wulf followed an invitation to perform measurements on top of the Eiffel tower on four days in April 1910. Assuming that the main part of $\gamma$-radiation comes from the area near to the ground, one would expect a reduction of ionisation at 300 m height by 27 % (see Table 2.2). In fact, Wulf measured a decrease of ionisation by 13 % compared to the ground (Wulf, 1910). This significant difference was in clear disagreement with his previous assumption that radioactive emanations in the atmosphere are negligible.

In Italy Domenico Pacini, a physicist at the Agency of Meteorology and Geodynamics, confirmed this result. Using electrometers of the Wulf-type, he performed measurements in 1910 and 1911 (Pacini, 1910, 1912) on board of a destroyer of the Italian Navy at more than 300 m distance from the coast, where the water was at least 4 m deep. Assuming that there is no influence of radiation from the Earth's solid ground, he estimated on the sea a fraction of 66 % of the ionisation measured in parallel on land. At the same time George Simpson, an English meteorologist, and Charles Wright, a Canadian physicist, investigated the 'atmospheric electricity over the ocean' (Simpson and Wright, 1911) on the way from England to New Zealand. Both were scientific members of Robert Scott's crew travelling in 1910 on board the 'Terra Nova' to Antarctica. They measured the ionisation also with a Wulf electroscope made by Günther & Tegetmeyer. Whereas Pacini's data showed strong fluctuations on land and on sea, Simpson and Wright measured in average 6–7 ions $cm^{-3}\,s^{-1}$ over the sea without variations during a day. They stated (Simpson and Wright, 1911):

> ... it was seen that near land a high radioactive-content of the air almost synchronised with a high natural ionisation. That this high ionisation is due to radioactive products deposited on the ship itself is highly probable from the fact that the ionisation persists for some time after the high air radioactivity had disappeared.

At the end of the 19th century balloon flights were very popular for military and scientific purposes. Especially meteorologists and geophysicists used balloons to study weather conditions, the electrical earth field and the electricity of the atmosphere at high altitudes. Probably Elster and Geitel were the first to suggest to use a balloon for ionisation measurements in the higher atmosphere. It was apparently forgotten or ignored that already between 1902 and 1903 the German meteorologist Franz Linke performed 12 balloon flights with interesting results (Linke, 1904). Starting in Berlin he reached altitudes up to 5 500 m and measured the electrical field of the Earth and the ionisation in the atmosphere. There was an agreement that

**Table 2.2** Ionisation measurements in the atmosphere at different altitudes. The measured values can be compared with the assumption that the radiation is concentrated close to the ground and is absorbed by the air corresponding to the exponential dependence on the distance (Wulf, 1910; Gockel, 1910)

| Scientist | Location | Date | Position | Measured ions, $cm^{-3}\,s^{-1}$ | Expected ions, $cm^{-3}\,s^{-1}$ |
|---|---|---|---|---|---|
| Th. Wulf | Eiffel Tower | 29.03.1910 | Ground | 17.5 | |
| | | 30.03.1910 | 300 m | 16.2 | 4.7 |
| | | 31.03.1910 | 300 m | 14.4 | |
| | | 01.04.1910 | 300 m | 15.0 | |
| | | 02.04.1910 | 300 m | 17.2 | |
| | | 03.04.1910 | Ground | 18.3 | |
| A. Gockel | Zürich | 11.12.1909 | Ground | 23.8 | |
| | | | 2 500 m | 16.2 | $4 \times 10^{-4}$ |
| | | | 4 000–4 500 m | 15.8 | $2 \times 10^{-7}$ |
| | Bern | 15.10.1910 | Ground | 11.4 | |
| | | | 2 000–2 800 m | 7–9 | $3 \times 10^{-4}$ |
| | Bern | 02.04.1911 | Ground | 14.7 | |
| | | | 1 900 m | 14 | $3 \times 10^{-3}$ |
| | | | 2 500 m | 11.3 | $3 \times 10^{-4}$ |

Elster and Geitel measured the ionisation in Wolfenbüttel at the same time for comparison. Linke observed at altitudes between 1 and 3 km about the same ionisation values as at ground and an increase by a factor of four at higher altitudes up to 5 km. Obviously, the existence of penetrating radiation in the higher atmosphere was detected too early to be recognised and appreciated by the physics community. A new series of balloon flights began in 1908 with Flemming and Bergwitz. Because of problems with their detectors, they did not achieve convincing results. In the end of 1909 Albert Gockel (Fig. 2.1) started the first of three balloon flights in Switzerland. With a Wulf electrometer he could establish previous observations that the ionisation of the atmosphere decreases slowly with altitude (Gockel, 1910). In Table 2.2 the results of Wulf and Gockel are summarised.

## 2.3 Discovery of Cosmic Rays by Victor F. Hess

V. F. Hess studied physics in Graz/Austria. From 1906 until 1910 he worked at University of Vienna under Franz Exner and Egon von Schweidler, both experts for electrical phenomena in the atmosphere. In 1910 the Institute for Radium Research of the Academy of Sciences was founded and Hess became the assistant of the first director, Stefan Meyer. The institute was for many years embedded in the international research of radioactivity and provided other institutes with gauged radioactive sources.

### 2.3.1 Calibration and Absorption Measurements

Inspired by the work of Wulf, Bergwitz and Gockel, Hess started with own measurements. First, he wanted to prove experimentally the absorption of $\gamma$-rays in air. With the strongest radium sources available in the institute Hess investigated the range of $\gamma$-rays at different distances to the detector (Hess, 1911, p. 999): "The sources were positioned at distances of 10, 20, 30, ... up to 90 m from the electrometer and then the saturation current was estimated as mean value of 5–10 single measurements." With an absorption coefficient for air of $\lambda = 0.0000447$, Hess confirmed the results of previous estimates of Eve and others and concluded (p. 1 000): "... that the penetrating radiation of the Earth must decrease rapidly with the altitude and at 500 m only few percent would be expected of the values on the ground." As Hess mentioned in the publication (Hess, 1913), Wulf proposed to him in winter 1911/1912 to calibrate two-string electrometers with different gauge radium sources. After some improvements of the electrometer construction carried out by the company Günther & Tegetmeyer, Hess performed the calibration with radioactive sources of different strengths. The accuracy to measure the radioactivity of a source with unknown strength could be improved to few per mille (Hess, 1913). In contrast, the same electrometer reached without this calibration a measurement accuracy of about 3 %.

### 2.3.2 Balloon Flights

In 1911 Hess planned balloon flights to repeat the investigations of penetrating radiation in the atmosphere. The Royal Imperial Austrian Aeronautical Club provided a balloon for two flights. Already with the first flight he confirmed the results of Gockel (Hess, 1911). The ionisation remained almost constant up to the maximum height of 1 000 m. The second flight in October 1911 was during the night. In general stable thermic conditions guarantee quiet flights at constant altitude, but in this case bad weather conditions did not allow to fly higher than 200–400 m above ground. Nevertheless, the observation of the identical ionisation at day and night became an important argument for later discussions.

That the Imperial Academy of Sciences in Vienna funded seven balloon flights in 1912 shows the high ranking of this research in Austria. To avoid the problems of Bergwitz and Goppel at higher altitudes, Hess had ordered at Günther & Tegetmeyer two pressure-sealed electrometers for $\gamma$-rays and a third one with thin zinc walls for $\beta$-rays. Six flights were launched from the area of the Aeronautical Club in Vienna's Prater. The balloons were filled with illuminating gas which did not allow one to reach very high altitudes. In Table 2.3 the characteristic data of these flights are summarised, and Fig. 2.3 shows the flight routes.

The results of the six flights at relatively low altitudes can be summarised as follows (Hess, 1912):

(i) All three electrometers showed identical variations with time and altitude.

**Table 2.3** Seven balloon flights of V.F. Hess in 1912 (Hess, 1912)

| Flight | Date | Time | Height above ground, m | Ions($\gamma-1$), $cm^{-3}\,s^{-1}$ | Ions($\gamma-2$), $cm^{-3}\,s^{-1}$ | Ions($\beta$-rays), $cm^{-3}\,s^{-1}$ |
|---|---|---|---|---|---|---|
| 1 | 17.4.1912 | 08:30–09:30 | 0 | 14.4 | 10.7 | |
| | | 11:00–12:15 | 1 700 | 13.7 | 11.1 | |
| | | 12:15–12:50 | 1 700–2 100 | 27.3 | 14.4 | |
| | | 12:50–13:30 | 1 100 | | 15.1 | |
| 2 | 26.–27.4.1912 | 16:00–22:30 | 0 | 17.0 | 11.6 | 20.2 |
| | | 23:00–09:35 | 140–190 | 14.9 | 9.8 | 18.2 |
| | | 06:35–09:35 | 800–1 600 | 17.6 | 10.5 | 20.8 |
| 3 | 20.–21.5.1912 | 17:00–21:30 | 0 | 16.9 | 11.4 | 19.8 |
| | | 22:30–02:30 | 150–340 | 16.9 | 11.1 | 19.2 |
| | | 02:30–04:30 | ~500 | 14.7 | 9.6 | 17.6 |
| 4 | 03.–04.5.1912 | 17:10–20:40 | 0 | 15.8 | 11.7 | 21.3 |
| | | 22:30–00:30 | 800–1 100 | 15.5 | 11.2 | 21.8 |
| 5 | 19.6.1912 | 15:00–17:00 | 0 | 13.4 | | |
| | | 17:30–18:40 | 850–950 | 10.3 | | |
| 6 | 28.–29.6.1912 | 20:10–23:10 | 0 | 15.5 | 12.2 | |
| | | 00:40–05:40 | 90–360 | 14.9 | 11.4 | |
| 7 | 7.8.1912 | 06:45–07:45 | 1 400 | 15.8 | 14.4 | 25.3 |
| | | 07:45–08:45 | 2 500 | 17.3 | 12.3 | 31.2 |
| | | 08:45–09:45 | 3 600 | 19.8 | 16.5 | 35.2 |
| | | 09:45–10:45 | 4 400–5 350 | 40.7 | 31.8 | |
| | | 10:45–11:15 | 4 200 | 28.1 | 22.7 | |
| | | 11:15–11:45 | 1 200 | 9.7 | 11.5 | |
| | | 11:45–12:10 | 150 | 11.9 | 10.7 | |
| | | 12:25–13:12 | 0 | 15.0 | 11.6 | |

(ii) The ionisation rate is not connected with Sun activities.
(iii) The observation of Wulf, Bergwitz and Goppel that the rate does not decrease significantly with distance to the Earth was established with high confidence.

But the main goal of Hess was the study of the ionisation rate at very high altitudes. With a hydrogen filled balloon provided by the German Aeroclub in Bohemia, he started in the morning of 7 August 1912 in Aussig (now Usti nad Labem, Czech Republic, close to the German border) together with the balloon pilot captain W. Hoffory and the meteorological observer E. Wolf. The maximum height of 5 350 m above ground was reached in the south of Brandenburg. The balloon landed in Bad Saarow/Pieskow, about 60 km south-east of Berlin. Figure 2.4 shows

Fig. 2.3 Routes of the seven balloon flights of V.F. Hess in 1912

the mean values of the observed ionisation for all three detectors. Unfortunately, the $\beta$-ray detector was damaged accidentally by Hess before the maximum height was reached. But between 500 and 3 000 m a continuous increase of ionisation was measured. Both $\gamma$-detectors registered an increase of ionisation by a factor four from 3 000 to 5 200 m. Summarising the results of the seven flights, Hess came to the following conclusion (Hess, 1912):

> The results of these observations seem to be explained by the assumption that a radiation of high penetration power hits our atmosphere from top, which causes also in their lower layers a fraction of the observed ionisation in the closed detectors. The intensity seems to underly variations which are visible in time intervals of one hour. Since I did not find a decrease of radiation during the night or during the sun eclipse, the sun cannot be the reason for this hypothetical radiation, at least if one assumes a direct $\gamma$-radiation with straight-line propagation.

The discovery of cosmic rays can be seen as a step-wise approach. First indications seen by Wulf, Bergwitz, Goppel and Pacini were convincingly established

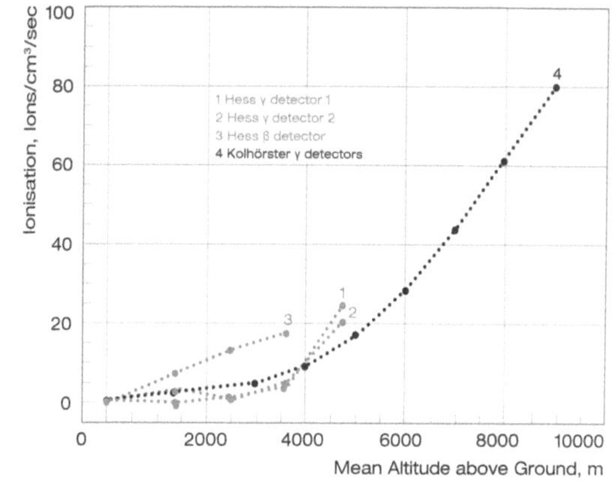


**Fig. 2.4** The number of ions measured with the three detectors (*1–3*) at the seventh high altitude flight of Hess (1912) and for the flight of W. Kolhörster up to 9 300 m altitude (*4*). The results of Kolhörster are the mean values of the results of his two $\gamma$-ray detectors (Kolhörster, 1914). For all values the ionisation measured at the surface of the Earth is subtracted

by the measurements of Hess. The essential step was the detection of the strong increase of penetrating radiation with growing altitude. Since $\gamma$-rays have the largest penetration power of the three known ionising radiations, it was natural to assume that also cosmic rays consist of energetic $\gamma$-rays.

### *2.3.3 Confirmation of the 'Höhenstrahlung'*

The discovery of the 'Höhenstrahlung' remained almost unnoticed. Probably nobody of the small number of scientists working in this field was sure that the measurements were correct. And even if there would have been no doubt, contrary to the discovery of X-rays, the possible consequences of the existence of extraterrestrial radiation were unknown. Therefore, it was necessary to establish the existence of the new radiation independently by other measurements.

Werner Kolhörster did this next step as an assistant in the Physics Institute of the University Halle in Germany. He had written his doctoral thesis about the radioactivity of mineral water coming from Karlovy Vary, the famous spa town in the western part of the Czech Republic. So, he was familiar with the problems of measuring small quantities of radioactivity. He knew the activities and publications of Hess and all others working on the penetrating radiation. Supported by the Aero-Physical Research Fund of Halle, Kolhörster could perform several balloon flights. First, he also improved the detector performance with help of the company Günther & Tegetmeyer (Fricke, 2011). Especially the measurement at very high altitude required good temperature and pressure stability of the electrometer corpus. Three flights in summer 1913, where altitudes up to 6 300 m were reached, showed the same behaviour of the ionisation rate (Kolhörster, 1913) as measured before by Hess. A new flight at 28 June 1914 demonstrated then, as stated by Kolhörster 'undoubtedly' (Kolhörster, 1914), the existence of radiation of cosmic origin. At an altitude of 9 300 m he measured an ionisation of 80 ions $cm^{-3}\,s^{-1}$. Figure 2.4 shows the ionisation in dependence of altitude for both, the measurements of Hess and of Kolhörster.

## 2.3.4 Doubts and Rediscovery

The World War I stopped most activities. Long-term measurements of Gockel and Hess in the Alps confirmed the balloon results for 2 500–3 500 m altitudes (Gockel, 1915; Hess, 1917). This was not important as regards corroborating the balloon results, but was convincing enough to think about research stations on high mountains. There was a group of physicists who had serious doubts that a new radiation of cosmic origin was discovered. Their main arguments ranged from a possible radiation in the upper atmosphere to measurement problems due to insulation leaks caused by the low temperatures at high altitudes.

A problematic role in these scientific debates played Robert A. Millikan at the California Institute of Technology. He received the Nobel Prize in 1923 for the measurement of the elementary charge of the electron, although his data analysis was challenged by experts. As director of the Norman Bridge Lab for Physics he started a cosmic ray research program. First results were presented by Russel M. Otis in 1923. He has measured the ionisation with a Kolhörster-like electrometer in balloons and airplanes up to 5 300 m altitude. A similar dependence was observed as before in Europe, although with a smaller increase. Another approach was tried by Millikan and Bowen, who used a simple and light electrometer with automated data recording. Their goal was to overcome the magic 10 000 m border with low-cost, unmanned sounding balloons. Two of four ascents in 1921 were successful and reached 11 200 and 15 500 m. But with the detector only one averaged ionisation value of 46.2 ions per $cm^2$ per second was measured above 5 500 m. This was about a factor three larger than at the surface. Nevertheless, Millikan concluded from this doubtful result in April 1926 that there is "complete disagreement" with the data of Hess and Kolhörster and one has therefore a "definite proof that there exists no radiation of cosmic origin having an absorption coefficient as large as 0.57 per meter of water" (Millikan and Bowen, 1926). A second publication from June 1926 summarised the experiments of Otis and measurements in the mountains. Also here no evidence for extraterrestrial radiation was observed. But five months later, measurements in snow-fed lakes at high altitudes (Millikan and Cameron, 1926) showed 'suddenly' that "This is by far the best evidence found so far for the view that penetrating rays are partially of cosmic origin."

An article appeared in the *New York Times* (NY-Times, 1925) at 12 November 1925 with the title "Millikan Rays" which referred to the sounding balloon measurements published five months later in April 1926 (Millikan and Bowen, 1926) with the conclusion given above. This is an interesting example for Millikan's 'abilities' in publicity and science marketing. Parts of this article, which was even reprinted in '*Science*' (Science, 1925), will be presented here:

> DR. R.A. MILLIKAN has gone out beyond our highest atmosphere in search for the cause of a radiation mysteriously disturbing the electroscopes of the physicists. ... The study had to be made out upon the edge of what the report of his discovery calls "finite space," many miles above the surface of the earth in balloons that carry instruments of men's devising where man himself cannot go. His patient adventuring observations through 20 years have at last been rewarded. He has brought back to earth a bit more of truth to add to what we

> knew about the universe. ... He found wild rays more powerful and penetrating than any that have been domesticated or terrestrialized, travelling toward the earth with the speed of light ... The mere discovery of these rays is a triumph of the human mind that should be acclaimed among the capital events of these days. The proposal that they should bear the name of their discoverer is one upon which his brother-scientists should insist. ... "Millikan rays" ought to find a place in our planetary scientific directory all the more because they would be associated with a man of such fine and modest personality.

The 'brother-scientists' in Europe insisted, but in the opposite way (Hess, 1926; Kolhörster, 1926). They made clear that what was called the discovery of 'Millikan rays' was nothing else than the radiation discovered in 1912 by Hess.

But Millikans aggressive campaign had a strong impact. In several scientific books of this time, also from European authors (see e.g. De Broglie and De Broglie, 1930, p. 130), Millikan was assigned as the discoverer of the extraterrestrial rays. Finally, with the Nobel Prize awarded in 1936 to V.F. Hess, the real development in this research field was put in perspective. Today it is assumed that Millikan created the terms 'cosmic radiation' and 'cosmic rays' for this radiation (Millikan and Cameron, 1926). But also this can be disputed. Gockel and Wulf used it in a paper from 1908 (Gockel and Wulf, 1908) summarising the results of their investigations in the Alps:

> An influence of the altitude on the ionisation could not be verified. This allows the conclusion that a cosmic radiation, if it exists at all, contributes with an inconsiderable fraction only.

In English, French and Russian publications the term 'cosmic rays' was the standard after Millikans paper in 1926, in German the terms 'Höhenstrahlung' and 'Ultrastrahlung' were in use until end of the 1940s. Afterwards 'cosmic rays' and 'cosmic particle physics' were commonly used.

## 2.4 Properties of Cosmic Rays

It was the higher penetration power which led to speculations that there could be something else than the known $\alpha$-, $\beta$- and $\gamma$-rays. As discussed before, it took years to isolate cosmic rays from background $\gamma$-radiations caused by radioactive impurities in the detector walls, in the environment of the detector, in the Earth crust and in the lower atmosphere. Until the early 1930s it was the general consensus that cosmic rays are $\gamma$-rays. In many long-term experiments the time variation of the ionisation was measured and correlations were investigated with temperature, velocity of the wind, air pressure, position of the sun and the stars. Most of these results were contradictory, and, finally, the only advantage was a better understanding of the used electrometers and the experimental conditions. Besides the discovery at high altitudes itself, the absorption measurements in water and ice as well as with lead shielding brought new insights. Real progress came with new detection methods like the cloud chamber, the Geiger–Müller counter and the possibility to measure coincident signals.

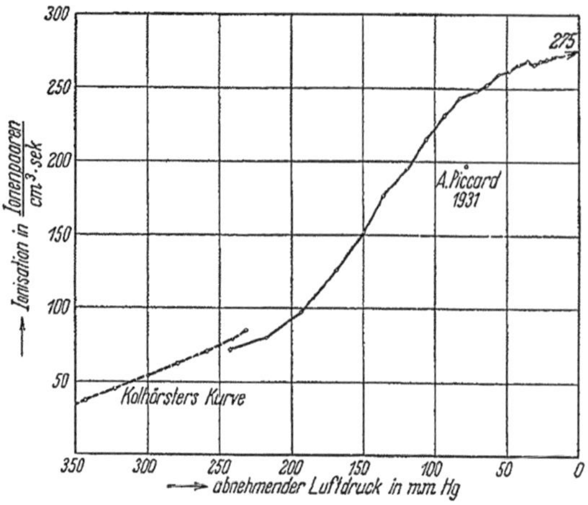


**Fig. 2.5** The measured ionisation in dependence on the air pressure. For comparison the results of Kolhörster (1914) and Piccard (1932) were shown (from Regener, 1932b)

### *2.4.1 Hardness of Cosmic Rays*

As discussed in Sect. 2.2.3 the absorption coefficient of $\gamma$-rays from radioactive sources was an important material parameter. Consequently, Kolhörster estimated the absorption for cosmic rays in air from his ionisation measurements up to 9 300 m altitude to $\lambda = 1 \times 10^{-5}\ \text{cm}^{-1}$. This is 4.4 times smaller than for $\gamma$-rays from radioactive sources (see Table 2.1), which means that cosmic rays are much harder. Later investigations by Kolhörster and Salis (1923) as well as by Millikan and Cameron (1926) in glacier ice and mountain lakes at altitudes between 1 400 and 3 900 m yielded absorption coefficients for cosmic rays in water of $2.2 \times 10^{-3}\ \text{cm}^{-1}$ and $1.8–3 \times 10^{-3}\ \text{cm}^{-1}$, respectively. The measurements showed also an inhomogeneity of the radiation, which was a first hint on secondary components caused by Compton scattering where a gamma-quant transfers its energy on an electron. Millikan and Cameron estimated the wave lengths of the radiation components to be $3.8–6.3 \times 10^{-12}$ cm. Secondary Compton electrons would then reach an energy of $1.5 \times 10^{7}$ eV (Kolhörster, 1928).

To complete these considerations the pioneering experiments of Erich Regener will be discussed. With an especially designed automatically recording electrometer enclosed in a compression-proof metal bomb he performed absorption measurements in Lake Constance up to a depth of 250 m (Regener, 1932a). The absorption coefficient of $1.8 \times 10^{-3}\ \text{cm}^{-1}$ was interpreted to mean that a very hard component of the cosmic radiation reached this depth. However, the relevance of these results was limited, since it was discovered that the primary cosmic radiation was not high energy $\gamma$-radiation (see Sect. 2.4.4). As will be seen, Regener came with his own investigations in the stratosphere to the same conclusion. Therefore, the absorption coefficients deduced so far for cosmic rays could not be considered as material constants.

With the same experimental accuracy Regener prepared sounding balloon experiments to measure the ionisation at altitudes where the primary cosmic radiation

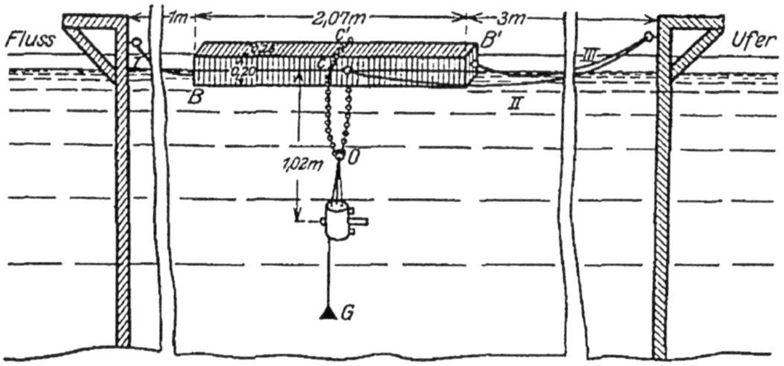


**Fig. 2.6** Schematic view of the experimental set-up to measure the ionisation variation with the air pressure in the Neva river (from Myssowsky and Tuwim, 1926)

hits the upper atmosphere (Regener, 1932b). The automatic photographic recording of ionisation, temperature and air pressure was adapted to the conditions at very high altitudes. The balloon flight at 12 August 1932 reached an altitude of 28 km. Figure 2.5 is an impressive demonstration of the continuation of Kolhörster's measurements into the stratosphere. The main conclusions of these investigations were (Regener, 1932b):

> ... 3. At pressures below 150 mmHg (above 12 km altitude) the curve becomes flatter, i.e. the intensity of the radiation increases more slowly approaching the end of the atmosphere. ... 6. If there would exist a $\gamma$-radiation of the known radioactive substances in the cosmos, then it would penetrate ... still 20 % of the corresponding air column. This would result in an increase of radiation intensity in the upper part of the curve. Since this is not the case, one can conclude that such kind of radiation does not exist with observable intensity.

## *2.4.2 Barometer Effect*

The influence of the air pressure on the ionisation rate was observed years before the discovery of cosmic particles. Simpson and Wright (see also Sect. 2.2.3) stated in their summary of atmospheric electricity measurements over the ocean in 1910 (Simpson and Wright, 1911):

> A slight dependence of the natural ionisation upon barometric pressure has been observed – a high barometer giving low value of ionisation.

The table of Wulf's measurement results on the Eiffel tower (Wulf, 1910) showed the same effect, but Wulf did not comment on it.

The effect was investigated by the Russian physicists L. Myssowsky and L. Tuwim in 1926 in the Neva river (Myssowsky and Tuwim, 1926). To reduce background radiation a Kolhörster electrometer was installed 1 m below the surface (see Fig. 2.6). Between 21 May and 11 June the ionisation and air pressure were registered. An increase by 1 mmHg (1.333224 HPa) reduced the ionisation by

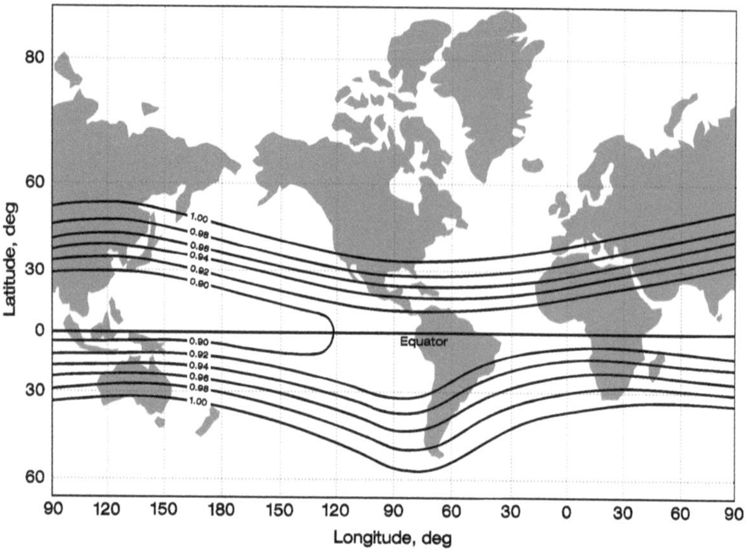


**Fig. 2.7** Schematic illustration of the latitude effect. The lines represent the intensity of cosmic particles in dependence on the latitude and longitude (from Johnson, 1938)

0.7 %. Their conclusion that the barometric effect has to be considered for precision measurements was the first result valid until today.

### 2.4.3 *Latitude Dependence*

The investigation of a possible latitude dependence of cosmic rays was proposed by Kolhörster in 1919 (Kolhörster, 1919). He interpreted the results of his solar eclipse observation in 21 August 1914, where no dependence on the sun intensity was detected. This was in agreement with earlier and later measurements by others. Kolhörster concluded that electrons could be emitted by the sun instead of $\gamma$-rays. They are then influenced by the Earth magnetic field before they hit the atmosphere to produce $\gamma$-rays. In this case one would expect a dependence on latitude.

First observations did not give clear results. Only J. Clay from the University of Amsterdam measured in 1926 on the way from Genua to Java a decrease of ionisation in the direction of the equator (Clay, 1927). Others, like Millikan along the American west coast as well as Bothe and Kolhörster on the way to Spitzbergen, did not observe an effect. Then, Clay and Compton initiated international campaigns where different groups measured the ionisation dependence on latitude and longitude with identical electrometers (Compton, 1932, 1933). As seen in Fig. 2.7, the ionisation follows the geomagnetic latitude dependency and not the geographical dependency.

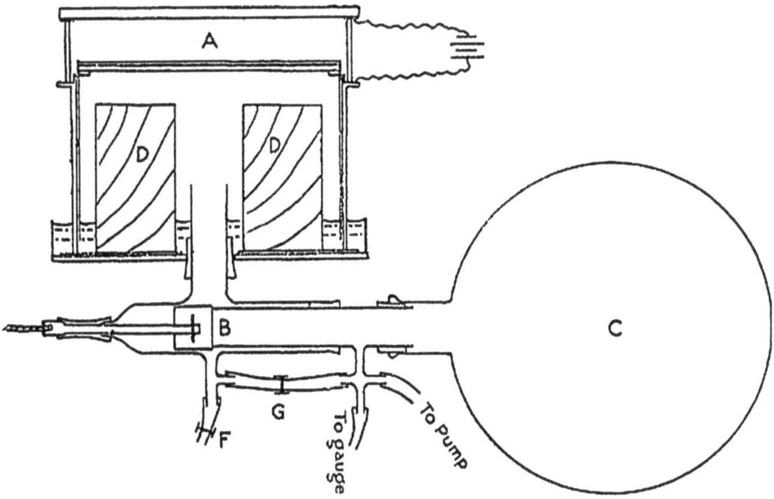


**Fig. 2.8** Schematic view of Wilson's cloud chamber from 1912 (from Wilson, 1912)

### *2.4.4 Particle Character*

For many years it was unquestioned that cosmic rays are highly energetic $\gamma$-rays. Radioactive decays were the only known source and $\gamma$-rays had the highest energy and penetration power. This view changed at the end of the 1920s when century when new detection methods came into operation. The old-fashioned electrometers were driven to high precision and stability. But they could not distinguish if a $\gamma$- or a $\beta$-ray ionised the air molecules. Only thicker detector walls could shield the lower energetic electrons.

The cloud chamber was not so new. Wilson made first studies in 1894 trying to understand the formation of clouds and fog. Motivated by his investigations on natural radioactivity and the conductivity of air by ionisation, he came back to this idea. In 1911 he published first results entitled "On a method of making visible the paths of ionising particles through a gas" (Wilson, 1911). In the following year Wilson produced, with an improved cloud chamber, impressive photographs of $\alpha$-, $\beta$- and X-rays (Wilson, 1912). The working principle of a cloud chamber is rather simple. A volume containing moist air reaches by fast expansion a supersaturated state. Irradiation with ionising rays produces air ions which then act as nuclei of condensation. Tiny water drops form the track of the ionising particle. Figure 2.8 presents a schematic view of this cloud chamber. Surprisingly on two photographs very straight tracks are visible. Cosmic rays had not been detected at this time. So Wilson misinterpreted these tracks. One of them is shown in Fig. 2.9.

There can be no question that the possibility to visualise the path of atomic particles revolutionised the research. The installation of the chamber between strong magnet coils opened for the first time the possibility of momentum and energy estimates by measuring the track curvature.

**Fig. 2.9** Photograph with a straight charged track, which is possibly the first cosmic ray electron. It was taken with Wilson's cloud chamber before June 1912 (from Wilson, 1912)

In 1927 Dmitri Skobeltzyn worked in Leningrad with a cloud chamber operating in a magnet (Skobeltzyn, 1927). He investigated Compton $\beta$-rays produced in the chamber gas by $\gamma$-rays of a Ra-C source. The photographs showed two straight tracks not related with Compton electrons. Because of their high energy of $>2 \times 10^7$ eV, Skobeltzyn concluded that these tracks were produced in the electric field of a thunderstorm. This demonstrates that even 17 years after their discovery the extraterrestrial origin of cosmic rays was not a common understanding. One year later Skobeltzyn found with a dedicated investigation in 600 pictures 36 electrons with energies larger than $1.5 \times 10^7$ eV (Skobeltzyn, 1929). This was the first 'visible proof' for the existence of charged secondary interaction products of cosmic rays.

A further important development for cosmic ray studies was the Geiger–Müller counter. In summer 1928 Hans Geiger and Walther Müller announced it in a half page article (Geiger and Müller, 1928), not knowing that nowadays it would be still an essential detection device in nuclear and particle physics. The counter consists of a metal tube with a radially spanned thin wire. The anode wire is on positive high voltage, the tube wall on ground. A charged particle traversing the tube ionises the counter gas and the electrons drift to the anode wire. First experiments used an electrometer to count the electrical signals on the wire. Walter Bothe and W. Kolhörster, working on cosmic rays in the 'Physikalisch-Technische Reichsanstalt' in Berlin, immediately saw new applications. Most interesting was the search for coincidences. Two counters give (within reachable accuracy) at the same time signals if both are crossed by a cosmic ray.

They designed a trend-setting experiment (see Fig. 2.10), whose results appeared in 1929 (Bothe and Kolhörster, 1929). Coincidence measurements were performed without and with a 4.1 cm thick gold absorber between the counters. The set-up was installed at two places: On the first floor with 3 m water-equivalent of concrete on top and for comparison below the roof of negligible material. At the first floor the coincident rates were identical, independent from the absorber. Below the roof the absorber reduced the rate by about 25 %. With these results Bothe and Kolhörster demonstrated:

(i) The cosmic rays measured in coincidence must be charged particles, $\gamma$-rays would not give coincidences.

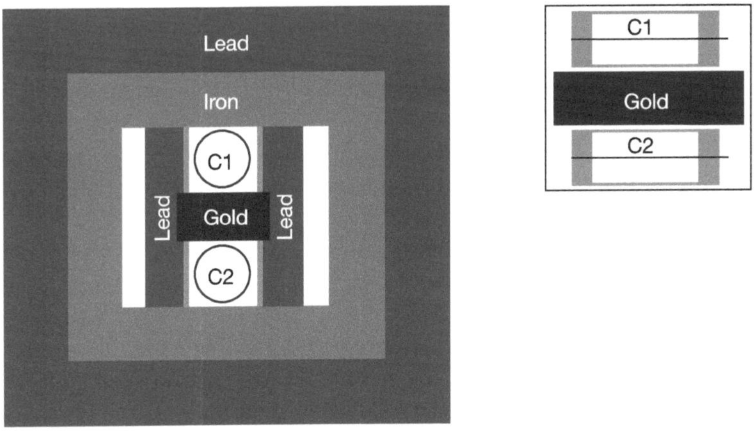


**Fig. 2.10** Layout of Bothe's and Kolhörster's coincidence experiment (from Bothe and Kolhörster, 1929). *C1* and *C2* are the Geiger–Müller counters. The coincidence condition requires the particles to cross the detector from *top* to *bottom*

(ii) These charged particles have a penetration power comparable with cosmic rays measured at high altitudes.

Therefore, it could be assumed that primary cosmic rays are also charged particles. The final answer to this question was given then in the following years by measuring the latitude dependence of the cosmic particle rate as discussed in Sect. 2.4.3.

Bothe and Kolhörster achieved the coincidence still with a photographic method. Analysing the registration film strips, which registered the electrometer string position, they looked for amplitudes appearing at the same time for both detectors. But the development of electronic components in the area of broadcast and telephony allowed new solutions. At the end of 1929 Bothe published his pioneering idea under the title 'Simplification of coincidence counting' (Bothe, 1929). With an electronic circuit and a two-grid vacuum tube he could realise an automatic coincidence counting. The circuit and the electronic components were improved in the following by Bruno Rossi and others. But the coincidence method is still an essential component in modern particle and astroparticle experiments.

## 2.5 Hypothesis About the Origin of Cosmic Rays

From the very beginning, it was the driving idea in all these research activities to identify possible sources of the penetrating radiation. In the following we will try to document the chronological development and then give a summary of the commonly accepted status end of the 1930s.

- 1901: The ionisation of air molecules was just discovered, but Wilson seemed to have visionary abilities, proposing to look also for sources outside the atmosphere (Wilson, 1901).

- 1906: O.W. Richardson studied the diurnal variation of ionisation in closed vessels (Richardson, 1906). He assumed that a correlation with the variation of the electric earth field near the surface could be "caused by radiation from extra-terrestrial sources".
- 1908: Gockel and Wulf used in their paper on high altitude measurements in the Alps (Gockel and Wulf, 1908) the term 'cosmic radiation' (kosmische Strahlung) many years before Millikan.
- 1912: Hess discovered the cosmic radiation 7 August 1912. With his previous balloon flights during the night and a solar eclipse, he concluded that the sun can be excluded as source.
- 1913: Kolhörster established the discovery. Why he favoured the sun to be the source is an open question. Perhaps he only wanted to distinguish himself from Hess. Especially in the first years he tried to convince the reader of his papers that Hess's results were not very confident.
- 1915: For the 'Elster-Geitel Festschrift' Egon von Schweidler (Univ. Vienna) performed theoretical estimates "about the possible sources of the Hess radiation" (von Schweidler, 1915). Based on the known knowledge about radioactivity, Schweidler could exclude most sources: The upper atmosphere, the moon, the sun, other planets and fixed stars. He concluded that "the less extreme requirements sets the hypothesis of radioactive substances distributed in the outer space."
- 1921: Walther Nernst, Nobel Prize laureate of 1920 and founder of physical chemistry, gave a public lecture on the status of newest research (Nernst, 1921). He also discussed the implications of the cosmic radiation: "... if many primordial matter is concentrated in the Milky Way, so this could be an area of stronger emissions. ... More detailed investigations should be done on high mountains. From here the fundamental question could be decided if it (the radiation) will be emitted uniformly in the space or stronger from the milky way." Subsequent investigations did not give conclusive answers. The reason became clear later with the discovery of the particle character of cosmic rays. The galactic magnet fields prevent a straight path from source to observer.
- 1926: In the publication, where Millikan and Cameron 'rediscovered' cosmic rays, they also presented their view on the origin of the radiation (Millikan and Cameron, 1926): "The cosmic rays are probably ... generated by nuclear changes having energy values not far from those recorded above. These changes may be (1) the capture of an electron by the nucleus of a light atom, (2) the formation of helium out of hydrogen, or (3) some new type of nuclear change, such as the condensation of radiation into atoms. The changes are presumably going on not in the stars but in nebulous matter in space, i.e., throughout the depths of the universe." It should be mentioned that Millikan was the last, giving up the $\gamma$-ray nature of cosmic rays.
- 1933: With the findings of Skobeltzyn, Bothe and Kolhörster and the proof of the latitude effect by Clay and Compton, the particle character of cosmic rays was established. This changed naturally the assumptions and requirements of their production.

- 1934: Fritz Zwicky, a Swiss, and Walter Baade, a German astrophysicist and astronomer, introduced the term supernova for short flaring, extremely bright objects (Baade and Zwicky, 1934a): "... the whole visible radiation is emitted during the 25 days of maximum brightness and the total thus emitted is equivalent to $10^7$ years of solar radiation of the present strength."
  But, more importantly, they demonstrated impressively that supernovae are sources of cosmic rays (Baade and Zwicky, 1934b): "The hypothesis that supernovae emit cosmic rays leads to a very satisfactory agreement with some of the major observations on cosmic rays." This concerns especially the energy release. They estimated the intensity of cosmic rays to be $\sigma = (0.8\text{–}8) \times 10^{-3}\ \text{ergs}\,\text{cm}^{-2}\,\text{s}^{-1}$, in rather good agreement with experimental results. Assuming that supernovae are the only source and knowing that very few appeared in our galaxy in the last 1 000 years, Baade and Zwicky argued: "The intensity of cosmic rays is practically independent of time. This fact indicates that the origin of these rays can be sought neither in the sun nor in any of the objects of our own Milky Way."
- 1942: The rebirth of the hypothesis that the sun is a source of cosmic rays came with observations of Scott Forbush, a USA geophysicist. He measured an increase of the cosmic ray rate during a strong solar flare in 1942 and concluded that at least a part of cosmic rays come from the sun (Forbush and Lange, 1942).

There were of course several publications discussing other ideas. Hannes Alfvén proposed in 1937 magnetic fields of double star systems as acceleration mechanism; Alfvén, Robert D. Richtmyer and Edward Teller discussed in 1949 the possibility that cosmic rays could have a solar origin. These and other suggestions were not mentioned here, since they did not have any relevance for future developments.

## 2.6 Begin of Particle Physics

At the beginning of 1930 three fundamental particles were known, the electron, the proton and the $\gamma$-quant. The atom was assumed to consists of a nucleus built by protons and electrons, surrounded by electrons on different orbits. To rescue the momentum conservation in the $\beta$-decay, Wolfgang Pauli had introduced a hypothetical neutral particle, later called neutrino. The neutron was detected by James Chadwick in 1932, which corrected then the picture of the atomic nucleus. On the theoretical side, quantum mechanics was developed and Paul A.M. Dirac had just formulated the theory of the electron, where another particle, the anti-electron, was postulated.

As described in Sect. 2.4.4, it was the time where new advanced particle detection devices came worldwide in operation. Their combination, i.e. the cloud chamber in a strong magnet field and, even better, the coincidence of Geiger–Müller counters to trigger a cloud chamber were the most efficient and successful methods to investigate cosmic rays. In the second half of the 1920s, cloud chambers were used to investigate charged particle tracks of radioactive sources. Skobeltzyn initiated a new area with his cosmic particle observations in a cloud chamber. This work was

**Table 2.4** Cloud chamber experiments for cosmic particle detection operating in a magnet field

| Author | Year | Chamber diameter, cm | Magnet field, tesla | Coincidence trigger counters | Discovery |
|---|---|---|---|---|---|
| Skobeltzyn (1927) | 1927 | 7.5 | 0.1 | no | first cosmic rays |
| Kunze (1933a) | 1933 | 16.0 | 2.0 | no | energy spectrum |
| Anderson (1933) | 1933 | 17.0 | 1.5 | no | positron |
| Blackett and Occhialini (1933) | 1933 | 13.0 | 0.3 | yes | $e^+e^-$-pair production, particle showers |
| Neddermeyer and Anderson (1938) | 1937 | 17.0 | 0.8 | yes | muon |

continued in 1931 by Paul Kunze in Germany, Patrick M.S. Blackett and Giuseppe Occhialini in Great Britain and by Carl D. Anderson in the USA (see Table 2.4).

But also other experimental approaches to study the properties of cosmic particles yielded important results. More advanced arrangements of Geiger–Müller counters in coincidence were used by Bruno Rossi, Bothe, Kolhörster and Erich Regener. Another photographic method, photographic emulsion, was brought to perfection by the efforts of Marietta Blau.

### *2.6.1 Discovery of the Positron*

Carl D. Anderson proposed at the end of his time as graduate student in 1929 a magnet cloud chamber experiment. The goal was to study electrons produced in a lead sheet within the chamber by 2.6 MeV $\gamma$-rays of a Th-C source. However, Millikan forced him to construct a cloud chamber with a very strong magnet for cosmic ray studies. First photographs taken in 1931 showed negatively and positively charged tracks. Mainly driven by Millikans view of the nature of cosmic rays, they were interpreted as electrons and protons produced by high energy cosmic $\gamma$-rays. But Anderson was in doubt, since for many positive particles the ionisation agreed with those of electrons. In August 1932 photographs with a 6 mm lead plate in the centre of the chamber were taken. A short announcement appeared in Science in September 1932. The more detailed publication from February 1933 (Anderson, 1933) presented the often-cited Fig. 2.11. It unambiguously demonstrated that the track must be a positively charged electron. A proton would have a ten times shorter track length.

At the same time, Blackett and Occhialini published a first analysis of photographs taken with their triggered cloud chamber (Blackett and Occhialini, 1933). The efficiency for taking cosmic track photographs was 80 % compared to 2 % for Anderson's untriggered chamber. The sketch in Fig. 2.12 shows the experimental set-up. Many photographs contained particle showers. To estimate momentum or

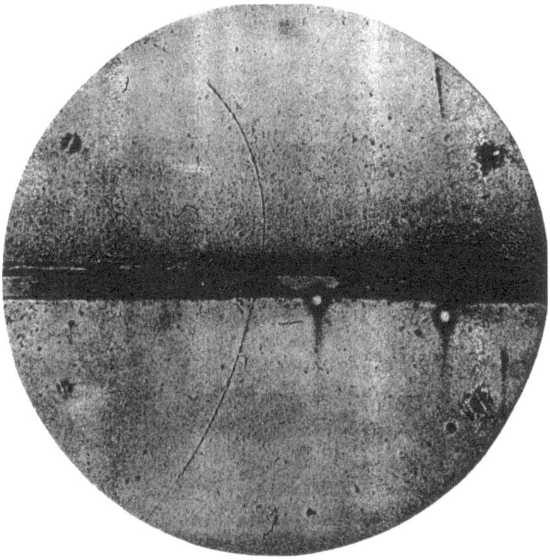

**Fig. 2.11** A positron track coming from *below* with 63 MeV energy. Passing a 6 mm lead plate, the remaining energy of the track in the *upper part* is 23 MeV (from Anderson, 1933)

energy of the tracks was difficult because of the small magnet field. But both particle charges were observed with almost identical fractions and ionisation values, which confirmed the assumption that electron–positron pairs were produced. Blackett and Occhialini discussed several hypotheses for the shower production and the properties of the positron:

> In this way one can imagine that negative and positive electrons may be born in pairs during the disintegration of light nuclei. If the mass of the positive electron is the same as that of the negative electron, such a twin birth requires an energy of $2mc^2 \sim 10^6$ eV, that is much less than the translationary energy with which they appear in general in the showers.
> The existence of positive electrons in these showers raises immediately the question of why they have hitherto eluded observation. It is clear that they can have only a limited life as free particles since they do not appear to be associated with matter under normal conditions. ... it seems more likely that they disappear by reacting with a negative electron to form two or more quanta. This latter mechanism is given immediately by Dirac's theory of electrons.

Anderson was aware of Dirac's prediction of the positron (Dirac, 1930). But as he stated in (Anderson, 1983), "... the discovery of the positron was wholly accidental. ... Dirac's relativistic theory ... played no part whatsoever in the discovery of the positron." For the paper of Blackett and Occhialini, Dirac computed the mean free path and the range of positrons in water for different energies. A positron with 1 MeV energy annihilates on average after 0.45 cm, and at 100 MeV the range is about 28 cm.

Probably, the visualisation of electron–positron particle showers initiated new theoretical activities. Heisenberg, Oppenheimer, Bethe, Heitler and others published models and theories. At the same time, experiments with cloud chambers and counter set-ups yielded new results. These important developments on particle showers will be discussed in the following article, by K.-H. Kampert and A. Watson.

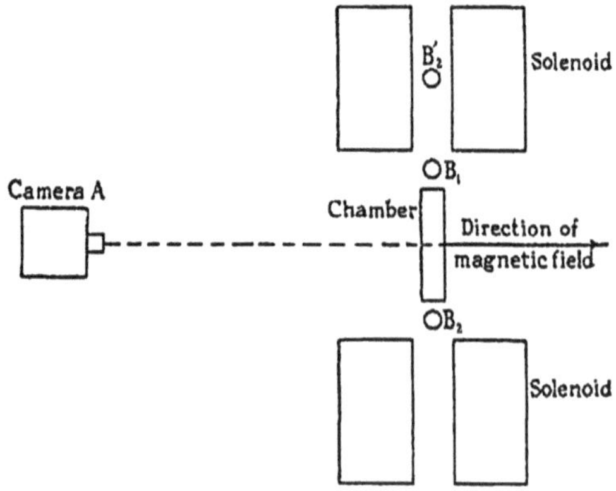


**Fig. 2.12** Set-up of Blackett's and Occhialini's cloud chamber experiment with three Geiger–Müller counters (*B*) in coincidence (from Blackett and Occhialini, 1933)

## *2.6.2 Discovery of the Muon*

After the discovery of the positron the main goal of the research with triggered cloud chambers and pure counter experiments was a better understanding of the particle properties. For theorists, the energy spectra of electrons, positrons and protons were important to verify and adjust their models. Cloud chambers in strong magnet fields and triggered by counters had clear advantages against other methods. The track visualisation, their momentum measurement and the mass estimate using the ionisation information allowed one to shed light on the complicated processes.

One of these paradoxes mentioned by Anderson appeared in photographs taken in 1934 with a 0.35 cm thick lead sheet in the centre of the chamber. S.H. Neddermeyer and Anderson found particles which were much less absorbed than electrons but had masses smaller than the proton mass. To solve this problem a new exposure of 6 000 photographs was performed with a 1 cm platinum plate in the chamber centre. Concerning electron absorption, this was more than a factor of five thicker than in the previous experiment. The data contained 55 events where the energy loss in platinum could be measured. Fourteen of them were identified as electrons and positrons with a considerable loss. For a large fraction the absorption was significantly smaller. Neddermeyer and Anderson announced the muon discovery in 1937 (Neddermeyer and Anderson, 1938) and concluded:

> ... that there exist particles of unit charge, but with a mass (which may not have a unique value) larger than that of a normal free electron and much smaller than that of a proton; this assumption would also account for the absence of numerous large radiative losses, as well as for the observed ionisation.

The name of the new particle should express that its mass is between those of electron and proton. In the first years the term 'mesotron' was used. After the discovery of the pion it was called '$\mu$-meson' and finally 'muon' to demonstrate that it is a lepton. Anderson later wrote about the history (Anderson, 1983):

> The discovery of the meson, unlike that of the positron, was not sudden and unexpected. Its discovery resulted from a two-year series of careful, systematic investigations all arranged

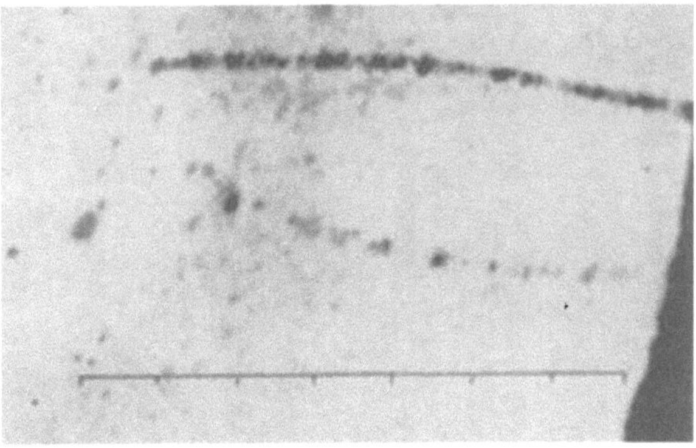

**Fig. 2.13** First photograph of a muon (*upper track*). The *lower track* is an electron (from Kunze, 1933b)

to follow certain clues and to resolve some prominent paradoxes which were present in the cosmic rays.

Paul Kunze published the first photograph of a probable muon in 1933 four years earlier (Kunze, 1933b) without knowing that he had missed a sensational discovery. He interpreted Fig. 2.13 as

> ... a thin electron track of 37 MeV and a considerably stronger ionising positive particle of smaller curvature. The nature of this particle is unknown; for a proton the ionisation is probably too small, and for a positive electron too large.

The existence of a hard, penetrating cosmic particle component consisting of muons was established later by many other experiments. Also the penetrating tracks measured in the pioneering experiment of Bothe and Kolhörster (see Sect. 2.4.4) were muons. The estimated muon mass varied still over a wide range with a mean value of $220 \pm 30$ electron masses, not so far away from the present value. The first photograph with a decaying muon was presented by E.J. Williams and G.E. Roberts in 1940 in a large cloud chamber at the picture at the University College of Wales (Williams and Roberts, 1940). Franco Rasetti was in 1941 the first who measured the muon lifetime. With a rather complex counter and absorber arrangement (Rasetti, 1941) he estimated a lifetime of $\tau_\mu = (1.5 \pm 0.3) \times 10^{-6}$.

Neddermeyer and Anderson did not know the publication of Hideki Yukawa, which appeared in the end of 1935 in a Japanese journal (Yukawa, 1935). Yukawa formulated a theory to describe the dense packing of protons and neutrons in the nucleus. In analogy to the electromagnetic theory, where the photon is the carrier of the force, here the short-ranged field needs a carrier with a mass inversely proportional to its range. Yukawa estimated a mass of about 200 electron masses and concluded: "The massive quanta may also have some bearing on the shower produced by cosmic rays." Thus, it is no surprise that the discovered muon was identified with the predicted Yukawa particle. This created new paradoxes which kept theoreticians and experimentalists busy. The main problem was that the muon with its penetration capability did not 'feel' the strong force. But the solution was found ten years later and will be discussed in the following section.

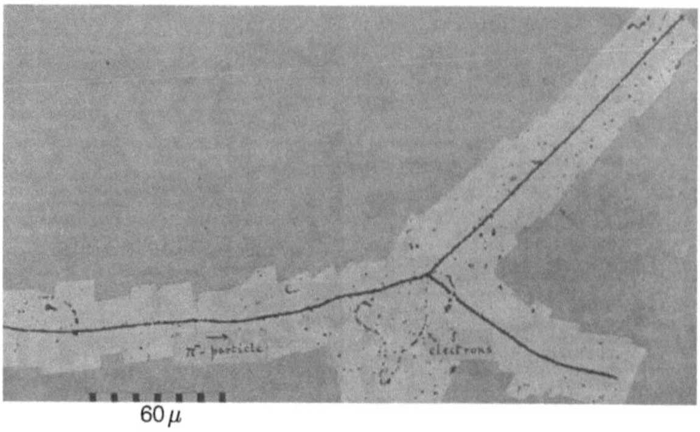


**Fig. 2.14** A pion entering the photographic emulsion from the left produces in the interaction two heavy particles and electrons (from Brown et al., 1949)

### *2.6.3 Discovery of the Pion*

In 1938 Yukawa and Sakata published a more detailed version of the theory. The lifetime of the Yukawa particle was predicted to be about $10^{-8}$ seconds, 100 times larger than the measured lifetime of the muon (Yukawa et al., 1938). This contradiction made it even more difficult to accept the possible identity of both particles. Almost ten years later, the mystery was finally solved with the discovery of the Yukawa-meson in a photographic emulsion plate.

This detection method was developed by Marietta Blau in the 1930s in Austria. Photographic emulsions accumulate the ionisation information of through-going tracks or interactions. The big advantage for the registration of rare processes is the long-term exposure from hours to months. Supported by Hess, Marietta Blau exposed an emulsion package in 1937 at the Hafelekar cosmic ray station in the Alps. One of the developed emulsion plates showed a 'star' of heavy particles (Blau and Wambacher, 1937). It was interpreted as the interaction of a cosmic particle with a nucleus of the emulsion material, leading to its disintegration into several parts. Because of her Jewish roots, Blau immigrated to Mexico, Her successful work was continued after the war, but, unfortunately, she had no possibility to participate.

In Great Britain Cecil Powell, Donald Perkins and others started in 1946 the development of photographic emulsions in cooperation with the Ilford company. Perkins, a graduate student at the Imperial College, performed an exposure of emulsion plates in an airplane at 10 km altitude (Perkins, 1947). He found about 20 'stars', one of them with an incoming particle track. From the measured ionisation and estimates for the elastic scattering of protons and lighter particles in the emulsion, Perkins concluded that the incoming particle is a meson of 100–300 electron masses.

Just a month later Occhialini and Powell published six events of the same signature (Occhialini and Powell, 1947), confirming Perkin's discovery. The group of the University of Bristol around Powell subsequently analysed 65 meson tracks, where 25 of them showed an interaction in the emulsion (Lattes et al., 1947). The estimated meson mass of $240 \pm 50$ electron masses agreed rather well with the pion mass. In Fig. 2.14 a pion interaction in a photographic emulsion is shown.

**Table 2.5** Results in elementary particle physics with cosmic rays and with experiments at the first particle accelerator, the 184 inch synchro-cyclotron at LBL Berkeley

| Year | Discovery with cosmic part. | Reference | Detector |
|---|---|---|---|
| 1929 | Charged secondaries | Skobeltzyn (1929) | Cloud chamber |
| 1929 | Charged secondaries | Bothe and Kolhörster (1929) | Counters and absorbers |
| 1932 | Charged primaries | Clay and Berlage (1932) | Electroscope |
| 1932 | Positron | Anderson (1933) | Cloud chamber |
| 1937 | Muon ($\mu$) | Neddermeyer and Anderson (1938) | Cloud chamber |
| 1947 | Pion ($\pi$) | Perkins (1947) | Photographic emulsion |
| | | Lattes et al. (1947) | Photographic emulsion |
| 1947 | Strange particles | Rochester and Butler (1947) | Cloud chamber |
| 1947 | $\mu$-absorption and decay | Conversi et al. (1945) | Counters and absorbers |
| 1949 | $K^o_L$-meson | Brown et al. (1949) | Photographic emulsion |
| 1951 | $\Lambda^o$-baryon | Armenteros et al. (1951) | Cloud chamber |
| 1952 | $\Xi$-hyperon | Armenteros et al. (1951) | Cloud chamber |
| 1953 | $\Sigma$-hyperon | York et al. (1953) | Cloud chamber |
| 1954 | $K^+$, $K^-$-meson | Menon and O'Ceallaigh (1954) | Photographic emulsion |

| Year | Discovery at accelerator | Reference | Detector/Accelerator |
|---|---|---|---|
| 1948 | $\pi^{\pm}$-lifetime | Richardson (1948) | Photogr. emulsion / 184" SC |
| 1949 | $\pi$-energy spectrum | Richman and Wilcox (1950) | Photogr. emulsion / 184" SC |
| 1950 | $\pi^{\pm}$- and $\mu^{\pm}$-mass | Barkas et al. (1951)) | Photogr. emulsion / 184" SC |
| 1950 | $\pi^o$-meson | Bjorklund et al. (1950) | Proportional counter / 184" SC |
| 1950 | $\pi^o$-mass | Panofsky et al. (1950) | Proportional counter / 184" SC |

### *2.6.4 Cosmic Particle Versus Accelerator Experiments*

Just about 50 years have passed since the first investigations on the conductivity of air and the search for the sources of radiation causing the ionisation of gases. With the discovery of cosmic rays, research activities have been started in many countries and over a wide range of scientific topics. Particle physics, one of the very strong and interesting branches since the beginning of the 1930s, began in 1948 with first steps into its own, autonomous life. Discoveries made with cosmic particles, being milestones for the development of elementary particle physics, are summarised in Table 2.5. Results are also shown from the worldwide first accelerator used since 1948 for particle physics investigations. Pions were produced with the 184 inch Berkeley synchro-cyclotron by accelerated $\alpha$-particles hitting a wire target. Most of the first small experiments used photographic emulsions as detector. Both the success in detecting new short living heavy mesons and baryons in cosmic particle experiments and the convincing first results at the Berkeley synchro-cyclotron triggered the construction of new accelerators and particle detectors. The table demonstrates in some degree the transition that particle physics performed within a few years. Yet in 1954 the 6.2 GeV Bevatron and the first hydrogen bubble chamber initiated a new area in elementary particle physics.

Finally, let some of the heroes in cosmic particle research present their view of this transition time in their own words.

> Carl D. Anderson (Anderson, 1983):
> ... the ever-encroaching larger and larger accelerators clearly indicated the end of the period when cosmic rays could be useful in studies of particle physics. ... However, undaunted by the irresistible encroachment of the accelerators, Cowan built a complex arrangement of eight flat ionisation chambers and 12 flat cloud chambers of a total height of 20 ft, designed for investigations at energies above those obtainable in any accelerator, and he continued his studies of cosmic-ray particle events until 1971.

> Cecil F. Powell (Powell, 1950):
> Even when the new machines have been brought successfully into operation, however, it will still be necessary to turn to natural sources in order to study the nuclear transmutations produced by particles of greatest energy. ... As a result of these developments there is today no line of division between nuclear physics and the study of cosmic radiation. The latter can be regarded as nuclear physics of the extreme high energy region.

> Bruno B. Rossi (Rossi, 1983):
> Today, thinking back to the work that produced these results and to the work in which other colleagues were engaged at that time, I am overtaken by a feeling of unreality. How is it possible that results bearing on fundamental problems of elementary particle physics could be achieved by experiments of an almost childish simplicity, costing a few thousand dollars, requiring only the help of one or two graduate students?
> In the few decades that have elapsed since those days, the field of elementary particles has been taken over by the big accelerators. These machines have provided experimentalists with research tools of a power and sophistications undreamed of just a few years before. All of us oldtimers have witnessed this extraordinary technological development with the greatest admiration; yet, if we look deep into our souls, we find a lingering nostalgia for what, in want of a better expression, I may call the age of innocence of experimental particle physics.

## References


Anderson, C.D.: The positive electron. Phys. Rev. **43**, 491–494 (1933)

Anderson, C.D.: Unraveling the particle content of cosmic rays. In: Brown, L.M., Hoddeson, L. (eds.) The Birth of Particle Physics, pp. 1–412. Cambridge University Press, Cambridge (1983)

Armenteros, R., et al.: Decay of V-particles. Nature **167**, 501–503 (1951)

Baade, W., Zwicky, F.: On super-novae. Proc. Natl. Acad. Sci. USA **20**, 254–259 (1934a)

Baade, W., Zwicky, F.: Cosmic rays from super-novae. Proc. Natl. Acad. Sci. USA **20**, 259–263 (1934b)

Barkas, W.H., et al.: Meson to proton mass ratios. Phys. Rev. **82**, 102–103 (1951)

Bjorklund, R., et al.: High energy photons from proton–nucleon collisions. Phys. Rev. **77**, 213–218 (1950)

Blackett, P.M.S., Occhialini, G.: Some photographs of the tracks of penetrating radiation. Proc. R. Soc. Lond. Ser. A **139**, 699–726 (1933)

Blau, M., Wambacher, H.: Disintegration processes by cosmic rays with the simultaneous emission of several heavy particles. Nature **140**, 585 (1937)

Bothe, W.: Zur Vereinfachung von Koinzidenzzählungen. Z. Phys. **59**, 1–5 (1929)

Bothe, W., Kolhörster, W.: Das Wesen der Höhenstrahlung. Z. Phys. **56**, 751–777 (1929)

Brown, R., et al.: Observations with electron-sensitive plates exposed to cosmic radiation. Nature **163**, 47–51 and 82–86 (1949)

Brown, L.M., Hoddeson, L. (eds.): The Birth of Particle Physics. Cambridge University Press, Cambridge (1983)

Clay, J.: Penetrating radiation. Proc. R. Soc. Amst. **30**, 1115–1127 (1927)

Clay, J., Berlage, H.P.: Variation der Ultrastrahlung mit der geographischen Breite und dem Erdmagnetismus. Naturwissenschaften **20**, 687–688 (1932)

Compton, A.H.: Variation of the cosmic rays with latitude. Phys. Rev. **41**, 111–113 (1932)

Compton, A.H.: A geographic study of cosmic rays. Phys. Rev. **43**, 387–403 (1933)

Conversi, M., Pancini, E., Piccioni, O.: On the decay process of positive and negative mesons. Phys. Rev. **68**, 232 (1945)

Cooke, H.L.: A penetrating radiation from the earth's surface. Philos. Mag. **34**, 403–411 (1903)

De Broglie, M., De Broglie, L.: Einführung in die Physik der Röntgen- und Gamma-Strahlen, pp. 1–205. Verlag Johann Ambrosius, Barth/Leipzig (1930)

Dirac, P.A.M.: A theory of electrons and protons. Proc. R. Soc. Lond. Ser. A **126**, 360–365 (1930)

Elster, J., Geitel, H.: Weitere Versuche über die Elektrizitätszerstreuung in abgeschlossenen Luftmengen. Phys. Z. **2**, 560–563 (1901)

Eve, A.S.: On the radioactive matter in the earth and the atmosphere. Philos. Mag. **12**, 189–200 (1906)

Forbush, S.E., Lange, I.: Further note on the effect on cosmic-ray intensity of the magnetic storm of March 1. Terr. Magn. Atmos. Electr. **47**, 331–334 (1942)

Fricke, R.: J. Elster & H. Geitel – Jugendfreunde, Gymnasiallehrer, Wissenschaftler aus Passion, pp. 1–159. Döring Druck, Druckerei und Verlag, Braunschweig (1992)

Fricke, R.: Günther & Tegetmeyer 1901–1958, pp. 1–299. AF-Verlag, Wolfenbüttel (2011)

Geiger, H., Müller, W.: Elektronenzählrohr zur Messung schwächster Aktivitäten. Naturwissenschaften **16**, 617–618 (1928)

Geitel, H.: Über die Elektrizitätszerstreuung in abgeschlossenen Luftmengen. Phys. Z. **2**, 116–119 (1900)

Gockel, A., Wulf, Th.: Beobachtungen über die Radioaktivität der Atmosphäre im Hochgebirge. Phys. Z. **9**, 907–911 (1908)

Gockel, A.: Luftelektrische Beobachtungen bei einer Ballonfahrt. Phys. Z. **11**, 280–282 (1910)

Gockel, A.: Beiträge zur Kenntnis der in der Atmosphäre vorhandenen durchdringenden Strahlung. Phys. Z. **16**, 345–352 (1915)

Hess, V.F.: Über die Absorption der $\gamma$-Strahlen in der Atmosphäre. Phys. Z. **12**, 998–1001 (1911)

Hess, V.F.: Über Beobachtungen der durchdringenden Strahlung bei sieben Freiballonfahrten. Phys. Z. **13**, 1084–1091 (1912)

Hess, V.F.: Über Neuerungen und Erfahrungen an den Radiummessungen nach der Gamma-Strahlenmethode. Phys. Z. **14**, 1135–1141 (1913)

Hess, V.F., Kofler, M.: Ganzjärige Beobachtungen der durchdringenden Strahlung auf dem Obir (2044 m). Phys. Z. **18**, 585–595 (1917)

Hess, V.F.: Über den Ursprung der Höhenstrahlung. Phys. Z. **18**, 159–163 (1926)

Johnson, Th.H.: Cosmic ray intensity and geomagnetic effects. Rev. Mod. Phys. **10**, 193–244 (1938)

Kolhörster, W.: Messungen der durchdringenden Strahlung im Freiballon in grösseren Höhen. Phys. Z. **14**, 1153–1155 (1913)

Kolhörster, W.: Messungen der durchdringenden Strahlungen bis in Höhen von 9300 m. Verh. Dtsch. Phys. Ges. **16**, 719–721 (1914)

Kolhörster, W.: Messungen der durchdringenden Strahlung während der Sonnenfinsternis vom 21 August 1914. Naturwissenschaften **7**, 412–415 (1919)

Kolhörster, W., Salis, G.v.: Intensitäts- und Richtungsmessung der durchdringenden Strahlung. Sitz.ber. Preuss. Akad. Wiss. **34**, 366–379 (1923)

Kolhörster, W.: Bemerkungen zu der Arbeit von R.A. Millikan: "Kurzwellige Strahlen kosmischen Ursprungs". Ann. Phys. **14**, 621–628 (1926)

Kolhörster, W.: Die durchdringende Strahlung in der Atmosphäre. In: Wegener, A. (ed.) Physik der Erde, pp. 565–580. Vieweg, Braunschweig (1928)

Kunze, P.: Magnetische Ablenkung der Ultrastrahlen in der Wilsonkammer. Z. Phys. **80**, 559–572 (1933a)

Kunze, P.: Untersuchung der Ultrastrahlung in der Wilsonkammer. Z. Phys. **83**, 1–18 (1933b)

Lattes, C.M.G., et al.: Processes involving charged mesons. Nature **159**, 694–697 (1947)

Linke, F.: Luftelektrische Messungen bei zwölf Ballonfahrten. Abh. Ges. Wiss. Göttingen **3**, 1–90 (1904)

Menon, M.G.K., O'Ceallaigh, C.: Observations on the decay of heavy mesons in photographic emulsions. Proc. R. Soc. A **221**, 295–318 (1954)

Miehlnickel, E.: Höhenstrahlung, pp. 1–313. Verlag von Theodor Steinkopf, Dresden und Leipzig (1938)

Millikan, R.A., Bowen, I.S.: High frequency rays of cosmic origin I. Sounding balloon observations at extreme altitudes. Phys. Rev. **27**, 353–363 (1926)

Millikan, R.A., Cameron, G.H.: High frequency rays of cosmic origin III. Measurements in Snow-Fed lakes at high altitudes. Phys. Rev. **28**, 851–869 (1926)

Myssowsky, L., Tuwim, L.: Unregelmässige Intensitätschwankungen der Höhenstrahlung in geringer Seehöhe. Phys. Z. **39**, 146–150 (1926)

Neddermeyer, S.H., Anderson, C.D.: Cosmic ray particles of intermediate mass. Phys. Rev. **54**, 88–89 (1938)

Nernst, W.: Das Weltgebäude im Lichte der Forschung, pp. 1–63. Springer, Berlin (1921)

NY-Times Editorial: Millikan rays. New York Times, November 12 (1925)

Occhialini, G.P.S., Powell, C.F.: Nuclear disintegrations produced by slow charged particles of small mass. Nature **159**, 186–190 (1947)

Pacini, D.: Penetrating radiation on the sea. Le Radium **8**, 307–312 (1910). See also De Maria, M., De Angelis, A.: arXiv:1101.3015v3 (2011)

Pacini, D.: Penetrating radiation at the surface of and in water. Nuovo Cimento **6**, 93–100 (1912). See also De Angelis, A.: arXiv:1002.1810v2 (2011)

Panofsky, W.K.H., et al.: The gamma-ray spectrum from the absorption of $\pi$-mesons in hydrogen. Phys. Rev. **78**, 825–826 (1950)

Perkins, D.H.: Nuclear disintegration by meson capture. Nature **159**, 126–127 (1947)

Piccard, A., Stahel, E., Kipfer, P.: Messung der Ultrastrahlung in 16000 m Höhe. Naturwissenschaften **20**, 592–593 (1932)

Powell, B.B.: Mesons. Rep. Prog. Phys., **13**, 350–424 (1950)

Rasetti, F.: Disintegration of slow mesotrons. Phys. Rev. **60**, 198–204 (1941)

Regener, E.: Über das Spektrum der Ultrastrahlung. Z. Phys. **74**, 433–454 (1932a)

Regener, E.: Messung der Ultrastrahlung in der Stratosphäre. Naturwissenschaften **20**, 695–699 (1932b)

Richardson, O.W.: Diurnal variation of ionisation in closed vessels. Nature **73**, 607 (1906)

Richardson, J.R.: The lifetime of the heavy meson. Phys. Rev. **74**, 1720–1721 (1948)

Richman, C., Wilcox, H.: Production cross sections for $\pi^+$ and $\pi^-$ mesons by 345 MeV protons on carbon at 90° to the beam. Phys. Rev. **78**, 496 (1950)

Rochester, G.D., Butler, C.C.: Evidence for the existence of new unstable elementary particles. Nature **160**, 855–857 (1947)

Rossi, B.B.: The decay of "Mesotrons" (1939–1943): experimental particle physics in the age of innocence. In: Brown, L.M., Hoddeson, L. (eds.) The Birth of Particle Physics, pp. 1–412. Cambridge University Press, Cambridge (1983)

von Schweidler, E.: Über die möglichen Quellen der Hessschen Strahlung. In: Bergwitz, K. (ed.) Elster- und Geitel Festschrift, pp. 1–719. Vieweg, Braunschweig (1915)

Science: Millikan rays. Science **62**, 461–462 (November 20, 1925)

Sekido, Y., Elliot, H. (eds.): Early History of Cosmic Ray Studies, pp. 1–408. Reidel, Dordrecht (1985)

Simpson, G.C., Wright, C.S.: Atmospheric electricity over the ocean. Proc. R. Soc. A **85**, 175–199 (1911)
Skobeltzyn, D.: Die Intensitätsverteilung in dem Spektrum der $\gamma$-Strahlen von RaC. Z. Phys. **43**, 354–378 (1927)
Skobeltzyn, D.: Über eine neue Art sehr schneller $\beta$-Strahlen. Z. Phys. **54**, 686–703 (1929)
Williams, E.J., Roberts, G.E.: Evidence for transformation of mesotrons into electrons. Nature **145**, 102–103 (1940)
Wilson, C.T.R.: On the leakage of electricity through dust-free air. Proc. Camb. Philos. Soc. **11**, 52 (1900)
Wilson, C.T.R.: On the ionisation of atmospheric air. Proc. R. Soc. Lond. **68**, 151–161 (1901)
Wilson, C.T.R.: On a method of making visible the paths of ionising particles through a gas. Proc. R. Soc. Lond. A **85**, 285–288 (1911)
Wilson, C.T.R.: On an expansion apparatus for making visible the tracks of ionising particles in gases and some results obtained by its use. Proc. R. Soc. Lond. A **87**, 277–297 (1912)
Wulf, Th.: Über die in der Atmosphäre vorhandene Strahlung von hoher Durchdringungsfähigkeit. Phys. Z. **10**, 152–157 (1909a)
Wulf, Th.: Über den Ursprung der in der Atmosphäre vorhandenen $\gamma$-Strahlung. Phys. Z. **10**, 997–1003 (1909b)
Wulf, Th.: Beobachtungen über Strahlung hoher Durchdringungsfähigkeit auf dem Eiffelturm. Phys. Z. **11**, 811–813 (1910)
York, C.M., Leighton, R.B., Bjonerud, E.K.: Direct experimental evidence for the existence of a heavy positive V particle. Phys. Rev. **90**, 167–168 (1953)
Yukawa, H.: On the interaction of elementary particles. Proc. Phys. Math. Soc. Jpn. **17**, 48–57 (1935)
Yukawa, H., Sakata, S., Taketani, M.: On the interaction of elementary particles IV. Proc. Phys. Math. Soc. Jpn. **20**, 720–745 (1938)

# Chapter 3
# Development of Cosmology: From a Static Universe to Accelerated Expansion

**Matthias Bartelmann**

## 3.1 A Static Universe, as Large as the Milky Way?

Modern cosmology has a birthdate: On November 25, 1915, after years of intense intellectual struggle, Albert Einstein published the final version of his field equations of General Relativity. This theory, which is entirely Einstein's achievement and which belongs to the most admirable accomplishments of human thought, superseded Newton's theory of gravity. It describes space-time as a ductile fabric shaped by the matter and the energy it contains. Its geometry is determined by a metric field whose dynamics is governed by Einstein's field equations. These equations quantify how the local curvature of space-time and the densities of matter and energy are related to each other.

It had been impossible to construct a model for the entire universe from Newton's theory of gravity. The essential reason is that infinitely extended matter distributions cannot be described by the field equation of Newtonian gravity, the Poisson equation, if the matter density is supposed to be non-zero. While Newton's theory could be applied to the spatially well-confined Solar System with remarkable success, it could not be used for cosmology.

At the time when Einstein published the final version of the field equations, the Universe was supposed to be static and eternal. Its physical dimensions were a matter of discussion. It had been known since William Herschel's meticulous observations that the Solar System is embedded in a disk-like structure of stars, the Milky Way galaxy, but it was unclear whether this was the only galaxy in the Universe. Faint, extended, diffuse objects were known in the sky. Charles Messier, a French astronomer and discoverer of comets, had published the final version of his catalogue of these so-called *nebulae* in 1781. To the present day, 109 objects of different kinds carry his name. The *New General Catalogue of Nebulae and Clusters of Stars*,

M. Bartelmann (✉)
Zentrum für Astronomie, Institut für Theoretische Astrophysik, Universität Heidelberg,
Albert-Ueberle-Str. 2, 69120 Heidelberg, Germany
e-mail: bartelmann@uni-heidelberg.de

B. Falkenburg, W. Rhode (eds.), *From Ultra Rays to Astroparticles*,
DOI 10.1007/978-94-007-5422-5_3, 

published in 1888 by the Danish astronomer Johan L.E. Dreyer, already contained 7840 objects which were evidently not individual stars, but of which many did not reveal their nature to the largest telescopes then available.

In 1920, the discussion about the nature of the nebulae culminated in the so-called Great Debate between Harlow Shapley and Heber D. Curtis, both renowned US-American astronomers. While Shapley was convinced that the nebulae were part of our own galaxy, Curtis took the view that they were extra-galactic. Both debaters had good arguments, and the debate was difficult to settle at the time. Only a few years later, in May 1925, Edwin Hubble announced his measurement of the distance to the nebula in the constellation of Andromeda, which settled the debate: This nebula turned out to be so far away that it had to be a galaxy of its own, much like the Milky Way itself (Hubble, 1925).

It is perhaps astounding in hindsight that the discovery is not even 100 years old that the Universe extends beyond our Galaxy. In any case, when Einstein published the field equations of General Relativity in 1915, it was not known whether the Universe consisted of anything else than the Milky Way. It is staggering how profoundly and quickly the picture changed thereafter.

In 1917, General Relativity was first applied to the Universe as a whole in two different papers. The first was by Einstein (1917). He discusses the difficulty Newton's theory has with static, extended mass distributions of constant density because of the boundary conditions to be set at infinity. To avoid having to set boundary conditions at all, he introduces a world model which is static in time and closed in space. Alexander Friedman later called this model Einstein's cylindric world. Since such a model does not satisfy Einstein's field equations of 1915, he extends them by introducing the cosmological constant. He closes the paper writing: "To arrive at this consistent interpretation, though, we had to introduce a new extension into the field equations of gravity, which is not justified by our actual knowledge of gravity. It is to be emphasised, however, that a positive curvature of space also results from the matter it contains if that additional term is not introduced; we require the latter only to enable a quasi-static matter distribution, as it corresponds to the fact of small stellar velocities" (Einstein, 1917, p. 152, my translation). Here it is: Spatially closed world models are possible with matter alone, but the conviction that the world is static forces Einstein to introduce the cosmological constant.

In the second paper, Willem de Sitter considered a Universe devoid of matter (de Sitter, 1917). This was an intriguing world model that he constructed to discuss the relation between gravity and inertia and Ernst Mach's hypothesis that the inertia of one body is caused by the presence of all others: He wanted to study the inertial motion of a single test particle in absence of any others. This model is, however, not globally static.

From today's point of view, it is hard to understand why Einstein overlooked that his model is unstable. Any small perturbation drives it to collapse or expand. This was a problem that had already disturbed Sir Isaac Newton. In 1693, he wrote to Bishop Bentley that if a static, infinitely extended universe was possible at all, it would be as unstable as a system of infinitely many needles standing on their points. It seems that Einstein was so firmly convinced that the Universe was static that he was satisfied to show that such a universe was in fact compatible with his theory, if only at the cost of introducing the cosmological constant.

## 3.2 An Expanding Universe, Extending Beyond the Milky Way

However, even in those years, there had already been observations of nebulae that revealed surprisingly high radial velocities. While it is possible only for very few astronomical objects, mostly bodies in the Solar System and some nearby stars, to measure the velocity transverse to the line-of-sight, the radial velocities along the line-of-sight can be spectroscopically measured by the Doppler shift of spectral lines. In 1917 already, the American astronomer Vesto Slipher summarised his spectroscopic observations of nebulae, writing (Slipher, 1917, p. 407): "The mean of the velocities with regard to sign is positive, implying that the nebulae are receding with a velocity of nearly 500 km [per second]." He speculates that "this result may still be considered as indicating that we [as inhabitants of the Milky Way] have some such drift through space. For us to have such motion and the stars not show it means that our whole stellar system moves and carries us with it. It has for a long time been suggested that the spiral nebulae are stellar systems seen at great distances. This is the so-called 'Island Universe' theory, which regards our stellar system and the Milky Way as a great spiral nebula which we see from within. This theory, it seems to me, gains favour in the present observations" (Slipher, 1917, p. 409). In a remarkable paper from 1922, Carl Wirtz, astronomer in Kiel, remarked: "Then again, the mean values [of measured radial velocities] formed with signs reveals an approximately linear trend in such a way as if the spiral nebulae near us had a tendency to approach, the distant ones a tendency to recede from our Milky Way system." He continues and then concludes: "All these statistical phenomena overlay the most prominent and main process, which can be described as the system of spiral nebulae driving apart from our position" (Wirtz, 1922, p. 352, my translation). Indications that the Universe may in fact expand, based on spectroscopic radial-velocity measurements and the speculation that the nebulae were extra-galactic, thus existed already before 1920.

In 1922, the Russian mathematician Alexander Friedman published a class of solutions to Einstein's field equations, with and without cosmological constant, that were only characterised by the two assumptions that the curvature of space be constant, but may depend on time, and that a time coordinate perpendicular to space could be constructed (Friedman, 1922). At this point, Friedman writes: "It seems to me that no physical or philosophical reasons can be given for this second assumption; it serves exclusively to simplify the calculations" (Friedman, 1922, p. 379, my translation). The Friedman models are generally unstable, expanding or shrinking with time. Friedman's paper can be seen as the unifying, mathematical construction of the modern, general-relativistic world models. As a mathematician, Friedman did not give any physical interpretation of his equations.

Friedman's solutions were independently obtained in 1927 by the Belgian physicist and Catholic priest Georges Lemaître (1927; see his portrait in Fig. 3.1). His

**Fig. 3.1** Georges Lemaître

paper is remarkably modern. Lemaître discusses homogeneous and isotropic cosmological solutions of Einstein's field equations, he re-derives Friedman's equations and discusses their time dependence, then he moves on to introduce cosmological redshift, and finally he concludes: "We have obtained a solution [of Einstein's field equations] that satisfies the following conditions: (...) (2) The radius of the Universe grows steadily from an asymptotic value $R_0$ at $t = -\infty$. (3) The distance to the extra-galactic nebulae is a cosmological effect due to the expansion of space (...)". Lemaître remarks: "This solution combines the advantages of those of de Sitter and Einstein. Note that the largest part of the Universe is forever beyond our reach" (Lemaître, 1927, p. 58, my translation). This paper, published in French in the Annals of the Scientific Society of Brussels, may be seen as the foundation of modern cosmology. Einstein created the theory, but insisted on a static world model despite its instability. Friedman constructed a class of cosmological solutions of Einstein's field equations, but did not comment on their physical meaning. Lemaître combined the inevitable time evolution of Friedman's solutions, which he derived on his own, with the growing indications that the Universe is in fact expanding.

Having established the extra-galactic nature of the nebulae by measuring the distance to the Andromeda galaxy, Edwin Hubble discovered that the radial velocities of distant galaxies grow approximately linearly with their distance. While this is an outstanding discovery in its own right, the recession of the galaxies had evidently been known long before. Hubble and Milton Humason, in their paper of 1931 (Hubble and Humason, 1931), even expressly refuse to interpret their data in a cosmological sense, writing: "The present contribution concerns a correlation of empirical data of observation. The writers are constrained to describe the 'apparent velocity-displacements' without venturing on the interpretation and its cosmological significance. Further observations are desirable and will be carried on, although it seems probable that the general features of the relation are already sketched in outline nearly to the limit of existing equipment" (Hubble and Humason, 1931, p. 80). Hubble must be credited with the discovery of the proportionality between radial velocities and distances of galaxies. That we may be living in an expanding Universe, however, was Lemaître's insight.

Einstein is often quoted as having called the introduction of the cosmological constant the "biggest blunder of his life". This quote was handed down to us by George Gamow, who writes in his autobiography *My Worldline*: "Much later, when I was discussing cosmological problems with Einstein, he remarked that the introduction of the cosmological term was the biggest blunder of his life" (Gamow, 1970, p. 44). Most certainly, Einstein felt that the cosmological constant severely impaired the beauty of his theory. However, in a short paper with Willem de Sitter in 1927 (Einstein and de Sitter, 1927, p. 51), the two authors laconically remark: "Historically, the term containing the 'cosmological constant' $\lambda$ was introduced into the field equations in order to enable us to account theoretically for the existence of a finite mean density in a static universe. It now appears that in the dynamical case this end can be reached without the introduction of $\lambda$." From the modern point of view of a classical field theory, the cosmological constant appears even quite natural. A theorem by David Lovelock states that the most general (second-rank, symmetric and divergence-free) tensor depending on derivatives of the metric up to second order must be a linear combination of the Einstein and metric tensors. The cosmological constant enters as the proportionality constant in front of the metric tensor. It thus appears, nowadays, that rather the absence of the cosmological constant would have to be justified than its presence.

## 3.3 Is the Earth Older than the Universe?

Hubble and Humason published a diagram showing the relation between the radial velocity and the distance to the galaxies. Seven years after Hubble's proof that the Andromeda nebula is extra-galactic, this diagram contains a galaxy with a radial velocity of 19 600 km $s^{-1}$, to which a distance of approximately 32 Mpc is assigned. The proportionality constant between the velocity and the distance, the so-called Hubble constant $H_0$, could thus be estimated to be $\approx$610 km $s^{-1}$ $Mpc^{-1}$. Innocent as it may seem at first glance, this value was extremely disturbing already at the time when it was published. Since the Hubble constant is a velocity divided by a distance, its dimension is $(\text{time})^{-1}$. The inverse of the Hubble constant is thus a time, which sets the typical time scale for the evolution of the Universe. With $H_0 \approx 610$ km $s^{-1}$ $Mpc^{-1}$, this time scale is approximately 1.6 billion years. Could the Universe be as young as that?

Most certainly, it could not. The decay of long-lived radioactive isotopes in minerals, such as uranium-235 and -238, had revealed already in 1918, only 22 years after Henri Becquerel had discovered radioactivity, that the Earth was more than a billion years old. In 1931, the most probable age of the Earth had already increased to three billion years. By the time Hubble and Humason published their result on the increase of the recession velocity with distance, the Universe seemed to be little more than half as old as the Earth. This was certainly a serious problem which remained unsolved for more than a decade.

A first and decisive step towards its resolution was taken in 1944 when Walter Baade realised that there are two distinct stellar populations in our Galaxy: a metal-rich Population I and a metal-poor Population II (Baade, 1944). In Baade's own words: "Although the evidence presented in the preceding discussion is still very fragmentary, there can be no doubt that, in dealing with galaxies, we have to distinguish two types of stellar population, one which is represented by the ordinary H–R [Hertzsprung–Russell] diagram (type I), the other by the H–R diagram of the globular clusters (type II) (...) Characteristic of the first type are highly luminous O- and B-type stars and open clusters; of the second, globular clusters and short-period Cepheids" (Baade, 1944, p. 145).

For cosmology, Baade's discovery was of paramount importance because Hubble had used a certain class of variable stars, the so-called Cepheids, to measure the distance first to the Andromeda galaxy and then to other distant galaxies. Cepheid stars pulsate with a period increasing with their absolute luminosity. From the measurable period of pulsation, their luminosity can be inferred, and by comparison with the observed flux the distance can be measured. Of course, this requires that this period–luminosity relation has been accurately calibrated. Baade found that Cepheids of population II have shorter periods than those of population I at the same luminosity. Hubble had mistaken the brighter Cepheids of population II with those of population I and thus underestimated their intrinsic luminosity. At the same flux, they could thus be much farther away. While Hubble had estimated the Andromeda galaxy to be 285 kpc away, its distance is now given as 765 kpc, higher by a factor of 2.7. Baade immediately remarked: "(...) it is now quite certain that Hubble's value of the distance modulus [a photometric expression for the distance, here to the Andromeda galaxy] is somewhat too small" (Baade, 1944, p. 141). With this correction of the distances, the Hubble constant shrank by the same factor, and the age of the Universe grew accordingly to approximately 4.3 billion years. Even though this was still not comfortably larger than the age of the Earth, it was reassuring that the Earth could now be younger than the Universe. After various further corrections, the value of the Hubble constant has today been measured to be $70.4^{+1.3}_{-1.4}\ \mathrm{km\,s^{-1}\,Mpc^{-1}}$, more than 8.5 times lower than the result published by Hubble and Humason. This illustrates impressively how difficult it is to measure cosmological distances reliably.

## 3.4 The Origin of the Elements

A cosmological model needs not only explain how the Universe is geometrically shaped and how it develops, but also how it could be filled with the structures it contains. We are surrounded by cosmological structures on all scales, ranging from planets to stars, star clusters, galaxies, galaxy clusters to the long filaments of matter surrounding huge voids. Perhaps the most obvious, if anthropocentric, question is how planets like the Earth and stars like the Sun could have been formed in the Universe we find ourselves in.

Life as we know it is based on carbon. As the lightest tetravalent element, it can form the long chains required by the complex biological molecules. How could carbon have been produced in large amounts?

Before we pursue this question, a word on terminology may be in order. What astronomers call "metals" are all elements heavier than helium. While baryons in physics are all particles composed of three quarks, "baryonic matter" is a more loosely defined term in astronomy, meaning essentially all forms of matter as we know it on Earth. In most circumstances, it is appropriate to read "electromagnetically interacting" when astronomers write "baryonic". Then the counterpart of baryonic matter is dark matter, a hypothetical form of matter which was first introduced to explain the kinematics in galaxy clusters and galaxies. How the arguments for dark matter developed will be exposed in the section on cold dark matter on p. 64 below.

The problem of the formation of elements more complex than hydrogen begins with helium. Helium makes up approximately 25 % of the baryonic mass in the Universe. Arthur Eddington proposed in 1920 that the energy of the Sun might be produced by the fusion of hydrogen to helium (Eddington, 1920). Remarkably, he already argues with the mass difference between four protons and the helium nucleus and concludes: "If 5 per cent of a star's mass consists initially of hydrogen atoms, which are gradually being combined to form more complex elements, the total heat liberated will more than suffice for our demands, and we need look no further for the source of a star's energy." Eddington asks: "But is it possible to admit that such a transmutation [of hydrogen into helium] is occurring?" and replies by himself: "Sir Ernest Rutherford has recently been breaking down the atoms of oxygen and nitrogen, driving out an isotope of helium from them; and what is possible in the Cavendish laboratory may not be too difficult in the Sun" (Eddington, 1920, p. 354).

Even though stars produce helium from hydrogen, it can quickly be estimated that they could by no means have enriched the baryonic matter in the Universe with helium up to 25 % by mass. Even if all stars gave away the helium they produce during their lives, their helium production could perhaps account for an abundance of approximately 5 % by mass rather than 25 %, but by far the most stars retain their helium in their cores.

If the Universe might have originated from a hot, early phase, could it perhaps as a whole be held responsible for the helium fusion? This question was studied in the 1940s by George Gamow and his collaborators Ralph Alpher and Robert Herman (see Fig. 3.2). In a sequence of papers that culminates with an article published by Alpher and Herman on April 1, 1949 (Alpher and Herman, 1949), they found among other things that the Universe could in fact have acted as a fusion reactor for the helium and some more light elements up to lithium-7.

This and the preceding studies led to a firm prediction. If the Universe had once been hot enough to brood helium in sizeable quantities, thermal radiation should be left over from this cosmic fusion process. This relic radiation should diffusely fill the entire Universe, cooled off substantially by the expansion of the Universe, but nonetheless ubiquitous. Presumably having originated under thermal-equilibrium

**Fig. 3.2** Robert Herman, George Gamow and Ralph Alpher (from *left* to *right*)

conditions, this radiation should have a Planck spectrum, which is fully characterised by its temperature. Applying and refining a very elegant argument originally due to Gamow, it was even possible for Alpher and Herman to predict this temperature. Going through different sets of parameters compatible with the constraints from helium production, they arrived at a remarkable conclusion. Alpher and Herman write: "(...) the temperature during the element-forming process must have been of the order of $10^8$–$10^{10}$ K. This temperature is limited, on the one hand, by photo-disintegration and thermal dissociation of nuclei and, on the other hand, by the lack of evidence in the relative abundance data for resonance capture of neutrons. For purposes of simplicity we have chosen (...) [a radiation density of 1 $\mathrm{g\,cm^{-3}}$] which corresponds to $T \simeq 0.6 \times 10^9$ K at the time when the neutron capture process became important (...) which corresponds to a temperature now of the order of 5 K. This mean temperature for the universe is to be interpreted as the background temperature which would result from the universal expansion alone" (Alpher and Herman, 1949, p. 1093). This was the prediction, in 1949, of the cosmic microwave background (CMB): Not only did the amount of helium in the Universe suggest a very hot and early phase of cosmic evolution, but the temperature of the remaining thermal radiation could even be estimated to lie around a few degrees Kelvin. Unfortunately, this remarkable insight seems to have gone utterly unnoticed when it was first published.

The problem posed by the existence of carbon and other "metals" in the astronomical sense was not solved, however. The main obstacle was that there are no stable elements with an atomic weight number of five. The first steps of cosmic nuclear fusion were the formation of deuterium from protons and neutrons, then the formation of helium-4 through tritium and helium-3, and finally of lithium-7 either directly through fusion of helium-4 with tritium, or indirectly by the decay of beryllium-7 formed by fusion of helium-3 with helium-4. In the rapidly diluting plasma in the early Universe, further fusion would have had to combine protons with helium-4, forming nuclei of atomic weight number five, but there is no such stable nucleus.

This problem was finally solved by Fred Hoyle in 1954. Quoting from the monumental review of 1957 on the *Synthesis of the elements in stars* (Burbidge et al., 1957, p. 565) by E. Margaret and Geoffrey Burbidge, William Fowler and Fred Hoyle: "Even though very small, the equilibrium concentration of $Be^8$ is sufficient to lead to considerable production of $C^{12}$ through radiative alpha-particle capture by the $Be^8$, and of $O^{16}$, $Ne^{20}$, etc., by succeeding alpha-particle captures. (…) Detailed consideration of the reaction rates and of the resulting relative abundances of $He^4$, $C^{12}$, and $O^{16}$ led Hoyle (…) to the prediction that the foregoing second reaction, in which $C^{12}$ is produced, must exhibit resonance within the range of energies at which the interaction between $Be^8$ and $He^4$ effectively occurs. Hoyle's predicted value for the resonance energy was 0.33 MeV, corresponding to an excited state in $C^{12}$ at 7.70 MeV. (…) The experiments reported (…) show (…) that the excitation energy of $C^{12*}$ is (…) $7.653 \pm 0.008$ MeV (…)."

In other words, Hoyle had recognised that the formation of carbon-12 in stars could be understood only if beryllium-8 and helium-4 could combine in such a way that a suitable resonance in the carbon-12 nucleus could accept the excess energy in the reaction. The discovery of this resonance at almost exactly the predicted energy marks one of the most outstanding masterpieces in astrophysics. The problem of the formation of carbon-12, necessary for our own existence, was thereby solved.

## 3.5 A Steady-State Universe?

Let us briefly recapitulate what has happened so far. We began with Einstein's theory of General Relativity in its form reported to the Prussian Academy of Sciences in November 1915. Within little more than a decade, a static universe the size of the Milky Way turned into an expanding Universe in which the Milky Way was one of very many other galaxies, separated by huge distances and driven apart by cosmic expansion. Lemaître's paper from 1927 already contained the essence of modern cosmology. About 20 years later, by the end of the 1940s, the grave problem of an old Earth in a young universe had been solved by correcting the distance scale, and the considerable amount of helium had been recognised as a strong piece of evidence for a hot, early phase in the evolution of the Universe. After a further decade, by the mid-1950s, the origin of carbon and heavier elements had essentially been solved.

Yet, there was substantial opposition against this emerging picture of the evolving Universe for which Fred Hoyle coined the intentionally derogatory term of a "Big Bang" conception. The age problem was still considered as potentially severe, and it was unclear how the cosmic structures could have formed against the expansion of space. At a more fundamental level, however, it seems that there were fierce objections against the idea of a Big Bang because it reminded one of an act of creation for which no scientific reason could be given. Yet, the recession of the galaxies was an undeniable observational fact. Could an alternative model for the

Universe be conceived that avoided an act of creation and could nonetheless account for the recession of the galaxies?

In 1948, two articles appeared in the same volume 108 of the Monthly Notices of the Royal Astronomical Society. The first, entitled *The steady-state theory of the expanding universe*, had been written by Herman Bondi and Thomas Gold (1948), the second, *A new model for the expanding universe*, by Fred Hoyle (1948). Bondi and Gold begin with a fundamental discussion of the conditions under which the laws of physics known on Earth can with some faith be applied to the Universe as a whole. They write: "As the physical laws cannot be assumed to be independent of the structure of the universe, and as conversely the structure of the universe depends upon the physical laws, it follows that there may be a stable position. We shall pursue the possibility that the universe is in such a stable, self-perpetuating state (...). We regard the reasons for pursuing this possibility as very compelling, for it is only in such a universe that there is any basis for the assumption that the laws of physics are constant; and without such an assumption our knowledge, derived virtually at one instant of time, must be quite inadequate for an interpretation of the universe (...)" (Bondi and Gold, 1948, p. 254). They pose an exciting epistemic problem: Can we with any reason believe that the laws of physics known to us could be extrapolated to the Universe? Their answer is that if this should at all be possible, then only in a universe that is as time-independent as we assume the physical laws to be. They postulate the "perfect cosmological principle", which is translation invariance not only in space, but also in time.

Intriguing as Bondi's and Gold's epistemic reasoning may be, how could it cope with the empirical fact that distant galaxies are receding the faster the farther they are? Bondi and Gold write: "If we considered that the principle of hydrodynamic continuity were valid over large regions and with perfect accuracy then it would follow that the mean density of matter was decreasing, and this would contradict the perfect cosmological principle. It is clear that an expanding universe can only be stationary if matter is continuously created within it. The required rate of creation, which follows simply from the mean density and the rate of expansion, can be estimated as at most one particle of proton mass per litre per $10^9$ years" (Bondi and Gold, 1948, p. 256). Even though matter had to be continuously created in a steady-state universe, the required rate of production was reassuringly low.

A severe initial problem of the Steady-State theory was that it had to violate local mass conservation, which is ensured by (the vanishing divergence of) Einstein's field equations. Bondi and Gold wish to retain a metric theory of gravity, though, and argue that the metric of the Universe then has to be of the exponentially expanding de Sitter type. Regarding the age problem, they notice: "The ages of the nebulae follow therefore a merely statistical law and there is no reason to suppose that a particular nebula (such as our Milky Way) is of some age rather than another" (Bondi and Gold, 1948, p. 264). Old and young galaxies should occur next to each other everywhere in time and space. From here, a suggestion already emerged for an observational test of the Steady-State model: In the Big-Bang model, young galaxies should all be distant, while they could also be nearby in the Steady-State model.

Bondi's and Gold's article remains tentative regarding the physics of creation. They acknowledge that the new matter must be created in such a way as to obey the observed recession velocity, that is, Hubble's law. Hoyle builds upon this insight and writes: "We now diverge from the usual procedure [of deriving Friedman's equations] by introducing at each point P of space-time a vector $C_\mu$ of fixed length directed along the geodesics from [a fixed space-time point] O to P. The sense of this vector is always taken as being away from O." Then he continues: "By differentiation, a symmetrical tensor field $C_{\mu\nu}$ is obtained. (...) The essential step in the present work is the introduction of the tensor $C_{\mu\nu}$ into the Einstein field equations. (...) The $C_{\mu\nu}$ term in (...) [the field equations] plays a rôle similar to that of the cosmological constant in the de Sitter model, with the important difference, however, that there is no contribution from the $C_{00}$ component. As we shall see, this difference enables a universe, formally similar to the de Sitter model, to be obtained, but in which [the matter density] $\rho$ is non-zero" (Hoyle, 1948, p. 376). Now the theory could be considered complete: Hoyle had specified a modification of Einstein's field equations by introducing the "creation field" $C$. The Steady-State model of the Universe quickly gained sympathy because it was undeniably elegant and seemed free of the difficulties that plagued the Big-Bang model.

The Steady-State model did not receive a severe blow until 1961, when Martin Ryle and Randolph Clarke published the results of their new survey for radio galaxies, undertaken with the Mullard Radio Astronomy Observatory at 178 MHz (Ryle and Clarke, 1961). (Note that the term "cycles per second", abbreviated c/s, was at that time used instead of "Hertz".) They derived the number density of faint radio galaxies as a function of radio flux. If the radio-galaxy population would not change in time, as required within the Steady-State model, the intrinsic distribution of radio galaxies with radio luminosity would have to be independent of time, and thus of distance. Their expected radio-flux distribution could therefore be predicted within the theory without any further assumptions on the radio-galaxy population itself. Ryle and Clarke found that faint radio galaxies were substantially more abundant than expected in the Steady-State model: "A comparison of the predicted with the observed curves [of the number density as a function of flux] shows a marked discrepancy, even when the smallest permissible source luminosity is adopted; the observed number of sources in the range $0.5 < S < 2 \times 10^{-26}$ watts $(c/s)^{-1}\,m^{-2}$ is $3 \pm 0.5$ times that predicted by the steady-state model. If a luminosity function similar to that of the identified sources is assumed, the discrepancy is $11 \pm 2$" (Ryle and Clarke, 1961, p. 361). Ryle and Clarke had discovered that radio galaxies were significantly farther away than they should have been in the Steady-State model; in other words, they had discovered that the radio-galaxy population must have undergone pronounced evolution over cosmic time scales. Evidently, cosmic evolution was not compatible with the idea of a steady state.

When another severe blow followed in 1965, the Steady-State model, elegant and compelling as it was, quickly disappeared in favour of the Big-Bang scenario. The pioneering discovery of 1965, however, needs to be described in a section of its own.

## 3.6 Relics from a Hot Beginning?

The story how Arno Penzias and Robert Wilson discovered the cosmic microwave background while testing a horn antenna (see Fig. 3.3) for the AT&T-Bell telephone company has frequently been told. In the course of their meticulous attempt at reducing the noise level of the antenna and its radiometer, they had even caught and deported the pigeons that used to soil the antenna, without success. In their paper of 1965 (Penzias and Wilson, 1965), they finally report: "Measurements of the effective zenith noise temperature of the 20-foot horn-reflector antenna (...) at the Crawford Hill Laboratory, Holmdel, New Jersey, at 4080 Mc/s have yielded a value about 3.5 K higher than expected. This excess temperature is, within the limits of our observations, isotropic, unpolarized, and free from seasonal variations. (...) The total antenna temperature measured at the zenith [i.e., overhead] is 6.7 K of which 2.3 K is due to atmospheric absorption. The calculated contribution due to ohmic losses in the antenna and back-lobe response is 0.9 K." Their error budget was thus significantly too low to explain the remaining noise. After a thorough discussion of the various noise contributions, Penzias and Wilson conclude: "From a combination of the above, we compute the remaining unaccounted-for antenna temperature to be $3.5 \pm 1.0$ K at 4080 Mc/s" (Penzias and Wilson, 1965, p. 419f). Without any word on a possible interpretation of this result, their brief report ends, filling little more than a single page in The Astrophysical Journal. In a *note added in proof*, the authors comment that additional measurements exclude the possibility that the observed radiation was due to radio sources of known type.

A possible explanation was presented in the same issue 142 of *The Astrophysical Journal* by Robert Dicke, Jim Peebles, Peter Roll and David Wilkinson (Dicke et al., 1965). Their paper begins with a discussion of possible cosmological models, oscillatory ones among them. The authors remark: "From this broader viewpoint we need not limit the discussion to closed oscillating models. Even if the universe had a singular origin it might have been extremely hot in the early stages. Could the universe have been filled with black-body radiation from this possible high-temperature state? If so, it is important to notice that as the universe expands the cosmological redshift would serve to adiabatically cool the radiation, while preserving the thermal character" (Dicke et al., 1965, p. 415). Later on, the paper declares: "Two of us (...) [Roll and Wilkinson] have constructed a radiometer and receiving horn capable of an absolute measure of thermal radiation at a wavelength of 3 cm. (...) While we have not yet obtained results with our instrument, we recently learned that Penzias and Wilson (...) have observed background radiation at 7.3-cm wavelength" (Dicke et al., 1965, p. 415f).

The paper moves on discussing the constraints on the matter density from the radiation temperature in several cosmological scenarios, mentioning their relation with the helium density, but without quoting the prediction by Alpher and Herman of a radiation background with a temperature of a few degrees Kelvin. The authors state: "we propose to present here the possible conclusion to be drawn if we tentatively assume that the measurements of Penzias and Wilson (...) do indicate black-body radiation at 3.5 K" (Dicke et al., 1965, p. 416) and conclude the paper by discussing possible cosmological constraints following from this interpretation.

**Fig. 3.3** Horn antenna of the AT&T-Bell Laboratories at Crawford Hill, New Jersey, with which Penzias and Wilson detected the cosmic microwave background

This summarises the winding route towards one of the most fundamental discoveries of modern cosmology: Well ahead of their time, Alpher and Herman had published a firm prediction of the microwave background and its temperature, based on the abundance of helium in the Universe. Dicke and collaborators searched for this radiation, apparently without knowing Alpher's and Herman's earlier temperature estimate. A suitable radiometer had already been constructed when Penzias and Wilson accidentally discovered the radiation as part of their noise budget without having searched for it. "Boys, you have been scooped" is the statement attributed to Robert Dicke when he put down the receiver, having been informed of Penzias' and Wilson's discovery by phone. 38 years after Lemaître had published his idea that we may be living in an expanding universe, the remains of its hot beginning had been found.

## 3.7 Structures in the Universe

Since we see ourselves surrounded by large cosmic structures, important questions are how these structures came into existence, what they consist of and how they have developed throughout the history of the Universe. The most natural assumption regarding their evolution was that primordial, small density fluctuations underwent gravitational collapse. This hypothesis raises three immediate new questions: Would gravitational collapse be fast enough? Should progenitors of the present-day structures be visible in the cosmic microwave background? And how could the primordial density fluctuations have been created?

Two years after the discovery of the cosmic microwave background, in 1967, Ray Sachs and Arthur Wolfe calculated the expected temperature fluctuations in the CMB on large angular scales (Sachs and Wolfe, 1967). They concluded: "We have estimated that anisotropies of order 1 per cent should occur in the microwave radiation if the radiation is cosmological. This figure is a reasonable lower limit provided even rather modest 10 per cent density fluctuations with a scale of 1/3 the Hubble radius occur at present. (...) Conversely, if isotropy to within 1 per cent or better could be established, this would be a quite powerful null result" (Sachs and Wolfe, 1967, p. 85). This estimate was necessarily rough because nobody knew the density-fluctuation level of large-scale structures. Nonetheless, the calculation by Sachs and Wolfe remained valid. On large scales, fluctuations in the gravitational potential cause temperature fluctuations in the CMB by gravitational redshift and time delay.

In 1968, Joseph Silk (1968) considered the effect of a finite mean-free path for the CMB photons prior to recombination and concluded that "primordial fluctuations may account for masses of the order of a typical galaxy; smaller fluctuations would not have survived to an epoch when condensation may occur. Primordial fluctuations of cosmogonic significance are found to imply anisotropy of the 3 K background radiation on an angular scale of between $10''$ and $30''$, depending on the cosmological model assumed" (Silk, 1968, p. 459). The damping by free streaming, aptly called Silk damping thereafter, suppresses small-scale fluctuations in the baryonic matter distribution and the corresponding CMB temperature fluctuations on angular scales of a few arc minutes.

Sachs and Wolfe had approached CMB temperature fluctuations from large angular scales, Silk from small angular scales. The treatment was completed by two pioneering studies in 1970, one by Rashid Sunyaev and Yakow Zeldovich (1970), the other by Peebles and Yu (1970). They went through the fairly complicated calculation of how temperature fluctuations could have been imprinted by density fluctuations during the formation of the CMB, which requires the solution of the collisional Boltzmann equation. As Peebles and Yu put it: "To obtain a more accurate description of the evolution through this complicated phase of recombination, we have resorted to direct numerical integration of the collision equation for the photon distribution function" (Peebles and Yu, 1970, p. 816). From their calculations, Sunyaev and Zeldovich concluded: "We note especially that perturbations corresponding to small masses in comparison with $10^{15}$ $M_\odot$ give quite a small contribution to $\delta T/T$; for example, for a single object with mass $M = 10^{11}$ $M_\odot$, in the case $\Omega = 1$ and $(\delta\rho/\rho) = 1$ for $z_0 = 2$ (...) we obtain (...) $\delta T/T = 10^{-8}$" (Sunyaev and Zeldovich, 1970, p. 15). Peebles and Yu arrived at compatible results and wrote, referring to larger angular scales: "Our result (...) yields characteristic angular scale (width at half-maximum) $\sim 7'$, and $\delta T/T \sim 1.7 \times 10^{-3}$ at this angular resolution" and added the cautionary note: "It is well to bear in mind that in this calculation the initial density fluctuations are invoked in an ad hoc manner because we do not have a believable theory of how they may have originated. (...) Our calculation thus is at best exploratory (...)" (Peebles and Yu, 1970, p. 834).

Even though the numbers had to be revised later for several reasons, the physical mechanisms leading to fluctuations in the CMB had now been put together. What we now call the Sachs–Wolfe effect is the imprint of gravitational-potential fluctuations on the largest angular scales. Silk damping removes fluctuations by photon diffusion on angular scales of a few arc minutes and smaller. In between, the interplay between gravity and radiation pressure gives rise to oscillations resembling sound waves in the cosmic plasma immediately prior to the release of the CMB.

## 3.8 Cosmological Inflation

Despite these detailed and pioneering calculations, the Big-Bang model posed a severe conceptual difficulty which appeared in different guises. It can perhaps best be highlighted in the following way. It is quite straightforward to estimate that the Universe must have been approximately 400 000 years old when it had cooled down sufficiently for atoms to form. Once electrons and nuclei combined to form mainly hydrogen and helium-4, free charges disappeared, the mean-free path of the photons increased abruptly, and the photons of the CMB were set free. This process is called recombination even though there had been no combination before.

This implies, however, that there is a firm upper limit for the size of causally connected regions at the end of recombination. During the first $\sim 400\,000$ years after the Big Bang, light could evidently travel by no more than $\sim 400\,000$ light years. But this length scale corresponds to a small patch on the sky, not very much larger than the Sun or the full Moon. How was it possible then that the CMB had a single temperature all over the sky? How could regions in the primordial plasma have adapted to the same temperature even though they must have been well outside any causal contact?

In 1981, Alan Guth (1981) pointed out this problem and wrote: "I have tried to convince the reader that the standard model of the very early universe requires the assumption of initial conditions with are very improbable for two reasons: (i) *The horizon problem*. Causally disconnected regions are assumed to be nearly identical; in particular, they are simultaneously at the same temperature." Guth added a second difficulty, called the flatness problem, and proposed a solution which sounded utterly speculative: "Both of these problems would disappear if the universe supercooled by 28 or more orders of magnitude below the critical temperature for some phase transition. (Under such circumstances, the universe would be growing exponentially in time.)" (Guth, 1981, p. 353f)

It was not at all clear what could be driving the exponential expansion of the Universe during such a phase of cosmological inflation, and how inflation could have ended. Initially, there was only the insight that Big-Bang cosmology had a severe causality problem that a period of inflation might remedy. However, it was recognised almost immediately by Viatcheslav Mukhanov and Gennady Chibisov in 1981 that an epoch of inflationary expansion might at the same time explain how structures could have been created (Mukhanov and Chibisov, 1981). They asked:

"Might not perturbations of the metric, which would be sufficient for the formation of galaxies and galactic clusters, arise in this stage?" (Mukhanov and Chibisov, 1981, p. 534). They carried out the quantum-theoretical calculations needed and concluded: "The fluctuation spectrum is thus nearly flat. (...) these perturbations can lead to the observed large-scale structure of the universe. The form of the spectrum (...) is completely consistent with modern theories for the formation of galaxies. (...) Thus we have one possible approach for solving the problem of the appearance of the original perturbation spectrum" (Mukhanov and Chibisov, 1981, p. 535).

## 3.9 Cold Dark Matter

At this time, 16 years after the discovery of the CMB, a detailed theory existed for the imprint of cosmic structures on temperature fluctuations in the CMB, and the inflationary hypothesis for how the causality problem could be avoided had turned out to provide at the same time a mechanism for seeding cosmic structures. One problem, however, remained and even intensified over the years: Even though radiometers became increasingly sensitive, the expected temperature fluctuations in the CMB were not found. The problem can be stated quite simply. Since the CMB was released by recombination, the Universe expanded by a factor of roughly $10^3$. The theory of gravitational instability on an expanding background shows that the amplitude of density fluctuations grows approximately by the same amount as distances in the Universe do, thus also by a factor of $\sim 10^3$. Today, structures are clearly non-linear on small scales, and just about at the onset of non-linearity when averaged over scales of a few Mpc. If such structures are assigned a relative density-fluctuation amplitude of $\sim 1$, their fluctuation amplitude should have been $\sim 10^{-3}$ at the time of recombination. Consequently, relative temperature fluctuations of $\sim 10^{-3}$ should be observed, but were not seen.

Stepping a few years forward in time, the difficulty was exacerbated in 1984 by a CMB measurement at 19.5 GHz undertaken by Juan Uson and David Wilkinson (1984). They found: "The final result (...) follows the common trend of these kinds of measurements: as observing techniques are improved, the experiments yield more stringent upper limits, but no anisotropy is detected. (...) The experiment is mostly sensitive at an angular scale of $4.5'$, but the fluctuations on scales between $1.5'$ and $15'$ must be less than 1 part in $10^4$" (Uson and Wilkinson, 1984, p. L3).

In 1982, Jim Peebles (1982) noticed: "The problem is relieved if the universe is dominated by massive, weakly interacting particles because density fluctuations on small scales can grow before decoupling" (Peebles, 1982, p. L1). This was a most remarkable suggestion. Cosmic structure formation could be reconciled with relative temperature fluctuations in the CMB much lower than $10^{-3}$ if the majority of the matter in the Universe could not interact electromagnetically, and would thus not leave a direct imprint in the cosmic background radiation. Peebles went on to calculate the relative temperature-fluctuation level expected in this case and concluded: "The rms fluctuation in [the temperature] $T$ smoothed over [angular scales of] $\theta = 10^\circ$ in a sample of size $\Theta = 100^\circ$ is $\delta T/T \sim 5 \times 10^{-6}$" (Peebles, 1982, p. L4).

That cosmic structures might be dominated by dark rather than visible matter had already been known since 1933. Examining the peculiar velocities of galaxies in the Coma cluster, Fritz Zwicky had found that this cluster could only be stable if the kinetic energy of its galaxies was balanced by the potential energy in the gravitational field of approximately ten times as much mass as could be seen. On the scale of galaxies, this was later confirmed in the 1970s by the measurements of stellar velocities by Vera Rubin and collaborators. The nature of this dark matter remained unclear, however. By Peebles' argument, the absence of temperature fluctuations in the CMB at the relative level of $10^{-3}$ was a strong hint at a form of dark matter that cannot interact electromagnetically. Asking for weakly interacting, potentially massive particles, neutrinos came to mind. Since the mass of the neutrinos was known to be finite but small, their velocities would be comparable to the speed of light. Such dark matter was called warm.

In 1984, George Blumenthal, Sandra Faber, Joel Primack and Martin Rees published an influential paper on the *Formation of galaxies and large-scale structure with cold dark matter* (Blumenthal et al., 1984). There, they remarked: "Although warm DM provides a natural (free streaming) scale for ordinary galaxies, it cannot account for massive haloes in dwarf spheroidals even though the warm DM mass is actually, if barely, consistent with the phase space constraint. This is because free streaming damps out fluctuations with mass $< 10^{11}\ M_\odot$, so dwarf galaxies with mass $\sim 10^7\ M_\odot$ can form in this picture only by fragmentation of much larger scale galactic masses" (Blumenthal et al., 1984, p. 519). After focusing on galaxy formation from cold dark matter, they concluded: "We have shown that a Universe with $\sim 10$ times as much cold dark matter as baryonic matter provides a remarkably good fit to the observed Universe. This model predicts roughly the observed mass range of galaxies, the dissipational nature of galaxy collapse, and the observed Faber–Jackson and Tully–Fisher relations [between galaxy luminosities and kinematic measures of their mass]. It also gives dissipationless galactic haloes and clusters. (...) Finally, the cold DM picture seems reasonably consistent with the observed large-scale clustering, including superclusters and voids. In short, it seems to be the best model available and merits close scrutiny and testing" (Blumenthal et al., 1984, p. 524).

In these years, the so-called "gang of four", Marc Davis, George Efstathiou, Carlos Frenk and Simon White, carried out a sequence of direct numerical simulations of cosmological structure formation in cold dark matter. These simulations achieved what was considered a very high resolution at the time, with $32^3 = 32\,768$ simulation particles. In a paper published in 1985 (Davis et al., 1985), they arrive at the conclusion: "A great virtue of the general theoretical framework investigated in this article is the fact that it makes very specific predictions on the initial conditions from which structure must form. (...) Conversely, if the properties of a universe filled with CDM are found to agree with observation, this must be considered a significant success, since there is very little freedom to adjust the theoretical predictions. (...) The major uncertainty in these implications comes from the fact that while we can predict the distribution of mass, what we see is the distribution of galaxies" (Davis et al., 1985, p. 393f).

By the mid-1980s, therefore, it had become clear that cold dark matter provided essentially the only way to reconcile the low, so far unseen, level of temperature fluctuations in the CMB with the existence of pronounced cosmic structures. The existence of dwarf galaxies argued against warm dark matter. The hypothesis of cosmological inflation was invoked to solve the horizon problem and at the same time provided a seed mechanism for cosmic structures. The spatial distribution of galaxies and galaxy clusters as well as their growth over cosmological time scales argued against a high cosmic matter density, while the lack of fluctuations in the CMB required that the matter density should be moderate, but not very low. In hindsight, it appears that all essential ingredients of what is now called the cosmological standard model had been in place around 1985. What was missing, however, was an experimental confirmation of temperature fluctuations in the CMB.

## 3.10 Fluctuations in the CMB and Supernovae Explosions

The rest of the story is now quickly told. In 1989, the satellite Cosmic Background Explorer was launched, COBE for short. This launch had been eagerly awaited by cosmologists world-wide, but had been delayed by the explosion of the space shuttle Challenger on January 28, 1986. COBE had three instruments on board, two of which are relevant for our discussion here. The Far Infrared Absolute Spectrometer, or FIRAS, had been built to measure the spectrum of the cosmic microwave background. According to the Big Bang theory, it was expected to be a thermal or Planck spectrum, but this had so far not been demonstrated. The Differential Microwave Radiometer, or DMR, had been constructed to finally detect the long-sought CMB temperature fluctuations. In a short paper published already in 1990 (Mather et al., 1990), John Mather and his team announced: "The spectrum is well fitted by a blackbody with a temperature of $2.735 \pm 0.06$ K, and the deviation from a blackbody is less than 1 % of the peak intensity over the range 1–20 $\text{cm}^{-1}$" (Mather et al., 1990, p. L37). The CMB was thus confirmed to be the thermal afterglow of the hot early Universe. The search for temperature fluctuations took somewhat longer because of the very careful data analysis required, but finally, in 1992 (Smoot et al., 1992), George Smoot and collaborators found: "The COBE DMR maps show structure with a characteristic anisotropy of $\Delta T/T \approx 6 \times 10^{-6}$. The structure is larger and of a different character than all identified systematic errors" (Smoot et al., 1992, p. L4). At the micro-Kelvin level, as expected in the cold dark matter scenario, the early progenitors of today's cosmic structures had been discovered. The excitement and the relief in the international cosmological community were overwhelming.

Further progress with CMB observations was made with balloons that carried radiometers into the stratosphere, where the microwave absorption by the water vapour in the Earth's atmosphere was tolerable. One of these balloon experiments, called BOOMERanG for *Balloon Observations Of Millimetric Extragalactic Radiation and Geophysics*, flew between December 29, 1998, and January 9, 1999, at an altitude of 37 km. With its bolometer array cooled to 0.28 K, it observed 1 800 square degrees of the sky in four frequency bands between 90 and 400 GHz with an angular resolution near 15′. Its results, announced by Paolo de Bernardis and his team in 2000 (de Bernardis et al., 2000), were summarised by: "We (...) find a peak [in the angular power spectrum] at Legendre multipole $\ell_{\rm peak} = (197 \pm 6)$, with an amplitude $\Delta T_{200} = (69 \pm 8)$ μK. This is consistent with that expected for cold dark matter models in a flat (Euclidean) Universe, as favoured by standard inflationary models" (de Bernardis et al., 2000, p. 955).

This statement, important as it is, may require a little more explanation. Together with the sound speed in the cosmic plasma prior to recombination, the time elapsed between the Big Bang and the recombination sets the characteristic length scale for CMB temperature fluctuations. The CMB originated ≈400 000 years after the Big Bang, the sound speed was very nearly $c/\sqrt{3}$. The largest wavelength of CMB temperature fluctuations is thus ≈230 000 light years, or ≈71 kpc. However, the angle that we see spanned by this length on the sky depends on the spatial curvature. From the angular scale of the peak in the power spectrum of the CMB fluctuations, first discovered by the BOOMERanG experiment, the spatial curvature could thus be directly inferred: it turned out to be compatible with zero. Within measurement uncertainties, the space in our Universe is flat.

This was a confirmation and a surprise at the same time. It was a confirmation of the expectation from inflationary cosmology that the brief period of exponential expansion should in fact have driven any finite curvature radius towards infinity. It was a surprise because spatial flatness requires all energy-density contributions in the Universe to add up to a critical value, the critical density. It was known from observations of cosmic structures as well as the CMB itself, however, that the density of baryonic and dark matter together should not amount to more than ~30 % of this critical density. The most obvious candidate for the missing ~70 % of the cosmic energy budget was the cosmological constant.

The verification that the remaining ~70 % of the present energy density could indeed be assigned to the cosmological constant formed the headstone in the edifice of the cosmological standard model. It came from observations of a certain class of stellar explosion, the so-called type-Ia supernovae. Supernovae of this type have a reasonably well-defined luminosity whose scatter can be substantially reduced by an empirical correction scheme. They form what has been called standardisable candles. By comparison of their luminosity with their measurable flux, their distance can be inferred. The relation between distance and redshift, however, depends on cosmology and thus allows the inference of cosmological parameters.

In September 1998, Adam Riess and the High-$z$ Supernova Search Team, among them Brian Schmidt, published measurements based on 16 distant and 34 nearby supernovae (Riess et al., 1998) from which they concluded: "We find the luminosity distances to well-observed SNe with $0.16 \leq z \leq 0.97$ measured by two methods to be in excess of the prediction of a low mass density ($\Omega_M \approx 0.2$) [$\Omega_M$ is the matter density in units of the critical density] universe by 0.25 to 0.28 mag [i.e., by a factor of 1.25 to 1.29]. A cosmological explanation is provided by a positive cosmological constant with 99.7 % ($3.0\sigma$) to more than 99.9 % ($4.0\sigma$) confidence using the complete spectroscopic SN Ia sample and the prior belief that $\Omega_M \geq 0$" (Riess et al.,

1998, p. 1034). Shortly thereafter, in June 1999, Saul Perlmutter and the members of The Supernova Cosmology Project announced the cosmological results obtained from a set of 42 type-Ia supernovae with redshifts up to 0.83 (Perlmutter et al., 1999). They found that: "A flat, $\Omega_\Lambda = 0$ cosmology is a quite poor fit to the data. The $(\Omega_M, \Omega_\Lambda) = (1, 0)$ line on Fig. 3.2b shows that 38 out of 42 high-redshift supernovae are fainter than predicted for this model" (Perlmutter et al., 1999, p. 580). This was the essential message from both teams: The distant supernovae appeared significantly fainter in reality than they should have appeared in a universe without cosmological constant. Assuming a spatially flat Universe, the cosmological constant should contribute $\approx$72 % of the critical energy density. The loop was closed.

## 3.11 Nobel Prizes

Of the scientists mentioned in this overview, ten received the Nobel Prize for their research, eight of whom for their research on cosmology. In chronological order, these were, quoting from the rationale given by the Nobel Committee (http://www.nobelprize.org/nobel_prizes/physics/laureates/, see also Appendix B): Arno A. Penzias and Robert W. Wilson in 1978 "for their discovery of cosmic microwave background radiation", William A. Fowler in 1983 "for his theoretical and experimental studies of the nuclear reactions of importance in the formation of the chemical elements in the universe", John C. Mather and George F. Smoot in 2006 "for their discovery of the blackbody form and anisotropy of the cosmic microwave background radiation", and Saul Perlmutter, Brian P. Schmidt and Adam G. Riess in 2011 "for the discovery of the accelerating expansion of the Universe through observations of distant supernovae".

Albert Einstein was awarded the Nobel Prize in 1921 not for cosmology and not specifically for General Relativity, but "for his services to Theoretical Physics, and especially for his discovery of the law of the photoelectric effect". Finally, Sir Martin Ryle received the Nobel Prize in 1974 "for his observations and inventions [in radio astrophysics], in particular of the aperture synthesis technique".

Empirical results, measurements and observations were clearly favoured by the Nobel Committee. Several of the fathers of modern cosmology were so far ahead of their time that their insights went almost unnoticed. Perhaps most notable are Georges Lemaître, who seems to have had the first clear vision of a Universe emerging from a hot beginning, as well as Ralph Alpher and Robert Herman, who first predicted the microwave background and its temperature.

## References

Alpher, R.A., Herman, R.C.: Remarks on the evolution of the expanding universe. Phys. Rev. **75**, 1089–1095 (1949). doi:10.1103/PhysRev.75.1089

Baade, W.: The resolution of Messier 32, NGC 205, and the central region of the Andromeda Nebula. Astrophys. J. **100**, 137 (1944). doi:10.1086/144650

Blumenthal, G.R., Faber, S.M., Primack, J.R., Rees, M.J.: Formation of galaxies and large-scale structure with cold dark matter. Nature **311**, 517–525 (1984). doi:10.1038/311517a0
Bondi, H., Gold, T.: The steady-state theory of the expanding Universe. Mon. Not. R. Astron. Soc. **108**, 252 (1948)
Burbidge, E.M., Burbidge, G.R., Fowler, W.A., Hoyle, F.: Synthesis of the elements in stars. Rev. Mod. Phys. **29**, 547–650 (1957). doi:10.1103/RevModPhys.29.547
Davis, M., Efstathiou, G., Frenk, C.S., White, S.D.M.: The evolution of large-scale structure in a Universe dominated by cold dark matter. Astrophys. J. **292**, 371–394 (1985). doi:10.1086/163168
de Bernardis, P., Ade, P.A.R., Bock, J.J., Bond, J.R., Borrill, J., Boscaleri, A., Coble, K., Crill, B.P., De Gasperis, G., Farese, P.C., Ferreira, P.G., Ganga, K., Giacometti, M., Hivon, E., Hristov, V.V., Iacoangeli, A., Jaffe, A.H., Lange, A.E., Martinis, L., Masi, S., Mason, P.V., Mauskopf, P.D., Melchiorri, A., Miglio, L., Montroy, T., Netterfield, C.B., Pascale, E., Piacentini, F., Pogosyan, D., Prunet, S., Rao, S., Romeo, G., Ruhl, J.E., Scaramuzzi, F., Sforna, D., Vittorio, N.: A flat Universe from high-resolution maps of the cosmic microwave background radiation. Nature **404**, 955–959 (2000). arXiv:astro-ph/0004404
de Sitter, W.: Einstein's theory of gravitation and its astronomical consequences. Third paper. Mon. Not. R. Astron. Soc. **78**, 3–28 (1917)
Dicke, R.H., Peebles, P.J.E., Roll, P.G., Wilkinson, D.T.: Cosmic black-body radiation. Astrophys. J. **142**, 414–419 (1965). doi:10.1086/148306
Eddington, A.S.: The internal constitution of the stars. Observatory **43**, 341–358 (1920)
Einstein, A.: Kosmologische Betrachtungen zur allgemeinen Relativitätstheorie, pp. 142–152. Sitzungsberichte der Königlich Preußischen Akademie der Wissenschaften, Berlin (1917)
Einstein, A., de Sitter, W.: On the relation between the expansion and the mean density of the Universe. In: Contributions from the Mount Wilson Observatory, vol. 3, pp. 51–52 (1927)
Friedman, A.: Über die Krümmung des Raumes. Z. Phys. **10**, 377–386 (1922). doi:10.1007/BF01332580
Gamow, G.: My world line: an informal autobiography (1970)
Guth, A.H.: Inflationary Universe: a possible solution to the horizon and flatness problems. Phys. Rev. D **23**, 347–356 (1981). doi:10.1103/PhysRevD.23.347
Hoyle, F.: A new model for the expanding Universe. Mon. Not. R. Astron. Soc. **108**, 372 (1948)
Hubble, E., Humason, M.L.: The velocity-distance relation among extra-galactic nebulae. Astrophys. J. **74**, 43 (1931). doi:10.1086/143323
Hubble, E.P.: Cepheids in spiral nebulae. Observatory **48**, 139–142 (1925)
Lemaître, G.: Un Univers homogène de masse constante et de rayon croissant rendant compte de la vitesse radiale des nébuleuses extra-galactiques. Ann. Soc. Sci. Brux. **47**, 49–59 (1927)
Mather, J.C., Cheng, E.S., Eplee, R.E. Jr., Isaacman, R.B., Meyer, S.S., Shafer, R.A., Weiss, R., Wright, E.L., Bennett, C.L., Boggess, N.W., Dwek, E., Gulkis, S., Hauser, M.G., Janssen, M., Kelsall, T., Lubin, P.M., Moseley, S.H. Jr., Murdock, T.L., Silverberg, R.F., Smoot, G.F., Wilkinson, D.T.: A preliminary measurement of the cosmic microwave background spectrum by the Cosmic Background Explorer (COBE) satellite. Astrophys. J. Lett. **354** L37–L40, (1990). doi:10.1086/185717
Mukhanov, V.F., Chibisov, G.V.: Quantum fluctuations and a nonsingular Universe. JETP Lett. **33**, 532 (1981)
Peebles, P.J.E.: Large-scale background temperature and mass fluctuations due to scale-invariant primeval perturbations. Astrophys. J. Lett. **263**, L1–L5 (1982). doi:10.1086/183911
Peebles, P.J.E., Yu, J.T.: Primeval adiabatic perturbation in an expanding Universe. Astrophys. J. **162**, 815 (1970). doi:10.1086/150713.
Penzias, A.A., Wilson, R.W.: A measurement of excess antenna temperature at 4080 Mc/s. Astrophys. J. **142**, 419–421 (1965). doi:10.1086/148307
Perlmutter, S., Aldering, G., Goldhaber, G., Knop, R.A., Nugent, P., Castro, P.G., Deustua, S., Fabbro, S., Goobar, A., Groom, D.E., Hook, I.M., Kim, A.G., Kim, M.Y., Lee, J.C., Nunes, N.J., Pain, R., Pennypacker, C.R., Quimby, R., Lidman, C., Ellis, R.S., Irwin, M., McMahon, R.G., Ruiz-Lapuente, P., Walton, N., Schaefer, B., Boyle, B.J., Filippenko, A.V., Math-

eson, T., Fruchter, A.S., Panagia, N., Newberg, H.J.M., Couch, W.J.: The supernova cosmology project: measurements of Ω and Λ from 42 high-redshift supernovae. Astrophys. J. **517**, 565–586 (1999). doi:10.1086/307221. arXiv:astro-ph/9812133

Riess, A.G., Filippenko, A.V., Challis, P., Clocchiatti, A., Diercks, A., Garnavich, P.M., Gilliland, R.L., Hogan, C.J., Jha, S., Kirshner, R.P., Leibundgut, B., Phillips, M.M., Reiss, D., Schmidt, B.P., Schommer, R.A., Smith, R.C., Spyromilio, J., Stubbs, C., Suntzeff, N.B., Tonry, J.: Observational evidence from Supernovae for an accelerating Universe and a cosmological constant. Astron. J. **116**, 1009–1038 (1998). doi:10.1086/300499. arXiv:astro-ph/9805201

Ryle, M., Clarke, R.W.: An examination of the steady-state model in the light of some recent observations of radio sources. Mon. Not. R. Astron. Soc. **122**, 349 (1961)

Sachs, R.K., Wolfe, A.M.: Perturbations of a cosmological model and angular variations of the microwave background. Astrophys. J. **147**, 73 (1967). doi:10.1086/148982

Silk, J.: Cosmic black-body radiation and galaxy formation. Astrophys. J. **151**, 459 (1968). doi:10.1086/149449

Slipher, V.M.: Nebulae. Proc. Am. Philos. Soc. **56**, 403–409 (1917)

Smoot, G.F., Bennett, C.L., Kogut, A., Wright, E.L., Aymon, J., Boggess, N.W., Cheng, E.S., de Amici, G., Gulkis, S., Hauser, M.G., Hinshaw, G., Jackson, P.D., Janssen, M., Kaita, E., Kelsall, T., Keegstra, P., Lineweaver, C., Loewenstein, K., Lubin, P., Mather, J., Meyer, S.S., Moseley, S.H., Murdock, T., Rokke, L., Silverberg, R.F., Tenorio, L., Weiss, R., Wilkinson, D.T.: Structure in the COBE differential microwave radiometer first-year maps. Astrophys. J. Lett. **396**, L1–L5 (1992). doi:10.1086/186504

Sunyaev, R.A., Zeldovich, Y.B.: Small-scale fluctuations of relic radiation. Astrophys. Space Sci. **7**, 3–19 (1970). doi:10.1007/BF00653471

Uson, J.M., Wilkinson, D.T.: New limits on small-scale anisotropy in the microwave background. Astrophys. J. Lett. **277**, L1–L3 (1984). doi:10.1086/184188

Wirtz, C.: Einiges zur Statistik der Radialbewegungen von Spiralnebeln und Kugelsternhaufen. Astron. Nachr. **215**, 349 (1922)

# Chapter 4
# Evolution of Astrophysics: Stars, Galaxies, Dark Matter, and Particle Acceleration

**Peter L. Biermann**

## 4.1 Evolution of Astrophysics

Astrophysics began when star-gazers began to ask already millenia ago, where comets came from, and what caused the observed novae and supernovae, how the planetary system began, what the Milky Way really was. Other galaxies were observed, but thought of as nebulae like any other such as around the Plejades or the sword of Orion. Even what we now know to be relativistic jets were already observed about 100 years ago in the case of the very large elliptical galaxy M87 without even knowing of other galaxies at the time.

Modern astrophysics began in the early 1900s with several developments, most importantly quantum mechanics, but also the realization by Oort that we live in a flattened distribution of stars far from the center. At this location optical emission by the stars near the center is obscured by interstellar dust. Quantum mechanics finally allowed one to quantitatively interpret the spectra of stars, to start studying their structure, and their evolution. Then using such understanding also the study of tenuous emission nebulae began with Woltjer (1958), considering the Crab nebula

P.L. Biermann (✉)
MPI for Radioastronomy, Bonn, Germany
e-mail: plbiermann@mpifr-bonn.mpg.de

P.L. Biermann
Institute for Nucl. Phys., Karlsruhe Inst. for Technology, Karlsruhe, Germany

P.L. Biermann
Dept. of Physics & Astronomy, Univ. of Alabama, Tuscaloosa, AL, USA

P.L. Biermann
Dept. of Physics, University of Alabama at Huntsville, Huntsville, AL, USA

P.L. Biermann
Dept. of Physics & Astron., Bonn, Germany

B. Falkenburg, W. Rhode (eds.), *From Ultra Rays to Astroparticles*,
DOI 10.1007/978-94-007-5422-5_4, © Springer Science+Business Media Dordrecht 2012

as a most famous example of a historical supernova of about a thousand years ago, which had been well observed.

First, with the discovery of X-ray sources outside the Solar System (Giacconi et al., 1962), the first detection of quasars (Schmidt, 1963), the first ultra-high-energy cosmic-ray event (Linsley, 1963), the cosmological microwave background (Penzias and Wilson, 1965), and pulsars (Hewish et al., 1968), the modern era of astrophysics began, with the most recent discovery of the existence of dark energy, after the detection and identification of Gamma-ray Bursts in the 1970s to 1990s.

There was a second, parallel strong development with the rapid expansion of observational opportunities, with imaging X-ray telescopes in space, intercontinental radio interferometry, infra-red and gamma-ray telescopes, and the unprecedented sensitivity of optical and ultraviolet space telescopes.

And thirdly, these opportunities are now shared by the entire world, with students and researchers mingling from all nationalities and ethnic origins, both men and women.

### *4.1.1 Historical Introduction*

In the following we will expand on the early developments in cosmology, stars, dark matter, and cosmic rays, and will add an outlook into the future:

- The big bang: The universe expands, and seems to do so for ever. As it expands it cools, and so the very early stages must have been very hot. The radiation of this hot phase was postulated to exist by Alpher et al. (1948). The residual emission from this hot phase was detected by Penzias and Wilson (1965), and immediately properly explained by Dicke et al. (1965). Today we use the very weak spatial wiggles in this radiation, and its polarization characteristics to learn about the very early universe. What is 99.5 percent of the universe made of?
- Dark energy: From 1998 the calibration of the light-curves of exploding white dwarfs (called Supernova type Ia) demonstrated that the universe expands with acceleration, most readily described by a cosmic equation of state with negative relativistic pressure; such a description defies a physical explanation up to now with predictive power and tests. Today the very precise measurements of the spatial wiggles in the cosmic microwave background give very precise information on the existence of dark energy. What can we say about the physics of dark energy?
- Magnetic fields: Polarization of star-light as well as the containment of the energetic charged particles (after 1945), called, for historical reasons, cosmic rays, require that magnetic fields permeate the interstellar medium, the gas between the stars in our Milky Way. Today we expect that magnetic fields run through the entire universe. Magnetic fields are readily produced in almost any rotating star consisting of ionized gas (referred to as Biermann-battery, 1950), and also in shock waves running through ionized gas (the Weibel instability, 1959). However, phase transitions in the very early universe can also give rise to magnetic fields.

Stars can distribute magnetic fields through their winds, or when they explode, and galaxies can distribute magnetic fields in their winds, or when the environment of their central super-massive black holes explodes. Where do magnetic fields come from?

- Cosmic rays: Cosmic rays were originally discovered through their ionizing properties by Hess in balloon flights (1912) and Kolhörster (1913), later due to their east–west asymmetry in arrival directions identified as charged particles, and today we have measured their entire spectrum accessible inside the magnetic Solar wind (discovered by Biermann, 1951). A model with predictions based on common explosions of stars into their stellar winds has passed a number of quantitative tests today (Biermann, 1993; Stanev et al., 1993; Biermann and de Souza, 2012), but we still are not completely sure whether this model is unique. Where do cosmic-ray particles come from?
- Massive black holes: In the 1960s it became apparent that super-massive back holes might help understand the huge luminosities seen in the center of some galaxies (often so bright that the galaxy is nearly invisible from a large distance). In the 1990s only did it become clear that all massive galaxies contain at their center a super-massive black hole, with a direct correlation between the mass of the spheroidal distribution of stars, and the mass of the central black hole. The mass distribution of these massive black holes starts essentially at a few million Solar masses, thus, far above stellar masses. What is the origin and evolution of these super-massive black holes?
- Dark matter: Dark matter implies that there is invisible matter which exerts gravitational attraction, holding a cluster of stars or galaxies (Zwicky, 1933) together, but we cannot identify this matter. The term was coined by Oort (1932) to explain the large perpendicular velocity dispersion of stars in the Galactic disk: these large motions have to be contained, otherwise the stellar system would fly apart. He argued correctly that the unseen component was due to stars. Zwicky used the same kind of reasoning to ask what is required of clusters of galaxies to hold them together. In this case we do need dark matter, which cannot be stars or even gas; it has to be of a form that can cluster, but does not participate in nuclear reactions in the early universe. What is dark matter?
- Ultra-high-energy cosmic-ray particles: From 1963 it became obvious (Linsley, 1963) that some particles of ultra-high energy exist that cannot be contained in the disk of our Galaxy by the interstellar magnetic fields. Their near isotropy as well as their energy implies that they originate far outside our own Galaxy, probably either from Gamma-ray Bursts (the formation of a stellar black hole probably), or from a feeding episode of a super-massive black hole, quite possibly after a merger of two host galaxies, and the subsequent merger of the two central massive black holes. These particles come in at an energy far above what we can attain in an accelerator such as the "Large Hadron Collider (LHC)", even when considering the center-of-mass frame of a collision with a particle at rest. Where do ultra-high-energy cosmic-ray particles come from, and what exactly are they?

The ever stronger mutual interaction between fundamental physics and astronomy can be safely predicted to grow.

## 4.2 Stars, Galaxies, Dark Matter, and Particle Acceleration

### *4.2.1 Background: Cosmology*

The universe started in a state of extremely high density and temperature and expanded almost explosively since then; this initial stage is called the big bang. This phase is very dense and very hot; emission from this phase was postulated in a famous paper by Alpher et al. (1948). The residual radiation from this background was detected by Penzias and Wilson (1965), and immediately properly explained by Dicke et al. (1965). – This radiation is called the microwave background (MWBG) or cosmic microwave background (CMB).

The background is slightly mottled by the weak gravitational disturbances due to dark matter at the epoch of last scattering, about redshift 1 100, corresponding to about 300 000 years after the big bang. The discovery of these spatial wiggles (Smoot, 2007; Mather, 2007) in the residual radiation (Penzias and Wilson, 1965) from the big bang seen in the sky allows us to measure the geometry of the universe to exquisite precision: The various contributing components add to unity, and so the universe is mathematically flat. However, the three main components, which we know about, are largely not understood, with about 72 percent dark energy, 23 percent dark matter, and 5 percent normal matter (WMAP seven year results, Komatsu et al., 2011):

(a) Dark energy drives the universe in ever faster expansion, and is independent of redshift (Riess et al., 1998; Goldhaber and Perlmutter, 1998, for a review see Frieman et al., 2008).
(b) Dark matter dominates the gravitational potential of galaxies, clusters of galaxies, and the soap-bubble-like filigree network of large scale structure.
(c) Normal matter participates in nuclear reactions, so is made up of hydrogen, helium, carbon, oxygen, iron and the like (see Burbidge et al., 1957). All of the hydrogen, most of the helium, and none of the heavy elements were made in the first few minutes of the big bang (e.g. Reeves, 1994). We now know (e.g. Woosley et al., 2002) that all the heavy elements are produced in nuclear reactions in massive stars; massive stars have powerful winds; and when such stars explode, most of their original mass has already been ejected and constitutes a wind with a shell from interacting with the surrounding interstellar medium: These stars blow up into their winds and expel much of their heavy element production. Considering then the gas and stars we observe in galaxies and groups and clusters of galaxies, one problem with the normal baryonic matter is that we have trouble finding it, even though we know from the helium abundance and several rare isotopes what the normal matter fraction is. Probably the best hypothesis is that this missing normal matter is in hot and warm gas (Ostriker et al., 2005), which is very difficult to detect. Obviously this missing baryonic mass could also be in cool extremely tenuous clouds. The first stars and galaxies were made from the baryonic soup, just hydrogen, helium and a trace of light elements like lithium. We do not yet know how early the first stars and the

first galaxies were made; suggestions run from about 40 million years to a few hundred million years. But we can expect the gamut of stellar phenomena very quickly after the first stars polluted the environment, making cooling instabilities and so the formation of a second generation of stars easier (e.g. Rollinde et al., 2009). Within just a few million years after the first massive stars one can expect heavy element enrichment, dust, magnetic fields, and energetic particles, also called, for historical reasons "cosmic rays".

The best published analyses of the spatial wiggles of the residual microwave-background radiation is based on the data from the WMAP satellite; these analyses are sequentially numbered: The WMAP7 analysis of the MWBG fluctuations (Komatsu et al., 2011) gives the latest determinations and error bars. $\Omega$ is the fraction of critical density, given by $\rho_{\rm crit} = 3H_0^2/(8\pi G_{\rm N})$, where $H_0$ is the Hubble expansion parameter, and $G_{\rm N}$ is Newton's constant of gravitation. If the universe were at the critical density it would expand parabolically, so expand forever, but with an ever slower expansion speed, just like a stone thrown out from Earth that would reach infinity with zero speed (which, obviously, would take forever), a mathematical limit.

For dark energy

$$\Omega_\Lambda = 0.725 \pm 0.016. \tag{4.1}$$

For dark matter

$$\Omega_{\rm dm} = 0.229 \pm 0.015. \tag{4.2}$$

For baryonic matter (stars, planets, gas)

$$\Omega_{\rm b} = 0.0458 \pm 0.0016. \tag{4.3}$$

Adding these three components yields

$$\Omega_{\rm k} = 1 - \Omega_\Lambda - \Omega_{\rm dm} - \Omega_{\rm b} = -0.0125^{+0.0064}_{-0.0067}. \tag{4.4}$$

As this quantity, $\Omega_{\rm k}$, approaches zero it measures whether the sum of the three angles in a triangle is 180 degrees – of course provided the path does not pass too closely to a black hole. It is useful to work out the angular mismatch due to black holes: Using a typical distance between black holes today of about 6 Mpc, and a typical mass of a black hole of, say, $10^7\ {\rm M}_\odot$ gives then a minimal angle of deviation of a light path of $\sim 3 \times 10^{-13}$ rad, which is 0.06 μarc-seconds. Of course in a random walk across the universe, this will increase to by a factor of order 30 and more, given the density of black holes, so it would become of order 2 μarc-seconds and more. One of the most precise astrometry measurements was in the μarc-seconds range (Marcaide et al., 1984). Therefore, the astrometry for high redshift objects ought to be limited to such a scale. In mathematical language the geometry is "flat", flat like a tabletop, but with this precision.

The expansion scale of the universe is the Hubble parameter,

$$H_0 = 70.2 \pm 1.4\ {\rm km/s/Mpc}. \tag{4.5}$$

For moderate distances the speed of a galaxy scales as $H_0$ times the distance; at very small distances peculiar velocities enter, at the scale of about 500 km/s.

This then enters into the distance scale and time integral (e.g. Frieman et al., 2008):

$$r(z) = \int_0^z \frac{c\,\mathrm{d}z'}{H(z')}, \tag{4.6}$$

which gives the various distance scales such as luminosity distance $d_L = (1+z)r(z)$, angular diameter distance $d_\mathrm{A} = r(z)/(1+z)$, and proper motion distance $d_M = r(z)$. These distances are required to determine the luminosity of a source from its apparent brightness, the size of a source from its angular diameter, and the motion of a source from its angular speed. If we were to recognize that some object could be thought of as a "standard candle", then we could obtain the luminosity distance, and so verify the cosmic distance scale. Supernova explosions of type Ia offer this possibility, based on an empirical law of their luminosity as a function of other observed properties, like color evolution, and time scales.

The time is

$$t(z) = \int_0^z \frac{\mathrm{d}z'}{(1+z')H(z')}. \tag{4.7}$$

Here and above the cosmic expansion rate $H(z)$ given by

$$H(z) = H_0\big(\{\Omega_\mathrm{dm} + \Omega_\mathrm{b}\}(1+z)^3 + \Omega_\Lambda\big)^{1/2}, \tag{4.8}$$

ignoring the radiation and neutrino component. We also ignore black holes, since they are probably made from baryonic matter (baryonic matter is matter that participates in nuclear reactions at sufficiently high density and temperature, such as inside stars, or in the very early universe at age of a few minutes), or perhaps even dark matter (see, e.g. Zhang et al., 2008). The contribution of super-massive black holes today can be estimated to $\Omega_\mathrm{BH} = 2 \times 10^{-6\pm0.40}$ (Caramete and Biermann, 2010). Concerning mini black holes, the mass range between $10^{15}$ g and $10^{26}$ g is not constrained and could contribute to total mass balance, while higher and lower mass black holes are strongly constrained (Abramowicz et al., 2009).

The baryonic gas in the universe starts off as hot and fully ionized, then recombines at a redshift near 1 000, and is later re-ionized by bright stars, active black holes, or in some other way (e.g. Wise and Abel, 2008a, 2008b; Mirabel et al., 2011). Assuming instantaneous re-ionization (corresponding to about $3 \times 10^8$ yr, counting from the big bang), we have

$$z_\text{re-ion} = 10.6 \pm 1.2. \tag{4.9}$$

Of course the re-ionization is very likely a very prolonged affair, with some arguments suggesting that it might start as early as redshift 80 (Biermann and Kusenko, 2006), then being initiated by massive stars. How the first stars may have formed has been reviewed by Ripamonti and Abel (2005).

The age of the universe is

$$t_0 = (13.76 \pm 0.11) \times 10^9 \text{ yr}. \tag{4.10}$$

In a fair approximation for $z \gg 0.4$ one can write the time since the big bang to redshift $z$: $t_0(1+z)^{-3/2}$.

The tally is that out there, say, in clusters of galaxies, where we believe we "see" everything, most of the gravitational mass is dark matter; the second most important component is hot gas, and only the third component is stars. A problem with these ingredients is that we do not know what dark energy and dark matter are, and even for the baryonic component we are confident only of order ten percent that we have detected in stars and visible gas. So we know only of $\sim 0.1\Omega_b \simeq 0.005$ where it is and what form it takes; of 0.995 we do not know this.

The conclusions discussed above derive from the measurements of the microwave-background fluctuations and spectrum, from large scale structure observations, from abundances of certain isotopes and elements, and – of course – from Supernova type Ia observations. But there are also other potential avenues to obtain information about the universe: There appears a possibility to measure directly the early expansion of the universe using frequency combs (Steinmetz et al., 2008). The precision using frequency combs is also approaching such a level that for instance in the not too distant future the cosmological change of some natural constants could be observed within a few weeks of time. Using atomic clocks one can also use time measurements to test relativity, geodesy and cosmology (Diddams et al., 2004; Rosenband et al., 2008; Chou et al., 2010).

## *4.2.2 Stars and Their Explosions*

It is only since relatively recently that quantum mechanics allowed us to interpret the observed absorption and emission line spectra of stars, their structure, their internal nuclear reactions, and their internal energy transport by radiation or convection. What we have learnt is that the lifetime of stars is a very strong function of their initial mass, with about two million years reached asymptotically for high masses, and a lifetime longer than the age of the universe at masses of about one Solar mass (depicted with a very ancient symbol as $\odot$, which looks similar in Chinese even).

Massive stars explode, and pollute their environment with heavy elements, dust, magnetic fields and cosmic rays. Lower mass stars live for times longer than the age of the universe. Stars in binary systems can gain and lose mass via mass exchange, and so show signs of extreme activity; the end point of moderate to low mass stars is a very compact star, called a "white dwarf", in which the electrons sit in a sort of lattice, commonly referred to as a "degenerate quantum state". Yet more compact stars are neutron stars, also a quantum lattice, this time for neutrons. Both these kinds of star have an upper limit as regards stability, and if adding too much mass for accretion they blow up. This limit is called the Chandrasekhar mass for white dwarfs (see below).

Cosmic rays are energetic particles that together make up as much pressure in the interstellar medium as the magnetic fields, and both components together again have an energy density almost as high as the thermal gas – which represents a famous stability limit due to Parker (1966). Exploding stars were the prime candidates to produce the cosmic rays as argued already by Baade and Zwicky (1934).

The cosmic-ray particles have a near power-law energy spectrum, with a highest energy of about $3 \times 10^{18}$ eV for heavy nuclei accelerated by the shocks of exploding stars. These highest energies can be reached in the case that a very massive star explodes into its own magnetic wind. An interesting alternative is the enhancement of the magnetic field strength by instabilities caused by the cosmic-ray particles themselves (Weibel, 1959; Bell and Lucek, 2001; Bell, 2004), which can also allow larger particle energies for a given space in the acceleration region. In this instability the magnetic field can approach in energy density a modest fraction of the ram pressure, produced by the stellar explosion. Cosmic rays will be discussed in more detail below.

Magnetic fields are also readily produced in stars. Whenever a star forms, and obtains angular momentum from shear or tidal forces, so rotates as usually – unless the thermodynamics of the star become extremely odd and simple –, surfaces of constant pressure and constant density do not coincide and so drive an electric current, which in turn produces a magnetic field. This magnetic field then immediately has the length scale characteristics of that star.

A simplified version of how this works is as follows (Biermann, 1950; Biermann and Schlüter, 1951): One starts by writing down separately the equation of motion for ions (charge $Z$, mass $M_i$, density $n_i$, velocity $\mathbf{v_i}$) and electrons (mass $m_e$, density $n_e$ and velocity $\mathbf{v_e}$). Then the electric current is given by

$$\mathbf{j} = e(n_i Z\mathbf{v_i} - n_e\mathbf{v_e}), \qquad (4.11)$$

where $e$ is the elementary charge. Then one can work out the equation of motion for the current from the difference of the equation of motion for the ions and electrons. One can renormalize the equation to isolate the electric field $\mathbf{E}$, and then take the curl. This finally yields one important term:

$$\mathrm{curl}\left(\frac{m_i}{eZ\rho}\nabla P_e\right) \sim \nabla\left(\frac{m_i}{eZ\rho}\right) \times \nabla P_e \sim \nabla\rho \times \nabla P_e. \qquad (4.12)$$

This is zero exactly when the surfaces of constant density of pressure coincide. It is a cross-product of two linear vectors and so is a rotational vector; it cannot be balanced by a linear vector, such as an electric field. In a freshly forming rotating star, as in an accretion disk, where density and temperature are governed by very different microphysics, this term, the cross-product of the gradients of density and pressure, is certainly non-zero, and drives an electric current. This driving of an electric current can be compensated by an existing magnetic field (Mestel and Roxburgh, 1962), so this works strictly only when the prior magnetic field is extremely close to vanishing or zero. In a second step, turbulence, shear and rotation enhance this seed field by the dynamo process and so attain the relatively strong magnetic fields we find ubiquitously in stars (work done by Steenbeck and Krause, 1965, 1966, 1966, 1967, and later papers; and in parallel by Parker, 1969, 1970a, 1970b, 1970c, and later papers). This can take very many rotation periods. Stars in turn eject their magnetic fields in winds and explosions; this in turn leads to the injection of magnetic fields into the interstellar medium (e.g., Bisnovatyi-Kogan et al., 1973), which are quite strong, possibly at the level of a few percent of equipartition. In the Galaxy shear, convection, and a cosmic-ray driven wind enhance the magnetic field to its stability limit (Hanasz et al., 2004, 2009; Otmianowska-Mazur et al., 2009; Kulesza-Zydzik et al., 2010; Kulpa-Dybel et al., 2011). These galactic winds eject magnetic fields possibly as far as hundreds of kpc (Breitschwerdt et al., 1991; Everett et al., 2008, 2010). As disk galaxies can be thought of as accretion disks (e.g. Y.-P. Wang and Biermann, 1998), the interstellar magnetic fields get accreted to the central super-massive black holes, which in turn during their activity episodes can eject powerful magnetic fields very far via their relativistic jets, typically 100 kpc, but occasionally up to 3 Mpc. In this way it is possible to distribute magnetic fields throughout the cosmos (for simulations see Ryu et al., 1998, 2008; Dolag et al., 2009; Vazza et al., 2011; for observations see Kronberg, 1999, 2005; Kronberg et al., 2008; Gopal-Krishna and Wiita, 2001; Ferrière, 2010; Oppermann et al., 2011). A review focusing on observations and theory alike is Beck et al. (1996); a recent review on cosmic magnetic fields is in Kulsrud and Zweibel (2008).

So exploding stars produce the bubbles of hot gas with lots of magnetic fields and energetic particles, shining at many electromagnetic wavelengths like radio, X-rays and gamma rays, as observed with the Cherenkov telescopes H.E.S.S., MAGIC and VERITAS today, as well as with the Fermi satellite (e.g. H.E.S.S.-Coll., 2011). These explosions can run through the immediate interstellar medium, or run through an extended stellar wind for the more massive stars.

Lower mass stars normally do not explode, except in rare cases, when their old-age stage of a white dwarf (Shapiro and Teukolsky, 1983) is pushed over an instability limit (Chandrasekhar, 1931; Landau, 1932) and creates what is called a supernova Ia. Amazingly it turns out that this class of supernova can be calibrated to provide a standard candle, so its distance can be measured reliably. This has convinced us that the universe expands with acceleration, thus providing evidence for dark energy (work by Riess et al., 1998, as well as Goldhaber and Perlmutter, 1998, and later papers; for a review see Frieman et al., 2008). Reviews of such supernovae are Branch (1998) and Hillebrandt and Niemeyer (2000).

However, we do not know what pushes these stars into an explosion (see, e.g. Mazzali et al., 2007). The common idea that accretion pushes them over the stability limit for white dwarfs, called the Chandrasekhar limit, has encountered observational problems (Gilfanov and Bogdan, 2010). The fact that stellar explosions, which we do not really understand, serve as stepping stones in an argument for the existence of the prime cosmological ingredient, dark energy, is disconcerting. On the other hand, the precision of the measurements of the microwave-background fluctuations, coupled with Large Scale Structure data, is so excellent nowadays that we now use the supernova data as a consistency check, and no longer completely depend on them.

Low mass stars in binary systems may include white dwarfs and neutron stars (Shapiro and Teukolsky, 1983); if such compact stars accrete, they display many of the same phenomena as also shown by accreting black holes, and they can be much easier to study (e.g. Ikhsanov, 2007). If the orbital periods are short and well determined, we can even identify planets in the data (e.g., DP Leonis, Biermann et al., 1985; Qian et al., 2010; Beuermann et al., 2011).

High mass stars explode, and make in rare cases a Gamma-ray Burst, probably giving birth to a black hole (Shapiro and Teukolsky, 1983; see, e.g., Woosley et al., 2002). These Gamma-ray Bursts constitute a narrow highly relativistic gas stream (Vietri, 1995; Waxman, 1995; e.g. Cenko et al., 2010; Mészáros and Rees, 2010, 2011); this means that the flow runs at a speed $\beta c$ very close to the speed of light, in fact so close that the difference to the speed of light is probably initially only about 1 in 100 000. We describe a velocity $\beta c$ very close to the speed of light by its Lorentz factor, given by $\gamma = (1 - \beta^2)^{-1/2}$. The difference with the speed of light can then be approximated by $1/(2\gamma^2)$. When such a relativistic jet flow happens to point at us, their emission is extremely enhanced, so very bright, and so such explosions become observable all across the universe – and have been seen from an epoch of the universe closer to the origin than one billion years (e.g., Cucchiara et al., 2011 identify a GRB at redshift 9.4). A beautiful summary of Gamma-ray Burst physics has been given by Piran (2004), and a recent book has been published by Mészáros (2010).

For high mass stars one idea is that the collapse of a rotating star stops at the angular momentum limit, when centrifugal forces balance the gravitational pull inwards, then ejects the rotational energy via torsional Alfvén waves, and finally collapses (Bisnovatyi-Kogan, 1970; Ardeljan et al., 1996; Moiseenko et al., 2006; Bisnovatyi-Kogan and Moiseenko, 2008; an early concept is in Kardashev, 1964). This picture has the advantage that it uses the observed magnetic fields in massive stars, and easily relates to the argument that such stars can also explode as a Gamma-ray Burst (Langer et al., 2010), given a substantial further mass loss in a binary system before the explosion. Another idea is that the ubiquitous neutrinos cause the explosion, with the very good reasoning that most of the potential energy is in fact transformed into neutrinos. However, despite considerable work by many (e.g. Bethe, 1990; Woosley and Weaver, 1986; Langer et al., 2010; Dessart et al., 2011) this has not yet yielded the final proof what the explosion mechanism is.

Most of high mass stars live in binary star systems, and when one of the stars explodes and makes a black hole, then tidal pull can attract gas from the other star onto the black hole. This accreting gas can form an accretion disk, given enough angular momentum (Lüst, 1952; Prendergast and Burbidge, 1968; Shakura and Sunyaev, 1973; Novikov and Thorne, 1973; Lynden-Bell and Pringle, 1974). This accretion disk lets the gas slowly spiral down into the black hole, getting very hot and emitting a lot of light. Up to 30 percent of the accretion energy can be dissipated into heat and emitted as light (Shakura and Sunyaev, 1973; Bardeen, 1970, e.g. Sun and Malkan, 1989); for stellar mass black holes this light is in the form of X-rays. In some cases, and we do not yet understand exactly how this happens, the accreting matter is funneled into a fast jet along the rotation axis of the black hole (Blandford and Znajek, 1977; Lovelace et al., 2002; e.g. Keppens et al., 2008). The flow along the jet usually is relativistic, with Lorentz factors possibly reaching up to of order 100, in Gamma-ray Bursts many hundreds quite possibly. Accretion and relativistic jet formation also occur for less compact objects like neutron stars, allowing us to study these phenomena in our own Galaxy. Active stars with such relativistic jets are called "micro-quasars" (Mirabel and Rodriguez, 1998, 1999, 2003; Mirabel et al., 2001, 2011; Mazzali et al., 2008).

Stars and their various activities present very many paradigmata that we can use for other phenomena as a stepping stone in our understanding, but we would better be sure about the steps (e.g. Plotkin et al., 2012). They teach us about the origin of the heavy elements, the acceleration of particles, the origin of magnetic fields, the birth of black holes, the phenomena of compact accretion disks and relativistic jets, through their motions about dark matter, and through their explosions about dark energy. And yet we still do not understand key aspects of stars, like exactly why they explode.

## *4.2.3 Galaxies and Black Holes*

### 4.2.3.1 Galaxies

Galaxies are visibly huge collections of stars, but in fact are gravitationally dominated by dark matter, which we of course do not see. In a very crude approximation many can be described as isothermal oblate spheres of older stars, but some are also tri-axial configurations; many galaxies have thin disks of a stellar population with gas (including magnetic fields and cosmic rays), like our Milky Way. A typical combination is a combination of a disk of old and young stars, with gas, and a broader spheroidal distribution of old stars, and all embedded in a near-spherical distribution of invisible dark matter. In our own Milky Way we observe the gas in the constellation Orion in the "sword" with our own eyes.

Galaxies come in many masses and sizes, and it appears from the data that there is a smallest mass, which is of order $5 \times 10^7$ $M_\odot$ (e.g. Gilmore et al., 2007); such galaxies are called dwarf ellipticals, and many think that they represent the original galaxies, the first ever to form. In this picture other galaxies grow out of merging small galaxies. There appears to be a clear separation in properties between star clusters and galaxies; star clusters never have any measurable dark matter, small galaxies are dominated by dark matter (Gilmore et al., 2007). Star clusters are also usually smaller. If we accept this picture, the scale of these small galaxies also represents the smallest scale in large scale structure, which is an indication of the dark matter particle and its property (here momentum) at the time when it was first produced in processes which we do not know. Galaxies grow by merging, and then initiate a starburst, a phase of very high temporary star formation (Sanders and Mirabel, 1996). In such a merger the central one or two black holes also get fed.

Galaxies come in many sizes, and have disk-like structures like spiral galaxies, and ellipsoidal structures like elliptical galaxies (Disney et al., 2008). Most galaxies are mixed, have a disk and an ellipsoidal distribution of old stars, usually called "bulges". The largest galaxies merge into groups of galaxies, and often dominate a region in a cluster of galaxies. Along the Hubble-sequence the so-called "early-type galaxies" have big bulges, and "late-type" galaxies have almost no bulges, only disks. Galaxies cluster easily, and usually occur in groups or clusters of galaxies; these groups and clusters form a soap-bubble-like large scale structure, with huge voids.

There are now observations and source counts for the early population of galaxies, such as Lagache et al. (2003, 2005), and Dole et al. (2006). The claimed detection of a universal radio background (Kogut et al., 2011) might be explainable by a very early activity of galaxies. The Planck satellite measurements will give strong further constraints (e.g. Planck-Coll., 2011).

There are now many detailed cosmological simulations, how galaxies, their first stars and their central black holes may form and evolve (e.g. Kim et al., 2011; Abel et al., 2009; Wise and Abel, 2008a, 2008b).

#### 4.2.3.2 Black Holes

The possibility of black holes was shown in a famous paper by Schwarzschild (1916) based on Einstein's General Theory of Relativity. The convincing proof of the existence of at least some black holes was only demonstrated by showing that a measurement of a certain mass within a very small radius did not allow any other known physical solution (e.g. Maoz, 1998), which was capable of surviving a large fraction of the age of the universe. Considering a dense star cluster allows the estimate the time scale of exchanging momentum between stars, which leads to a growing density contrast between the inner and outer zone, and finally, after reaching a critical density contrast, to a "gravothermal catastrophe", which then may lead to a massive black hole.

Upwards of a certain mass galaxies all seem to have a central super-massive black hole; the mass of this black hole is tightly correlated with the mass in the old stellar population in the bulge (Kormendy and Richstone, 1995; Faber et al., 1997; Häring and Rix, 2004). Since black holes grow, either by merging with other black holes, or by accretion – observable as very bright active galactic nuclei – the growth process of the black holes should be connected to the development of the old stellar population. This tight correlation can be understood in the frame-work of a galaxy evolving, and merging, with infall from the same gas, that supports star formation, accreting towards the central black hole and feeding it (Wang and Biermann, 1998; Silk and Rees, 1998). Just like galaxies seem to have a smallest mass, these central super-massive black holes seem to have a lowest mass in their mass distribution (Greene et al., 2006, 2008; Greene and Ho, 2007a, 2007b; Caramete and Biermann, 2010). To within our understanding of the errors, this mass coincides with the mass of our own central black hole, in our Galaxy, at $3 \times 10^6\ M_\odot$ (e.g. Schödel et al., 2009; Genzel et al., 2010).

There are some well established ideas about how these first black holes have have formed, and a typical concept is the formation out of a very massive star, which explodes and makes a super-massive black hole. This super-massive star in turn may form out of agglomerating massive stars (Sanders, 1970; Portegies Zwart et al., 2004). However, since massive stars have powerful winds, driven by radiative interaction with heavy element line transitions in the wind, the ensuing mass loss is thought to prevent this formation of a very massive stars (Yungelson et al., 2008), with the exception of the very early stage in the universe, when the heavy element abundance was still close to zero. Very massive stars live just a few million years; they suffer from an instability as their mass approaches about $10^6$ $M_\odot$, and so can blow up (Appenzeller and Fricke, 1972a, 1972b). Fall-back of additional material may increase the mass of the resulting black hole beyond that of the exploding star, thus possibly explaining the lower mass limit of the distribution of super-massive black holes. So in this way the first generation of super-massive black holes may have formed. Again we are not yet sure whether this is the key process, although it seems plausible.

The original number density of black holes can be derived (Caramete and Biermann, 2010) in two approximations: If their growth is all by baryonic accretion, then their number is conserved and their mass is above $3 \times 10^6$ $M_\odot$,

$$6 \times 10^{-3\pm0.40}\ \text{Mpc}^{-3}. \tag{4.13}$$

If they grow mostly by mergers, then their mass is conserved, and the original number is

$$8 \times 10^{-2\pm0.40}\left(\frac{M_{\text{BH,min}}}{3 \times 10^6\ \text{M}_\odot}\right)^{-1}\ \text{Mpc}^{-3}. \tag{4.14}$$

Black holes can grow by accretion and by merging. If they grow by accretion the accretion disk and the relativistic jet almost always seem to be there, in various strengths, and provide enormous power to the environment, and great visibility, all across the universe. The highest luminosities we observe currently reach about $10^{13}$ times the luminosity of our Sun. If they grow by merging, then to current observations things are less visible, because we cannot yet detect gravitational waves. The mass distribution of these super-massive black holes can be well interpreted by the merging process, using a classical merger scenario (e.g. Smoluchowski, 1916; Silk and Takahashi, 1979; Berti and Volonteri, 2008; Gergely and Biermann, 2009). But we do not know for sure which of these processes is the more important. Observations show that black hole growth predates final galaxy assembly (Woo et al., 2010), but this does not unequivocally decide which growth mechanism is preferred.

The creation of stellar black holes from merging binary stars, perhaps from white dwarfs or neutron stars, may provide an opportunity to be directly observed by gravitational wave interferometers (e.g. Hild et al., 2011; Zhang et al., 2008; Abadie et al., 2010).

Merging super-massive black holes produce ubiquitous gravitational waves, which may also be observable. If black holes start at a few million solar masses, then merge, and do this at some high redshift, then for each specific merger the integral gravitational wave spectrum rises steeply to a maximum frequency given by the merger itself (e.g. Wyithe and Loeb, 2003). The best current limits are given by pulsar timing (e.g. Kramer, 2010; Liu et al., 2011). There is reasonable hope that within a few years this observational method may detect the gravitational wave background produced when black holes merged probably prodigiously in the early universe. Due to the low mass cut in the black hole mass distribution there ought to be a rather significant spike in the spectrum of gravitational waves, probably somewhere near the mass of the Galactic Center black hole of about $3 \times 10^6$ $M_\odot$, probably also corresponding to the mass of the first generation of massive black holes, modified by the redshift factor $(1 + z)$, which could be large (giving a lower frequency).

#### 4.2.3.3 Active Black Holes

Almost all black holes studied with sufficient sensitivity show signs of activity, often in the form of a gaseous jet flow, with relativistic velocities. This jet flow is detectable via the emission of energetic electrons with a power-law spectrum, giving what is called non-thermal emission.

Many of the phenomena of relativistic jet flow can be seen already in highly supersonic flow, as studied by E. Mach in the 19th century. He already determined the geometry of the basic shockwave patterns.

The production of a relativistic jet by a super-massive black hole leads to some of the most spectacular sights in the universe. Since the emission from the jet is boosted by the relativistic motion, the emission makes these jets (and by inference their hosting black holes) so eminently detectable that of a random selection of radio sources at 5 GHz (or 6 cm wavelength) half of all sources show a jet pointed at us (Gregorini et al., 1984). Looking at these jets in gamma rays, the percentage goes up to nearly 100. The activity of such relativistic jets is almost certainly the origin of the highest energy particles observed to date. Copious neutrino emission is expected to accompany this (e.g. Becker, 2008).

For sources that are selected to show a relativistic jet pointing at Earth, there is a spectral sequence commonly referred to as the blazar sequence, which shows a double-bump spectrum, for which many observations and tests exist (Fossati et al., 1998; Ghisellini et al., 1998, 2009a, 2009b, 2009c, 2010; Ghisellini, 2004; Guetta et al., 2004; Celotti et al., 2007; Celotti and Ghisellini, 2008; Ghisellini and Tavecchio, 2008, 2009; Cavaliere and D'Elia, 2002). One common interpretation is that we observe at radio to X-ray frequencies synchrotron emission from relativistic electrons, and at TeV energies the corresponding inverse Compton emission (Kellermann and Pauliny-Toth, 1969; Maraschi et al., 2008). An alternative interpretation is (see, e.g., Biermann et al., 2011) that the lower frequency peak is indeed primarily synchrotron emission from relativistic electrons. But following some speculations in Biermann and Strittmatter (1987), the higher frequency peak is synchrotron emission from protons (or nuclei). Obviously, under such a hypothesis many secondary emissions will also exist. On the other hand, under this hypothesis, the frequency ratio between the primary emissions is fixed to $(m_p/m_e)^3$, the proton to electron mass ratio to the third power; the frequencies both scale with the black mole mass as the inverse square root $M_{BH}^{-1/2}$.

There are a variety of interesting effects that arise from emission from a relativistic jet. First of all, and perhaps most famous, is the effect of apparent superluminal motion: When an emitting region moves at a speed very close to the speed of light, the energy of any particle moving along with this emitting region is enhanced in the observer by a factor called the Lorentz factor, explained above. When this motion is very close to the line of sight to the observer, then the apparent motion can reach $\Gamma c$. Second, the variability seen can be shortened by up to $1/(2\Gamma^2)$. And third, the emission is boosted into a narrow cone of angular width $1/\Gamma$, while each photon is increased in its energy by $\Gamma$, so as a consequence the visible emission is enhanced by $\Gamma^3$. This leads to enormous selection effects in source surveys, as noted above: Of all the isotropically distributed sources, only those within an angle of about $1/\Gamma$ to the line of sight to the observer have their apparent emission enhanced by at least $\Gamma^3$ (or more depending on spectrum, and on whether the jet is intermittent or steady); this means that we can detect such sources to much larger distances, and so for a given observational sensitivity the number of sources visible to us is very much greater if we see this boosted emission.

#### 4.2.3.4 Black Hole Entropy

Finally, since the early 1970s arguments by Bardeen, Bekenstein and Hawking have shown that black holes have entropy. For a non-rotating black hole the entropy is

$$\frac{S_{\mathrm{BH}}}{k_{\mathrm{B}}} = \frac{1}{4}\frac{A}{l_{\mathrm{Planck}}^2}, \tag{4.15}$$

where $A$ is the area of a black hole:

$$A = 16\pi \left(\frac{G_{\mathrm{N}} M_{\mathrm{BH}}}{c^2}\right)^2, \tag{4.16}$$

and

$$l_{\mathrm{Planck}}^2 = \frac{G_{\mathrm{N}} \hbar}{c^3}, \tag{4.17}$$

which boils down to

$$\frac{S_{\mathrm{BH}}}{k_{\mathrm{B}}} = 1.15 \times 10^{91} \left(\frac{M_{\mathrm{BH}}}{10^7\,\mathrm{M}_{\odot}}\right)^2. \tag{4.18}$$

For maximal spin the entropy is lower by a factor of 2 (see also Bardeen, 1970; Bekenstein, 1973, 2004; Bardeen et al., 1973; Hawking, 1971, 1973, 1976; Wald, 1997).

The characteristic energy of a graviton is of order

$$h\nu = \frac{ch}{2\pi r_{\mathrm{g}}}. \tag{4.19}$$

There is another equivalence, in that

$$\frac{M_{\mathrm{BH}} c^2}{h\nu} = \left(\frac{G_{\mathrm{N}} M_{\mathrm{BH}}}{c^2}\right)^2 \frac{c^3}{G_{\mathrm{N}} \hbar} = \frac{S_{\mathrm{BH}}}{k_{\mathrm{B}}}, \tag{4.20}$$

which says that to within a factor of order unity the number of gravitons possibly produced with a black hole is equal to its entropy in units of $k_{\mathrm{B}}$.

Let us consider the 6D phase space in making a black hole.

The characteristic length scale is

$$r_g = \frac{G_N M_{BH}}{c^2}, \tag{4.21}$$

and so we obtain

$$r_g^3 \left( \frac{h\nu}{c} \right)^3 = \hbar^3. \tag{4.22}$$

Any thermodynamic system fills a certain volume in 6D phase space, by multiplying the volume in 3D real space with the volume in 3D momentum space; as a first example one can take the Sun, with a radius of about 700 000 km, and made up mostly of hydrogen atoms, with a typical velocity of about 300 km/s, giving a 6D phase-space volume of about $10^{-16}$ erg$^3$ s$^3$. As a second example we take a star cluster of a million Suns, which fills a certain volume in real space, say, a pc cube; and in momentum phase space the typical momentum of a star like the Sun, taken in all directions because there are many stars, so that we have a momentum cube. The product of the two separate volumes is the 6D phase-space volume, therefore we now have $10^{176}$ erg$^3$ s$^3$. For black holes this result implies that making a black hole results in $\hbar^3$ in 6D phase space, pure and simple, to within a factor of order unity, independent of the BH mass. And this enormous number of states – the entropy divided by $k_B$ – are in that 6D Planck volume.

Such a line of reasoning also leads to the idea that black holes can evaporate, giving "Hawking radiation" (Hawking, 1975; Gibbons and Hawking, 1977; Parikh and Wilczek, 2000). Only very small black holes, of mass below $10^{15}$ g, evaporate within the lifetime of the universe. The expected evaporation signatures lead to a very strong limit concerning a possible contribution of $\sim 10^{15}$ g primordial black holes to dark matter (a fraction of $< 7 \times 10^{-9}$: Barrau et al., 2003). The range above $10^{15}$ g and up to $10^{26}$ g is fully unexplored so far (Abramowicz et al., 2009). At higher masses, the contribution to dark matter is constrained to less than $\sim 10$ percent from microlensing effects.

We are still lacking a deeper understanding of what "entropy" really means in this context. Bekenstein noted (1973) that the entropy of a black hole must contribute to the entropy of the entire universe. We still do not grasp the significance of such a statement.

### *4.2.4 Dark Matter*

Dark matter – as confirmed today – was discovered by observing that galaxies in clusters of galaxies moved much faster than their combined masses would allow (work by Zwicky, 1933, 1937). Originally it just meant mass which acts gravitationally (Oort, 1932), so exerts a gravitational pull on its surroundings, but which cannot be detected in any other way (e.g., Alcock et al., 1998). Today, we differentiate the unseen mass into baryonic matter that participates in nuclear reactions, so contributes to helium and some rare isotope formation in the early universe, and other kinds of mass, which have no relation to these nuclear reactions in the first few minutes of the universe. As noted above, most of the normal matter is not seen either. But dark matter, as defined today, is just that gravitating mass that is not participating in nuclear reactions; in its spatial and velocity fluctuations in the early universe it could limit the spatial scale of these fluctuations from below. Some think, as also noted already, that this scale corresponds to the smallest dwarf elliptical galaxies (Gilmore et al., 2007; Wyse and Gilmore, 2008; Donato et al., 2009). If we assume that this is true, then we can estimate the mass of these particles very simply by assuming that the population of these particles behaves thermodynamically just as any normal gas. So we have three constraints: the velocity dispersion in the halo of a galaxy gives us the typical speed of these unseen particles; they must go at this speed just to match the radial profile of the gravitational well. We have the density of these particles just from the volume and the mass to generate the potential well. These two numbers give us "density" and "pressure". And, finally, assuming that this "gas" of dark matter particles behaves just as a normal gas, we also find that the "pressure" has to scale as density to the $5/3$ power. This last relationship allows us then to obtain an estimate of the mass of these particles, and the result is in the keV range, so at the level of one percent of the mass of an electron. Obviously, this quasi-gas of dark matter might behave very differently from a normal gas, but simplicity suggests this line of reasoning is worth pursuing (work done by Hogan and Dalcanton, 2000; Dalcanton and Hogan, 2001; Boyanovsky et al., 2008a, 2008b; de Vega and Sanchez, 2010, 2011; de Vega et al., 2011).

The simple version of the argument runs as follows (Hogan and Dalcanton, 2000; Weinberg, 1972, Eq. (15.6.29)): Consider the phase-space density $f(p)$ of some particle of momentum $p$, which is produced in some unknown interaction and decay chain, and which will give today the dark matter particles of mass $m$:

$$f(p) \simeq \frac{1}{e^{(pc-\mu)/(k_B T_D)} \pm 1}, \tag{4.23}$$

where $\mu$ is the chemical potential, $T_D$ is the temperature at decoupling, while $\pm$ applies to fermions and bosons. We assume that we start in an equilibrium situation, which is initially highly relativistic. Expansion of the universe just lowers the momenta of the particles without changing their distribution function. Using the temperature of the distribution in its diluted form from decoupling to today, $T_0$, we then obtain simple integrals. In the following we will assume that the particles are today non-relativistic and that the chemical potential can be neglected.

Then we can work out density $n$ and pressure $P$, using the number of spin degrees of freedom $g$:

$$n \sim \frac{g}{(2\pi)^3 \hbar^3} \int f(p)\, \mathrm{d}^3 p, \tag{4.24}$$

$$P \sim \frac{g}{3m(2\pi)^3 \hbar^3} \int p^2 f(p)\, \mathrm{d}^3 p. \tag{4.25}$$

The "phase density" $Q = \rho/\{\langle v^2\rangle\}^{3/2}$ is a measure of the inverse specific entropy, and so is conserved except in shock waves or "violent relaxation" (Lynden-Bell, 1967); numerical experiments show that regions of enhanced entropy move to the outer parts of halos, while the lowest entropy material sinks to the center. One may conjecture that the cores of galaxies conserve a memory of the original "phase density". This is our key assumption here, and the data seem to confirm that cores of smaller galaxies all show similar properties (Gilmore et al., 2007; Strigari et al., 2008; Gentile et al., 2009; de Vega and Sanchez, 2011). So today's quasi-temperature $T_0$ can be used to rewrite the two expressions for density and pressure:

$$n \sim \frac{g(k_B T_0)^3}{(2\pi)^3 \hbar^3 c^3} \int f(x)\, d^3x, \tag{4.26}$$

$$P \sim \frac{g(k_B T_0)^5}{3m(2\pi)^3 \hbar^3 c^5} \int x^2 f(x)\, d^3x, \tag{4.27}$$

where $x$ is the dimensionless momentum, scaled with $k_B T_0/c$. Noting that $\langle v^2\rangle = 3P/(nm)$, we combine the two expressions eliminating $T_0$ and write

$$Q_X = q_X g_X m_X^4. \tag{4.28}$$

In the thermal case $q \simeq 2 \times 10^{-3}$, just given by the ratio of two simple integrals; in a highly degenerate case of fermions this coefficient can be much larger. Translating this into astronomical units for a thermal neutrino-like relic particle gives

$$Q_T = 5 \times 10^{-4} \left(\frac{\mathrm{M}_\odot}{\mathrm{pc}^3}\right) \left(\frac{\mathrm{km}}{\mathrm{s}}\right)^{-3} \left(\frac{m_X}{\mathrm{keV}}\right)^4. \tag{4.29}$$

The application is that we determine the density of dark matter in a core of a galaxy, and its velocity dispersion, calculate $Q$ using the same units as above and compare. For instance, 0.1 $\mathrm{M}_\odot\ \mathrm{pc}^{-2}$ over a core radius of 100 pc with a velocity dispersion of 15 km/s (Gilmore et al., 2007) yields a value for $Q$ lower than the reference value above, suggesting a dark matter particle mass of slightly less than 1 keV, based on our simple argument.

Going any further requires a specific model for this particle, and one model typically explored is that of a right-handed sterile neutrino, which can decay into a left-handed standard model neutrino and photons; these photons give a background, and for each galaxy a specific emission line at a photon energy at half the mass of the particle. Such models can be compared with data and then suggest that such a particle is actually sub-thermal (Kusenko, 2005, 2006, 2009; Biermann and Kusenko, 2006; Loewenstein and Kusenko, 2010), and so our equations need to be modified; however, the principal route remains, and for each specific model this can be worked out (see Boyanovsky et al., 2008a, 2008b; de Vega and Sanchez, 2010, 2011; de Vega et al., 2011). This suggests in the end a somewhat higher mass, a few keV. There are some speculations now on how such particles could naturally arise (Kusenko et al., 2010; and Bezrukov et al., 2010).

This kind of reasoning runs directly counter to other arguments, also based on simplicity, that derive from particle physics (for an observational argument derived from simulations, see, e.g., Font et al., 2011). Along this line of reasoning it is noted that in supersymmetry there ought to be a lightest particle, which cannot decay, and which would be a natural candidate to explain dark matter. Such a particle is very massive, of order 100 GeV or even more. The particle physics detector experiments based on this hypothesis have not yielded a definitive positive detection of anything yet, but have indicated numerous hints at seeing evidence (e.g. Bernabei et al., 2011; Frandsen et al., 2011; Savage et al., 2011), which has the correct behavior, like an oscillation with the yearly phase. This would be plausible, because in one part of the year the Earth moves into the dark matter particle population, and so increases the flux slightly, and in the opposite phase of the year it moves away from the population, and so the flux is expected to decrease. There is some evidence that the detections do have this pattern. But no consistent signal across all experiments has emerged yet. Obviously, this interpretation runs directly counter to the phase-space argument described above, which is based on detailed astronomical data.

Some others have proposed to modify the theory of gravity (Milgrom and Bekenstein, 1987), but this has not yet yielded any prediction that could be independently verified; on the other hand, this approach has also passed many tests of consistency (e.g. Sanders, 2008; Milgrom, 2009).

There is no direct and fully convincing proof of any of these hypotheses yet – or any other –, on what dark matter really is made of. A real detection of the dark matter particles would very much help to clinch the case.

### *4.2.5 Extragalactic and Galactic High Energy Particles*

Already in 1912 it was noted by V. Hess (Hess, 1912; Kolhörster, 1913) that the ionization of enclosed gas containers with two charged electrodes inside decreased rapidly with time, suggesting an outside ionizing invisible source. Going up in a balloon demonstrated that this effect increased with height above ground, and so the term "cosmic rays" was coined. Today we know that these cosmic rays are charged particles, and contain all the normal chemical elements, but with many isotopic variations and enhancements, as well as a component of electrons and positrons. Since 1934 (Baade and Zwicky) it has been thought that supernova explosions accelerate particles to cosmic-ray energies. The basic concept is acceleration in flow discontinuities, originally proposed by Fermi (1949, 1954), and then applied to shock waves by Axford et al. (1977), Krymskii (1977), Bell (1978a, 1978b), Blandford and Ostriker (1978), with a thorough review by Drury (1983). All the heavier elements in cosmic rays are strongly increased in abundance, and it is thought that this arises because massive stars expose their inner layers through their powerful winds, and so the subsequent explosion rips through these layers of dominant helium, or even carbon and oxygen, with significant variations in isotopes (work by Prantzos, 1984, 1991, and later papers), and so provide the particle injection for a shock accelerating cosmic rays. Textbooks are, e.g., Berezinsky et al. (1990), Gaisser (1991), Schlickeiser (2002), Aharonian (2004), Mészáros (2010), and Stanev (2010).

Cosmic-ray particles are charged and so they react strongly to magnetic fields; this was already used very early to estimate the strength of the then unmeasured magnetic fields in interstellar space (Biermann and Schlüter, 1951). The numbers determined then still hold (Beck et al., 1996). It was also realized that cosmic-ray particles can get reflected as they spiral along magnetic flux tubes that narrow as they approach a magnetic pole structure. Such reflections were then used in almost all cosmic-ray acceleration models from Fermi (1949, 1954) on.

Considering the explosion into the interstellar medium, most relevant for the lower mass range of normal and most common supernova explosions, the shock accelerates particles as described above, in a naive analogy to a tennis game. In a tennis game the tennis racket moves towards the other player, when it encounters the tennis ball, and considering the two sides the two tennis rackets constitute a compressing system. The two sides of a shock in an ionized gas with magnetic fields (called "plasma") also constitute a compressing system, since the helical motion of energetic particles in a magnetic field connects the particles to the plasma; irregularities in the magnetic field then change the direction of the particle's motion, and so the stronger the irregularities, the faster the particle forgets where it came from, and so has a high probability to move back through the shock. As the speed of the plasma towards the shock is much faster than the speed at which it moves away from the shock downstream, the difference is negative, and so effectively we do have compression. Every time a particle goes through it gets a kick, gains momentum and energy, and as long as it does not just escape, keeps gaining energy. Balancing energy gain with how many particles escape at each cycle gives in a simple limit for normal strong shock waves a particle spectrum of $E^{-2}$ (see Drury, 1983). The energy contained in this population is the integral of the spectrum times the energy; this integral diverges, and so the acceleration of the population naturally hits a limit, when the total energy in particles becomes comparable to the other energy densities in the shock system, like the magnetic field, or upstream flow energy (called the "ram pressure").

Following Drury (1983) the argument runs as follows: We consider a population of particles of mass $m$, velocity v, and momentum $p$, injected near the shock, that get scattered back and forth. Occasionally some particle gets lost from the system. This means that we have to follow the momentum gains and the probability that particles get lost.

The relative momentum gain per crossing is $\mu(U_1 - U_2)/\mathrm{v}$, with $\cos\theta = \mu$. Averaging then over the probability of crossing the shock, using a given surface area of the shock surface, gives a factor $\mu$. Scaling with integral of $\mu$ from 0 to 1 gives another factor of 2. This means that the weighting factor is $2\mu$. The average momentum gain is then $\frac{4}{3}p(U_1 - U_2)/\mathrm{v}$, going back and forth once each.

In the flow downstream particles escape net with $nU_2$. At shock, however, just downstream, the particles have an isotropic distribution in momentum phase space. Averaging over the isotropic distribution, we have the average of $\cos\theta n d(\cos\theta) = n\mathrm{v}/4$ with a factor $1/2$ downstream from isotropy and a factor $1/2$ from the average of $\cos\theta$, where $\theta$ is the angle between the direction downstream and the momentum vector of some particle with velocity v. The fraction of particles lost downstream therefore have ratio $4U_2/\mathrm{v}$, in a standard shock for relativistic particles equal to $U_1/c$.

After $n$ cycles the momentum is $p_n \sim p_0 \Pi(1 + \frac{4}{3}\{U_1 - U_2\}/\mathrm{v}_i)$, or in logarithmic form $\ln(p_n/p_0) \sim \frac{4}{3}(U_1 - U_2)\Sigma\frac{1}{\mathrm{v}_i}$. Here we denote with index $i$ the various particles. The probability to get there is $P_n \sim \Pi(1 - 4U_2/\mathrm{v}_i)$, or again in logarithmic form $\ln P_n \sim -4U_2\Sigma\frac{1}{\mathrm{v}_i}$.

Now we have to remember that losses might intervene and so the momentum gain may be reduced, say by a factor of $\alpha_{\mathrm{acc}}$, and the probability to get lost might also be different, by a factor of $\beta_{\mathrm{acc}}$.

Combining now the two expressions, and inserting the correction factors we get for the integral spectrum

$$P_n = (p_n/p_0)^{-(3U_2\beta_{\mathrm{acc}})/(\alpha_{\mathrm{acc}}(U_1-U_2))}. \tag{4.30}$$

Taking the derivative gives the spectral index

$$-\frac{3\beta_{\mathrm{acc}}U_2 + \alpha_{\mathrm{acc}}(U_1 - U_2)}{\alpha_{\mathrm{acc}}(U_1 - U_2)}. \tag{4.31}$$

For a strong shock in a normal gas of adiabatic gas index 5/3 we have $U_1/U_2 = 4$, and so for $\alpha_{\mathrm{acc}} = \beta_{\mathrm{acc}} = 1$ we obtain a spectral index of $-2$, corresponding in synchrotron emission by relativistic electrons to a flux density spectrum of $\nu^{-1/2}$. As an alternative one can also consider a strong non-relativistic shock in a relativistic gas of adiabatic gas index 4/3 with $U_1/U_2 = 7$; the spectrum then has $-3/2$, and the spectrum of synchrotron emission is $\nu^{-1/4}$. Now using smaller $\alpha_{\mathrm{acc}} < 1$, so less energy gain per cycle, and also a higher probability to get lost, so $\beta_{\mathrm{acc}} > 1$, yields a steeper spectrum; for example, for $\alpha_{\mathrm{acc}} = 1/2$ and $\beta_{\mathrm{acc}} = 2$, and $U_1 = 4U_2$ again we obtain a very steep spectrum of $-5$, corresponding to a synchrotron spectrum of $\nu^{-2}$.

The configuration normally considered is for the shock normal to be parallel to the magnetic field, for an isotropic random distribution of angles a very unlikely situation. Jokipii (1987) has introduced a concept to deal with highly oblique shocks by introducing a specific limiting scattering coefficient; this does not change the derivation of the spectrum, as sketched here, but it does modify the rate of acceleration, and the maximal particle energy attainable. Another aspect is that radio polarization observations (e.g. Dickel et al., 1991), and now also theory, show that the shock region is highly unstable (Bell, 2004; Caprioli et al., 2010; Bykov et al., 2011), and therefore quite turbulent, and perhaps extremely unsteady. This is very difficult to deal with, and has been done only in certain limiting situations. However, it has been used as a starting point to derive scaling relations for the unsteadiness and turbulent transport (Biermann, 1993), following Prandtl (1925).

Since charged particle also can constitute an electric current, it appears possible that the magnetic fields get enhanced (Weibel, 1959; Bell and Lucek, 2001), and not just the particle population. So both particle population and magnetic field may reach their limit of some fraction of order 0.1 of the kinetic energy of the flow going into the shock. Considering the spherical shock around an exploding star, the shock system also includes adiabatic losses, as the system expands, and then the spectrum is slightly steeper. When these particles encounter normal interstellar matter outside the expanding shock shell they interact, and produce other emissions; these emissions verify the spectrum. The material of the gas the expanding shock moves through then determines the abundances of the energetic particles being picked up. The mix of massive stars that explode into the interstellar medium (Lagage and Cesarsky, 1983), and the very massive stars that explode into their own wind (Völk and Biermann, 1988), with the chemical composition enhanced in heavy elements allows us to understand the abundances of cosmic rays (Biermann, 1993, and later papers; Binns et al., 2008). Gamma-ray Bursts exploding into Wolf–Rayet stars offer an almost isomorphic alternative (Wang et al., 2008). Since these winds are very massive they in turn form a thick shell in the interstellar medium around them, where most of the former mass of the star is waiting for the later supernova shock to hit. When this shock, loaded with energetic particles, does encounter the massive shell a lot of nuclei get broken up, or are "spallated" (Silberberg and Tsao, 1990), and this may explain the many secondary nuclei in cosmic rays. The wandering of these very fast particles through the shocked shell is dominated by magnetic irregularities given by the cosmic rays themselves. This allows a ready understanding of the spectra of these secondary nuclei.

The model for cosmic rays using a more detailed view of the stars that explode, by invoking the magnetic stellar wind, was developed quantitatively by Biermann et al. (Biermann, 1993; Biermann and Cassinelli, 1993; Biermann and Strom, 1993; Stanev et al., 1993), then developed further in Biermann et al. (1995, 2001), Wiebel-Sooth et al. (1998), and in Meli and Biermann (2006). The new tests began with accounting for the cosmic-ray positrons and electrons (Biermann et al., 2009), then for the high-frequency radio emission excess near the Galactic Center (called the WMAP excess) as well as the 511 keV annihilation emission also near the Galactic Center (Biermann et al., 2010a). This entails the model that used the galactic cosmic rays as seeds for ultra-high-energy cosmic rays (Gopal-Krishna et al., 2010), the spectral curvature predicted in 1993 confirmed by the CREAM mission (Biermann et al., 2010b). Finally, in 2012 we had the spectral test of the quantitative 1993 model in Biermann and de Souza (2012), now accounting for the entire cosmic-ray spectrum in a first simple test. This especially used the cosmic-ray data only available in 2011 from the Kaskade-Grande experiment, in the energy range $10^{15}$ eV to $10^{18}$ eV, and with the Kaskade array on the lower energy side, and the Auger array on the higher energy side covering $10^{14}$ eV through $10^{20.5}$ eV. The key for this development was:

(a) Massive stars explode into their winds.
(b) Cosmic rays interact with the matter in the shell snow-plowed together by that stellar wind.
(c) These galactic cosmic rays can be seeds for ultra-high-energy cosmic-ray acceleration by a relativistic jet, preserving basically the spectral shape and abundances. A relativistic shock as pushed through the region of enhanced star formation (Achterberg et al., 2001) can do this, as noted below.

There is also evidence about the origin of cosmic rays from observations of stellar absorption lines of elements which are produced mainly in spallation of cosmic-ray nuclei. Spallation is the destruction of nuclei in nucleus–nucleus collisions, giving rise to different elements and isotopes, usually secondary. This shows that the local environment was already enriched when the spallation happened, arguing for massive star origin (see, e.g. Ramaty et al., 2000, 2001).

However, we have as yet no fully convincing observational confirmation of all these plausible concepts. The pathway to detect such a confirmation could be the unequivocal discovery of neutral pion decay in the gamma-ray spectra of supernova remnants. Obviously charged pion decay gives rise to abundant neutrinos, not yet detected. Another possibility is to look for molecular transitions that require energetic protons for excitation, for instance the discovery of molecular hydrogen molecular transitions, of $H_2^+$ or $H_3^+$, which indicate the low end of the proton spectrum (see Nath and Biermann, 1994; Goto et al., 2008, 2011; Indriolo et al., 2010; Becker et al., 2011).

These cosmic-ray particles then meander through the interstellar medium in a diffusive process; this meandering (Rickett, 1977; Matthaeus and Zhou, 1989; Spangler and Gwinn, 1990; Goldstein et al., 1995) or diffusion is faster for more energetic particles, and so the spectrum becomes steeper for a normal observer (e.g. Garcia-Munoz et al., 1987). These cosmic rays together with the magnetic fields can be thought of as a very hot gas that tries to "bubble" out of the interstellar medium disk (Parker, 1966; Everett et al., 2008, 2010, Nath and Silk, 2009). This process then also drives a magnetic wind from the Galaxy, as noted above already, rather akin to the Solar wind, discovered (Biermann, 1951) by the wind's action on comet tails. These winds then distribute magnetic fields through the universe, in parallel to the action of the relativistic jets emanating from the central supermassive black holes (e.g. Ryu et al., 1998, 2008). The scale height at which we have a transition from simple diffusion of cosmic-ray particles to convection in the wind is of order kpc near the Sun (e.g. Beuermann et al., 1985), but it could be much smaller in those regions of a galactic disk where star formation and supernova activity is much stronger, perhaps as small as 0.1 kpc (Biermann et al., 2010a).

Considering the expansion of a supernova shock through a magnetic stellar wind, it can easily be shown that for heavy nuclei we reach energies close to $3 \times 10^{18}$ eV, just where the overall spectrum of cosmic rays shows a dip (Völk and Biermann, 1988). Also, at these energies the disk itself can no longer contain these particles, as the radius of the helical Larmor motion of a charged proton is of order 0.6 kpc, too close to the scale height itself to ensure isotropy from scattering even if the magnetic irregularities were maximized. However, the data do not show any anisotropy, suggesting that magnetic irregularities are strong enough to ensure a very near isotropy situation. The magnetic wind, with lower magnetic field, but larger length scale, also contributes to scattering of these particles.

Much of the acceleration, interaction and propagation of cosmic-ray particles can be tested with gamma-ray data (for a comprehensive textbook see Aharonian, 2004; for recent results see, e.g., Casanova et al., 2011) and CR and neutrino data (e.g. IceCube Coll., 2011a, 2011b; Halzen, 2010).

The transition between galactic and extragalactic cosmic-ray particles is a challenging question, and has been discussed by, e.g., Hillas (2006) and Berezinsky

(2009), who have proposed some solutions such as a pair-production dip under the assumption that the main particles are protons.

Since 1963 (Linsley, 1963) particles of much higher energy have been detected, up to $3 \times 10^{20}$ eV today, clearly not easily explained as originating from a source in our Galaxy. Already before Linsley's publication Ginzburg and Syrovatski (also 1963, 1964) suggested the famous radio galaxies Centaurus A (Cen A = NGC5128), Virgo A (Vir A = M87 = NGC4486), and Fornax A (For A = NGC1316) as the best candidates outside our own Galaxy for producing cosmic-ray particles in abundance. Further development of radio galaxies as sources of ultra-high-energy cosmic rays using observational cues was, e.g., in Lovelace (1976), Biermann and Strittmatter (1987), and Rachen et al. (1993). These radio galaxies could be excluded to explain the normal lower energy cosmic rays (readily understood via supernovae, as suggested first by Baade and Zwicky, 1934), but easily could explain the extra-galactic cosmic rays observed up to now. The most recent reviews are Kotera and Olinto (2011), and Letessier-Selvon and Stanev (2011); classical textbooks include Berezinsky et al. (1990) and Gaisser (1991), and the most recent textbook is by Stanev (2010).

The sites where ultra-high-energy cosmic-ray particles are accelerated are also sites where many interactions and secondary particle production can occur. Since even in the center-of-mass frame the energies are higher than at the Large Hadron Collider (LHC) at CERN in Geneva, at least in principle we could learn more about the nature of particles and their physics from these sources. However, the extreme flux arriving at Earth makes this very difficult. One of the first attempts to discuss various interactions that are likely to happen was made by Gould and Burbidge (1967). Others are described in the various textbooks and reviews mentioned.

The Auger experiment (Auger-Coll., 2011a, 2011b, 2011c, 2011d, 2011e) gives a lot of information on particle character, energy and direction. There is a strong hint suggesting that these particles at energies beyond $3 \times 10^{18}$ eV are usually heavy nuclei, and the composition seems to vary in this energy range. Their arrival directions seem to cluster around the radio galaxy Cen A. The data do not unambiguously show whether all particles come from Cen A or only a small fraction of all; the data allow both, depending on various models for the spectrum and the magnetic scattering.

After the discovery of the cosmic microwave background by Penzias and Wilson (1965) it was quickly realized that energetic particles interact with this cosmic background, and so suffer considerable losses. Assuming as a reference example that sources are homogeneously distributed in space, source spectra are $E^{-2}$ to very high energy, and particles propagate in straight lines outside galaxies. Then the observed spectrum turns down significantly at an energy now commonly referred to as the GZK-energy, after the authors who made its discovery (Greisen, 1966; Zatsepin and Kuz'min, 1966). For protons this energy is $6 \times 10^{19}$ eV; for nuclei this can be at lower energy (Allard et al., 2008). It follows that far below this energy the cosmos is effectively transparent and we should be able to "see" sources at very large distances. So far there is no evidence that this "sub-GZK population" of sources has been detected. All that we observe is compatible with some source in the radio galaxy Cen A that also produces heavy nuclei. There are a number of ways to explain these findings.

One option is that a very massive star explodes as a Gamma-ray Burst, and the relativistic shock plowing through the enriched wind of the former star picks up heavy nuclei, pushes them up in energy and then sends these particles out into space (Vietri, 1995; Waxman, 1995; Wang et al., 2008). Another option is that the black hole observed near the center of Cen A pushes through the region of the massive star formation inside the radio galaxy, triggered by the merger between an elliptical and a spiral galaxy some 50 million years ago (Gopal-Krishna et al., 2010). In this picture the relativistic shock of the fresh jet picks up the existing population of galactic cosmic rays, and pushes it up in energy (Achterberg et al., 2001). The second possibility is specific enough to allow for a test with the observed spectrum, and shows full consistency (Biermann and de Souza, 2012). However, the spectral fit does suggest that all these particles basically go straight between Cen A and us, and then get scattered into near isotropy in the magnetic wind of our Galaxy. This may be possible, but sets a serious constraint on our Galactic wind and its magnetic fields.

One can also use the concepts of relativistic jets to explore which sources out there could contribute even in principle. Using a complete sample of radio galaxies just confirms the 1963 arguments of Ginzburg and Syrovatski, with Cen A predicted to be the strongest contributor (Caramete et al., 2011), just as suggested by the Auger data. These new calculations suggest that the radio galaxies Cen A, Vir A, and For A contribute as 10:1:0.6; so at the present level of statistics Cen A is the only source discernible. Using a complete sample of starburst galaxies to allow for a contribution from Gamma-ray Bursts suggests many possible sources, with no obvious preference for some single source (Caramete et al., 2011). The radio galaxy Cen A is a prime candidate (see, also, Anchordoqui et al., 2011; Fargion and D'Armiento, 2011).

The collisions of the charged particles inside Cen A occur at much higher energy than at the LHC at CERN (Geneva), even when considering the center-of-mass frame. But of course at a distance of order 3–4 Mpc ($\simeq 10^{25}$ cm) the opportunities for a high flux observations are nil. Any particle physics experiment would have to be done very differently from an experiment on the ground. The fluxes at the source are very high, but we have no control over what exactly happens, and will have to infer the "experimental set-up" in detail before learning anything really new and having any significance.

The final conclusion and confirmation is still pending, but it appears possible that our highest energy interaction experiment accessible to observations of charged as well as neutral particles may be the radio galaxy Cen A. If that were the case, and with a few years of statistics, we might be able to verify this possibility. Then Cen A could be thought of as an extension of the LHC at CERN, with higher particle energies, higher luminosities, but lower fluxes per given area, and beyond any human control. The combination of insights gathered at the LHC with measurements of such a distant particle interaction laboratory such as Cen A might help us push the frontier.

## 4.3 From the Past to the Future of Astrophysics

Over the last 100 years the interaction between physics and astronomy has first spawned astrophysics, and now has given rise to what is commonly called astroparticle physics, or sometimes also particle astrophysics. We learn from the cosmos, and in the future we can expect that this interaction will get yet stronger, as not only radio galaxies might turn into particle interaction laboratories at energies far beyond what we have achieved on Earth, but at the very highest energies only the very early universe in combination with laboratories on Earth might teach us more about the fundamental properties of matter and nature.

**Acknowledgements** The author wishes to thank his many collaborators, friends and partners for intense discussions over many years, among them F. Aharonian, J.K. Becker, V. Berezinsky, G. Bisnovatyi-Kogan, H. Blümer, S. Britzen, L. Caramete, S. Casanova, J. Cassinelli, L. Clavelli, A. Donea, R. Engel, T. Enßlin, J. Everett, H. Falcke, T.K. Gaisser, L.A. Gergely, G. Gilmore, Gopal-Krishna, F. Halzen, B. Harms, A. Haungs, N. Ikhsanov, G.P. Isar, S. Jiraskova, R. Jokipii, K.-H. Kampert, H. Kang, A. Kogut, P.P. Kronberg, A. Kusenko, N. Langer, R. Lovelace, K. Mannheim, I. Mariş, S. Markoff, G. Medina-Tanco, A. Meli, S. Moiseenko, F. Munyaneza, B. Nath, A. Obermeier, K. Otmianowska-Mazur, R. Protheroe, G. Pugliese, J. Rachen, W. Rhode, M. Romanova, D. Ryu, N. Sanchez, M.M. Shapiro, V. de Souza, E.-S. Seo, R. Sina, Ch. Spiering, T. Stanev, J. Stasielak, P.A. Strittmatter, R.G. Strom, H. de Vega, H.-J. Völk, Y.-P. Wang, T. Weiler, P. Wiita, A. Witzel, Ch. Zier, et multae aliae et multi alii.

## References

Abadie, J., et al.: Phys. Rev. D **81**, id. 102001 (2010)
Abel, T., Bryan, G., Teyssier, R.: In: Chabrier, G. (ed.) Structure Formation in Astrophysics. Cambridge Contemporary Astrophysics, p. 159. Cambridge University Press, Cambridge (2009)
Abramowicz, M.A., et al.: Astrophys. J. **705**, 659–669 (2009). arXiv:0810.3140
Achterberg, A., et al.: Mon. Not. R. Astron. Soc. **328**, 393 (2001)
Aharonian, F.A.: Very High Energy Cosmic Gamma Radiation: A Crucial Window on the Extreme Universe. World Scientific, Singapore (2004)
Alcock, C., et al.: Astrophys. J. Lett. **499**, L9 (1998)
Allard, D., et al.: J. Cosmol. Astropart. Phys. **10**, 33 (2008)
Alpher, R.A., Bethe, H., Gamow, G.: Phys. Rev. **73**, 803–804 (1948)
Anchordoqui, L.A., et al.: arXiv:1103.0536 [astro-ph] (2011)
Appenzeller, I., Fricke, K.: Astron. Astrophys. **18**, 10 (1972a)
Appenzeller, I., Fricke, K.: Astron. Astrophys. **21**, 285 (1972b)
Ardeljan, N.V., et al.: Astron. Astrophys. Trans. **10**, 341–355 (1996)
Auger-Coll.: eprint arXiv:1107.4809 (2011a)
Auger-Coll.: eprint arXiv:1107.4804 (2011b)
Auger-Coll.: eprint arXiv:1107.4805 (2011c)
Auger-Coll.: eprint arXiv:1107.4806 (2011d)
Auger-Coll.: eprint arXiv:1107.4807 (2011e)
Axford, W.I., Leer, E., Skadron, G.: In: Proc. 15th Intern. C. R. Conf., Plovdiv, Bulgaria, vol. 11, pp. 132–137 (1977)
Baade, W., Zwicky, F.: Proc. Natl. Acad. Sci. **20**, 259 (1934)
Bardeen, J.M.: Nature **226**, 64–65 (1970)
Bardeen, J.M., Carter, B., Hawking, S.W.: Commun. Math. Phys. **31**, 161–170 (1973)

Barrau, A., et al.: Astron. Astrophys. **398**, 403 (2003)
Beck, R., et al.: Annu. Rev. Astron. Astrophys. **34**, 155–206 (1996)
Becker, J.K.: Phys. Rep. **458**, 173–246 (2008)
Becker, J.K., et al.: Astrophys. J. Lett. **739**, L43 (2011)
Bekenstein, J.D.: Phys. Rev. D **7**, 2333–2346 (1973)
Bekenstein, J.D.: Contemp. Phys. **45**, 31–43 (2004)
Bell, A.R.: Mon. Not. R. Astron. Soc. **182**, 147–156 (1978a)
Bell, A.R.: Mon. Not. R. Astron. Soc. **182**, 443–455 (1978b)
Bell, A.R.: Mon. Not. R. Astron. Soc. **353**, 550–558 (2004)
Bell, A.R., Lucek, S.G.: Mon. Not. R. Astron. Soc. **321**, 433 (2001)
Berezinsky, V.: Nucl. Phys. B, Proc. Suppl. **188**, 227 (2009)
Berezinsky, V.S., Bulanov, S.V., Dogiel, V.A., Ginzburg, V.L., Ptuskin, V.S.: Astrophysics of Cosmic Rays. North-Holland, Amsterdam (1990)
Bernabei, R., et al.: Can. J. Phys. **89**, 141–152 (2011)
Berti, E., Volonteri, M.: Astrophys. J. **684**, 822–828 (2008)
Bethe, H.A.: Rev. Mod. Phys. **62**, 801–866 (1990)
Beuermann, K., Kanbach, G., Berkhuijsen, E.M.: Astron. Astrophys. **153**, 17–34 (1985)
Beuermann, K., et al.: Astron. Astrophys. **526**, id. A53 (2011)
Bezrukov, F., Hettmansperger, H., Lindner, M.: Phys. Rev. D **81**, id. 085032 (2010)
Biermann, L.: Z. Naturforsch. **5a**, 65–71 (1950)
Biermann, L.: Z. Astrophys. **29**, 274 (1951)
Biermann, L., Schlüter, A.: Phys. Rev. **82**, 863–868 (1951)
Biermann, P.L.: Astron. Astrophys. **271**, 649 (1993)
Biermann, P.L., Cassinelli, J.P.: Astron. Astrophys. **277**, 691 (1993). astro-ph/9305003
Biermann, P.L., de Souza, V.: Astrophys. J. **746**, 72 (2012). arXiv:1106.0625
Biermann, P.L., Kusenko, A.: Phys. Rev. Lett. **96**, id. 091301 (2006)
Biermann, P.L., Strittmatter, P.A.: Astrophys. J. **322**, 643 (1987)
Biermann, P.L., Strom, R.G.: Astron. Astrophys. **275**, 659 (1993). astro-ph/9303013
Biermann, P.L., Strom, R.G., Falcke, H.: Astron. Astrophys. **302**, 429 (1995). astro-ph/9508102
Biermann, P.L., Langer, N., Seo, E.-S., Stanev, T.: Astron. Astrophys. **369**, 269–277 (2001)
Biermann, P.L., et al.: Astrophys. J. **293**, 303–320 (1985)
Biermann, P.L., et al.: Phys. Rev. Lett. **103**, id. 061101 (2009). arXiv:0903.4048
Biermann, P.L., et al.: Astrophys. J. Lett. **710**, L53–L57 (2010a)
Biermann, P.L., et al.: Astrophys. J. **725**, 184–187 (2010b). arXiv:1009.5592
Biermann, P.L., et al.: In: Proc. "Neutrino Oscillations 2010" Meeting, Near Lecce, Italy, September 2010. Nucl. Phys. B, Proc. Suppl. **217**, 284–286 (2011). arXiv:1012.0204
Binns, W.R., et al.: New Astron. Rev. **52**, 427–430 (2008)
Bisnovatyi-Kogan, G.S.: Astron. ž. **47**, 813 (1970). Transl. in Sov. Astron. **14**, 652 (1971)
Bisnovatyi-Kogan, G.S., Moiseenko, S.G.: Chin. J. Astron. Astrophys. Suppl. Ser. **8**, 330–340 (2008)
Bisnovatyi-Kogan, G.S., Ruzmaikin, A.A., Syunyaev, R.A.: Astron. ž. **50**, 210 (1973). Transl. in Sov. Astron. **17**, 137 (1973)
Blandford, R.D., Ostriker, J.P.: Astrophys. J. Lett. **221**, L29–L32 (1978)
Blandford, R.D., Znajek, R.L.: Mon. Not. R. Astron. Soc. **179**, 433–456 (1977)
Boyanovsky, D., de Vega, H.J., Sanchez, N.G.: Phys. Rev. D **77**, id. 043518 (2008a)
Boyanovsky, D., de Vega, H.J., Sanchez, N.G.: Phys. Rev. D **78**, id. 063546 (2008b)
Branch, D.: Annu. Rev. Astron. Astrophys. **36**, 17–56 (1998)
Breitschwerdt, D., McKenzie, J.F., Völk, H.J.: Astron. Astrophys. **245**, 79–98 (1991)
Burbidge, E.M., Burbidge, G.R., Fowler, W.A., Hoyle, F.: Rev. Mod. Phys. **29**, 547–650 (1957)
Bykov, A.M., Osipov, S.M., Ellison, D.C.: Mon. Not. R. Astron. Soc. **410**, 39 (2011)
Caprioli, D., Amato, E., Blasi, P.: Astropart. Phys. **33**, 307 (2010)
Caramete, L.I., Biermann, P.L.: Astron. Astrophys. **521**, id. A55 (2010)
Caramete, L.I., Tascau, O., Biermann, P.L., Stanev, T.: eprint arXiv:1106.5109 (2011)
Casanova, S., et al.: eprint arXiv:1104.1109 (2011)

Cavaliere, A., D'Elia, V.: Astrophys. J. **571**, 226–233 (2002). arXiv:astro-ph/0106512
Celotti, A., Ghisellini, G.: Mon. Not. R. Astron. Soc. **385**, 283–300 (2008). arXiv:0711.4112
Celotti, A., Ghisellini, G., Fabian, A.C.: Mon. Not. R. Astron. Soc. **375**, 417–424 (2007). arXiv:astro-ph/0611439
Cenko, S.B., et al.: Astrophys. J. **711**, 641–654 (2010)
Chandrasekhar, S.: Astrophys. J. **74**, 81 (1931)
Chou, C.W., et al.: Science **329**, 1630 (2010)
Cucchiara, A., et al.: Astrophys. J. **736**, id. 7 (2011)
Dalcanton, J.J., Hogan, C.J.: Astrophys. J. **561**, 35–45 (2001)
de Vega, H.J., Sanchez, N.G.: Mon. Not. R. Astron. Soc. **404**, 885–894 (2010)
de Vega, H.J., Sanchez, N.G.: Int. J. Mod. Phys. A **26**, 1057–1072 (2011)
de Vega, H.J., et al.: eprint arXiv:1109.3452 (2011)
Dessart, L., et al.: Mon. Not. R. Astron. Soc. **414**, 2985–3005 (2011)
Dicke, R.H., Peebles, P.J.E., Roll, P.G., Wilkinson, D.T.: Astrophys. J. **142**, 414–419 (1965)
Dickel, J.R., van Breugel, W.J.M., Strom, R.G.: Astron. J. **101**, 2151–2159 (1991)
Diddams, S.A., et al.: Science **306**, 1318 (2004)
Disney, M., et al.: Nature **455**, 1082–1084 (2008)
Dolag, K., et al.: In: Cosmic Magnetic Fields: From Planets, to Stars and Galaxies. Proc. IAU Symp., vol. 259, pp. 519–528 (2009)
Dole, H., Lagache, G., Puget, J.: In: Armus, L., Reach, W.T. (eds.) The Spitzer Space Telescope: New Views of the Cosmos. ASP Conf. Ser., vol. 357, p. 290 (2006)
Donato, F., et al.: Mon. Not. R. Astron. Soc. **397**, 1169–1176 (2009)
Drury, L.O'C.: Rep. Prog. Phys. **46**, 973–1027 (1983)
Everett, J.E., et al.: Astrophys. J. **674**, 258–270 (2008)
Everett, J.E., et al.: Astrophys. J. **711**, 13–24 (2010)
Faber, S.M., et al.: Astron. J. **114**, 1771 (1997)
Fargion, D., D'Armiento, D.: eprint arXiv:1101.0273 (2011)
Fermi, E.: Phys. Rev. **75**, 1169–1174 (1949)
Fermi, E.: Astrophys. J. **119**, 1 (1954)
Ferrière, K.: Astron. Nachr. **331**, 27 (2010)
Font, A.S., et al.: Mon. Not. R. Astron. Soc. **417**, 1260–1279 (2011)
Fossati, G., et al.: Mon. Not. R. Astron. Soc. **299**, 433–448 (1998)
Frandsen, M.T., et al.: Phys. Rev. D **84**, id. 041301 (2011)
Frieman, J.A., Turner, M.S., Huterer, D.: Annu. Rev. Astron. Astrophys. **46**, 385–432 (2008)
Gaisser, T.K.: Cosmic Rays and Particle Physics. Cambridge University Press, Cambridge (1991)
Garcia-Munoz, M., et al.: Astrophys. J. Suppl. **64**, 269–304 (1987)
Gentile, G., et al.: Nature **461**, 627–628 (2009)
Genzel, R., Eisenhauer, F., Gillessen, St.: Rev. Mod. Phys. **82**, 3121–3195 (2010)
Gergely, L., Biermann, P.L.: Astrophys. J. **697**, 1621–1633 (2009)
Ghisellini, G.: New Astron. Rev. **48**, 375–380 (2004). arXiv:astro-ph/0306429
Ghisellini, G., Tavecchio, F.: Mon. Not. R. Astron. Soc. **387**, 1669–1680 (2008). arXiv: 0802.1918
Ghisellini, G., Tavecchio, F.: Mon. Not. R. Astron. Soc. **397**, 985–1002 (2009). arXiv:0902.0793
Ghisellini, G., et al.: Mon. Not. R. Astron. Soc. **301**, 451–468 (1998)
Ghisellini, G., et al.: Mon. Not. R. Astron. Soc. **393**, L16–L20 (2009a). arXiv:0810.5555
Ghisellini, G., Maraschi, L., Tavecchio, F.: Mon. Not. R. Astron. Soc. **396**, L105–L109 (2009b). arXiv:0903.2043
Ghisellini, G., Tavecchio, F., Ghirlanda, G.: Mon. Not. R. Astron. Soc. **399**, 2041–2054 (2009c). arXiv:0906.2195
Ghisellini, G., et al.: Mon. Not. R. Astron. Soc. **402**, 497–518 (2010). arXiv:0909.0932
Giacconi, R., et al.: Phys. Rev. Lett. **9**, 439–443 (1962)
Gibbons, G.W., Hawking, S.W.: Phys. Rev. D **15**, 2752–2756 (1977)
Gilfanov, M., Bogdan, A.: Nature **463**, 924–925 (2010)
Gilmore, G., et al.: Astrophys. J. **663**, 948–959 (2007)

Ginzburg, V.L., Syrovatskii, S.I.: Astron. ž. **40**, 466 (1963). Transl. in Sov. Astron., A.J. **7**, 357 (1963)
Ginzburg, V.L., Syrovatskii, S.I.: The Origin of Cosmic Rays. Pergamon Press, Oxford (1964). Orig. Russ. edn. (1963)
Goldhaber, G., Perlmutter, S.: Phys. Rep. **307**, 325–331 (1998)
Goldstein, M.L., Roberts, D.A., Matthaeus, W.H.: Annu. Rev. Astron. Astrophys. **33**, 283 (1995)
Gopal-Krishna, Wiita, P.J.: Astrophys. J. Lett. **560**, L115–L118 (2001)
Gopal-Krishna, et al.: Astrophys. J. Lett. **720**, L155–L158 (2010)
Goto, M., et al.: Astrophys. J. **688**, 306–319 (2008)
Goto, M., et al.: Publ. Astron. Soc. Jpn. Lett. **63**, L13–L17 (2011)
Gould, R.J., Burbidge, G.R.: Handb. der Physik, Band 46/2, pp. 265–309. Springer, Heidelberg (1967)
Greene, J.E., Ho, L.C.: Astrophys. J. **670**, 92–104 (2007a)
Greene, J.E., Ho, L.C.: In: Proc. the Central Engine of Active Galactic Nuclei. ADSP Conf., vol. 373, p. 33 (2007b)
Greene, J.E., Barth, A.J., Ho, L.C.: New Astron. Rev. **50**, 739–742 (2006)
Greene, J.E., Ho, L.C., Barth, A.J.: Astrophys. J. **688**, 159–179 (2008)
Gregorini, L., et al.: Astron. J. **89**, 323–331 (1984)
Greisen, K.: Phys. Rev. Lett. **16**, 748–750 (1966)
Guetta, D., et al.: Astron. Astrophys. **421**, 877–886 (2004). arXiv:astro-ph/0402164
H.E.S.S. Coll.: Astron. Astrophys. **531**, id. A81 (2011)
Halzen, F.: In: Proc. Exotic Nuclei and Nuclear/Particle Astrophysics (III): From Nuclei to Stars. AIP Conf., vol. 1304, pp. 268–282 (2010)
Hanasz, M., et al.: Astrophys. J. Lett. **605**, L33–L36 (2004)
Hanasz, M., et al.: Astron. Astrophys. **498**, 335–346 (2009)
Häring, N., Rix, H.-W.: Astrophys. J. Lett. **604**, L89–L92 (2004)
Hawking, S.W.: Phys. Rev. Lett. **26**, 1344–1346 (1971)
Hawking, S.W.: In: Hegyi, D.J. (ed.) Proc. Sixth Texas Symp. on Rel. Astrop. Ann. N.Y. Acad. Sci., vol. 224, p. 268 (1973)
Hawking, S.W.: Commun. Math. Phys. **43**, 199–220 (1975)
Hawking, S.W.: Phys. Rev. D **13**, 191–197 (1976)
Hess, V.F.: Physik. Z. **13**, 1084 (1912)
Hewish, A., et al.: Nature **217**, 709–713 (1968)
Hild, S., et al.: Class. Quantum Gravity **28**, id. 094013 (2011)
Hillas, A.M.: eprint astro-ph/0607109 (2006)
Hillebrandt, W., Niemeyer, J.C.: Annu. Rev. Astron. Astrophys. **38**, 191–230 (2000)
Hogan, C.J., Dalcanton, J.J.: Phys. Rev. D **62**, id. 063511 (2000)
IceCube Coll.: eprint arXiv:1109.1017 (2011a)
IceCube Coll.: Astrophys. J. **740**, id. 16 (2011b)
Ikhsanov, N.R.: Mon. Not. R. Astron. Soc. **375**, 698–704 (2007)
Indriolo, N., et al.: Astrophys. J. **724**, 1357–1365 (2010)
Jokipii, J.R.: Astrophys. J. **313**, 842–846 (1987)
Kardashev, N.S.: Astron. ž. **41**, 807 (1964)
Kellermann, K.I., Pauliny-Toth, I.I.K.: Astrophys. J. Lett. **155**, L71–L78 (1969)
Keppens, R., et al.: Astron. Astrophys. **486**, 663–678 (2008)
Kim, J.-H., et al.: Astrophys. J. **738**, id. 54 (2011)
Kogut, A., et al.: Astrophys. J. **734**, id. 4 (2011)
Kolhörster, W.: Physik. Z. **14**, 1153 (1913)
Komatsu, E., et al.: Astrophys. J. Suppl. **192**, id. 18 (2011)
Kormendy, J., Richstone, D.: Annu. Rev. Astron. Astrophys. **33**, 581 (1995)
Kotera, K., Olinto, A.V.: Annu. Rev. Astron. Astrophys. **49**, 119–153 (2011)
Kramer, M.: In: Relativity in Fundamental Astronomy: Dynamics, Reference Frames, and Data Analysis. Proc. IAU Symp., vol. 261, pp. 366–376 (2010)

Kronberg, P.P.: In: Wielebinski, R., Beck, R. (eds.) Cosmic Magnetic Fields. Lect. Notes in Phys., vol. 664, p. 9 (2005)
Kronberg, P.P., Lesch, H., Hopp, U.: Astrophys. J. **511**, 56–64 (1999)
Kronberg, P.P., et al.: Astrophys. J. **676**, 70–79 (2008)
Krymskii, G.F.: Dokl. Akad. Nauk **234**, 1306–1308 (1977). Transl. in Sov. Phys. Dokl. **22**, 327–328 (1977)
Kulesza-Zydzik, B., et al.: Astron. Astrophys. **522**, id. A61 (2010)
Kulpa-Dybel, K., et al.: Astrophys. J. Lett. **733**, id. L18 (2011)
Kulsrud, R.M., Zweibel, E.G.: Rep. Prog. Phys. **71**, id. 046901 (2008)
Kusenko, A.: New Astron. Rev. **49**, 115–118 (2005)
Kusenko, A.: Phys. Rev. Lett. **97**, id. 241301 (2006)
Kusenko, A.: Phys. Rep. **481**, 1–28 (2009)
Kusenko, A., Takahashi, F., Yanagida, T.T.: Phys. Lett. B **693**, 144–148 (2010)
Lagache, G., Dole, H., Puget, J.-L.: Mon. Not. R. Astron. Soc. **338**, 555–571 (2003)
Lagache, G., Puget, J.-L., Dole, H.: Annu. Rev. Astron. Astrophys. **43**, 727–768 (2005)
Lagage, P.O., Cesarsky, C.J.: Astron. Astrophys. **125**, 249–257 (1983)
Landau, L.D.: Phys. Z. Sov. **1**, 285 (1932). Transl. in ter Haar, D. (ed.) (1965). "Collected papers of L.D. Landau", p. 60. Gordon & Breach, New York. Quoted on p. 505 and 716 of Binney, J., Tremaine, S., Galactic Dynamics. Princeton University Press, Princeton (1987)
Langer, N., van Marle, A.-J., Yoon, S.-C.: New Astron. Rev. **54**, 206–210 (2010)
Letessier-Selvon, A., Stanev, T.: Rev. Mod. Phys. **83**, 907–942 (2011)
Linsley, J.: Phys. Rev. Lett. **10**, 146–148 (1963)
Liu, et al.: Mon. Not. R. Astron. Soc. **417**, 2916–2926 (2011)
Loewenstein, M., Kusenko, A.: Astrophys. J. **714**, 652–662 (2010)
Lovelace, R.V.E.: Nature **262**, 649 (1976)
Lovelace, R.V.E., et al.: Astrophys. J. **572**, 445 (2002)
Lüst, R.: Z. Naturforsch. **7**, 87 (1952)
Lynden-Bell, D.: Mon. Not. R. Astron. Soc. **136**, 101 (1967)
Lynden-Bell, D., Pringle, J.E.: Mon. Not. R. Astron. Soc. **168**, 603–637 (1974)
Maoz, E.: Astrophys. J. Lett. **494**, L181–L184 (1998)
Maraschi, L., et al.: Mon. Not. R. Astron. Soc. **391**, 1981–1993 (2008). arXiv:0810.0145
Marcaide, J.M., et al.: In: Fanti, R., Kellermann, K., Setti, G. (eds.) VLBI and Compact Radio Sources. Proc. IAU Symp., vol. 110, p. 361 (1984)
Mather, J.C.: Rev. Mod. Phys. **79**, 1331–1348 (2007)
Matthaeus, W.H., Zhou, Y.: Phys. Fluids B **1**, 1929–1931 (1989)
Mazzali, P.A., et al.: Science **315**, 825 (2007)
Mazzali, P.A., et al.: Science **321**, 1185 (2008)
Meli, A., Biermann, P.L.: Astron. Astrophys. **454**, 687–694 (2006). astro-ph/0602308
Mestel, L., Roxburgh, I.W.: Astrophys. J. **136**, 615 (1962)
Mészáros, P.: The High Energy Universe: Ultra-High Energy Events in Astrophysics and Cosmology. Cambridge University Press, Cambridge (2010)
Mészáros, P., Rees, M.J.: Astrophys. J. Lett. **715**, 967–971 (2010)
Mészáros, P., Rees, M.J.: Astrophys. J. Lett. **733**, id. L40 (2011)
Milgrom, M.: Mon. Not. R. Astron. Soc. **398**, 1023–1026 (2009)
Milgrom, M., Bekenstein, J.: In: Proc. Dark Matter in the Universe. IAU Symp., pp. 319–330 (1987)
Mirabel, I.F., Rodriguez, L.F.: Nature **392**, 673–676 (1998)
Mirabel, I.F., Rodriguez, L.F.: Annu. Rev. Astron. Astrophys. **37**, 409–443 (1999)
Mirabel, I.F., Rodrigues, I.: Science **300**, 1119–1121 (2003)
Mirabel, I.F., et al.: Nature **413**, 139–141 (2001)
Mirabel, I.F., et al.: Astron. Astrophys. **528**, id. A149 (2011)
Moiseenko, S.G., Bisnovatyi-Kogan, G.S., Ardeljan, N.V.: Mon. Not. R. Astron. Soc. **370**, 501–512 (2006)
Nath, B.N., Biermann, P.L.: Mon. Not. R. Astron. Soc. **267**, 447 (1994)

Nath, B.N., Silk, J.: Mon. Not. R. Astron. Soc. **396**, L90–L94 (2009)
Novikov, I.D., Thorne, K.S.: In: Black Holes (Les astres occlus), pp. 343–450 (1973)
Oort, J.H.: B.A.N. **6**, 249 (1932)
Oppermann, N., et al.: eprint arXiv:1111.6186 (2011)
Ostriker, J.P., Bode, P., Babul, A.: Astrophys. J. **634**, 964–976 (2005)
Otmianowska-Mazur, K., et al.: In: Proc. Magnetic Fields in the Universe II: From Laboratory and Stars to the Primordial Universe. Rev. Mex. de Astron. y Astrof. (Ser. de Conf.), vol. 36, pp. CD266–CD271 (2009)
Parikh, M.K., Wilczek, F.: Phys. Rev. Lett. **85**, 5042–5045 (2000)
Parker, E.N.: Astrophys. J. **145**, 811 (1966)
Parker, E.N.: Astrophys. J. **157**, 1129 (1969)
Parker, E.N.: Annu. Rev. Astron. Astrophys. **8**, 1 (1970a)
Parker, E.N.: Astrophys. J. **160**, 383 (1970b)
Parker, E.N.: Astrophys. J. **162**, 665 (1970c)
Penzias, A.A., Wilson, R.W.: Astrophys. J. **142**, 419–421 (1965)
Piran, T.: Rev. Mod. Phys. **76**, 1143–1210 (2004)
Planck Coll.: eprint arXiv:1101.2028 (2011)
Plotkin, R.M., et al.: Mon. Not. R. Astron. Soc. **419**, 267–286 (2012)
Portegies Zwart, S.F., et al.: Nature **428**, 724–726 (2004)
Prandtl, L.: Z. Angew. Math. Mech. **5**, 136 (1925)
Prantzos, N.: Adv. Space Res. **4**, 109–114 (1984)
Prantzos, N.: In: van der Hucht, K.A., Hidayat, B. (eds.) Wolf-Rayet Stars and Interrelations with Other Massive Stars in Galaxies. Proc. IAU Symp., vol. 143, p. 550 (1991)
Prendergast, K.H., Burbidge, G.R.: Astrophys. J. Lett. **151**, L83–L88 (1968)
Qian, S.-B., et al.: Astrophys. J. Lett. **708**, L66–L68 (2010)
Rachen, J.P., Stanev, T., Biermann, P.L.: Astron. Astrophys. **273**, 377 (1993)
Ramaty, R., et al.: Astrophys. J. **534**, 747 (2000)
Ramaty, R., Lingenfelter, R.E., Kozlovsky, B.: Space Sci. Rev. **99**, 51 (2001)
Reeves, H.: Rev. Mod. Phys. **66**, 193–216 (1994)
Rickett, B.J.: Annu. Rev. Astron. Astrophys. **15**, 479–504 (1977)
Riess, A.G., et al.: Astron. J. **116**, 1009–1038 (1998)
Ripamonti, E., Abel, T.: eprint arxiv:astro-ph/0507130. Lecture Notes for the Spring 2003 SIGRAV Doctoral School The Joint Evolution of Black Holes and Galaxies (2005, to be published)
Rollinde, E., et al.: Mon. Not. R. Astron. Soc. **398**, 1782–1792 (2009)
Rosenband, T., et al.: Science **319**, 1808 (2008)
Ryu, D., Kang, H., Biermann, P.L.: Astron. Astrophys. **335**, 19–25 (1998)
Ryu, D., et al.: Science **320**, 909 (2008)
Sanders, R.H.: Astrophys. J. **162**, 791 (1970)
Sanders, R.H.: Mon. Not. R. Astron. Soc. **386**, 1588–1596 (2008)
Sanders, D.B., Mirabel, I.F.: Annu. Rev. Astron. Astrophys. **34**, 749 (1996)
Savage, Ch., et al.: Phys. Rev. D **83**, id. 055002 (2011)
Schlickeiser, R.: Cosmic Ray Astrophysics. Springer, Berlin (2002)
Schmidt, M.: Nature **197**, 1040 (1963)
Schödel, R., Merritt, D., Eckart, A.: Astron. Astrophys. **502**, 91–111 (2009)
Schwarzschild, K.: Sitz.-berichte der Königl. Preuss. Akad. d. Wiss. zu Berlin, Phys.-Math. Kl., 189–196 (1916)
Shakura, N.I., Sunyaev, R.A.: Astron. Astrophys. **24**, 337–355 (1973)
Shapiro, S.L., Teukolsky, S.A.: Black Holes, White Dwarfs, and Neutron Stars: The Physics of Compact Objects. Wiley-Interscience, New York (1983), 663 pp.
Silberberg, R., Tsao, C.H.: Phys. Rep. **191**, 351–408 (1990)
Silk, J., Rees, M.J.: Astron. Astrophys. Lett. **331**, L1–L4 (1998)
Silk, J., Takahashi, T.: Astrophys. J. **229**, 242–256 (1979)
Smoluchowski, M.: Z. Phys. **17**, 557 (1916)
Smoot, G.F.: Rev. Mod. Phys. **79**, 1349–1379 (2007)

Spangler, St.R., Gwinn, C.R.: Astrophys. J. Lett. **353**, L29 (1990)
Stanev, T.: High Energy Cosmic Rays. Springer, Berlin/Heidelberg (2010)
Stanev, T., Biermann, P.L., Gaisser, T.K.: Astron. Astrophys. **274**, 902 (1993). astro-ph/9303006
Steenbeck, M., Krause, F.: Monatsb. Dtsch. Akad. Wiss. **7**, 335–340 (1965). Transl. in NCAR-TN/IA-60, The turbulent dynamo, edited by P.H. Roberts, M. Stix, 1971, p. 21
Steenbeck, M., Krause, F.: Z. Naturforsch. **21a**, 1285–1296 (1966). Transl. in NCAR-TN/IA-60, The turbulent dynamo, edited by P.H. Roberts, M. Stix, 1971, p. 49
Steenbeck, M., Krause, F., Rädler, K.-H.: Z. Naturforsch. **21a**, 369–376 (1966). Transl. in NCAR-TN/IA-60, The turbulent dynamo, edited by P.H. Roberts, M. Stix, 1971, p. 29
Steenbeck, M., et al.: Monatsb. Dtsch. Akad. Wiss. **9**, 714–719 (1967). Transl. in NCAR-TN/IA-60, The turbulent dynamo, edited by P.H. Roberts, M. Stix, 1971, p. 97
Steinmetz, T., et al.: Science **321**, 1335 (2008)
Strigari, L.E., et al.: Nature **454**, 1096–1097 (2008)
Sun, W.-H., Malkan, M.A.: Astrophys. J. **346**, 68–100 (1989)
Vazza, F., et al.: Mon. Not. R. Astron. Soc. **418**, 960–985 (2011), eprint arXiv:1106.2159
Vietri, M.: Astrophys. J. **453**, 883 (1995)
Völk, H.J., Biermann, P.L.: Astrophys. J. Lett. **333**, L65 (1988)
Wald, R.M.: Phys. Rev. D **56**, 6467–6474 (1997)
Wang, Y.-P., Biermann, P.L.: Astron. Astrophys. **334**, 87–95 (1998)
Wang, X.-Y., Razzaque, S., Meszaros, P.: Astrophys. J. **677**, 432–440 (2008)
Waxman, E.: Phys. Rev. Lett. **75**, 386 (1995)
Weibel, E.S.: Phys. Rev. Lett. **2**, 83–84 (1959)
Weinberg, St.: Gravitation and Cosmology: Principles and Applications of the General Theory of Relativity. Wiley-VCH, New York (1972), 688 pp.
Wiebel-Sooth, B., Meyer, H., Biermann, P.L.: Astron. Astrophys. **330**, 389 (1998). astro-ph/9709253
Wise, J.H., Abel, T.: Astrophys. J. **684**, 1–17 (2008a)
Wise, J.H., Abel, T.: Astrophys. J. **685**, 40–56 (2008b)
Woltjer, L.: Bull. Astron. Inst. Neth. **14**, 39 (1958)
Woo, J.-H., et al.: In: Proc. Co-Evolution of Central Black Holes and Galaxies. IAU Symp., vol. 267, pp. 183–188 (2010)
Woosley, S.E., Weaver, T.A.: Annu. Rev. Astron. Astrophys. **24**, 205–253 (1986)
Woosley, S.E., Heger, A., Weaver, T.A.: Rev. Mod. Phys. **74**, 1015–1071 (2002)
Wyithe, J.St.B., Loeb, A.: Astrophys. J. **590**, 691–706 (2003)
Wyse, R.F.G., Gilmore, G.: In: Proc. Dark Galaxies and Lost Baryons. IAU Symp., vol. 244, pp. 44–52 (2008)
Yungelson, L.R., et al.: Astron. Astrophys. **477**, 223–237 (2008)
Zatsepin, G.T., Kuz'min, V.A.: Zh. Exp. Th. Fis. Pis'ma **4**, 114 (1966). Transl. in J. Exp. Th. Phys. Lett. **4**, 78 (1966)
Zhang, W., Woosley, S.E., Heger, A.: Astrophys. J. **679**, 639–654 (2008)
Zwicky, F.: Helv. Phys. Acta **6**, 110–127 (1933)
Zwicky, F.: Astrophys. J. **86**, 217 (1937)

# Chapter 5
# Development of Ultra High-Energy Cosmic Ray Research

**Karl-Heinz Kampert and Alan A. Watson**

## 5.1 Introduction and General Overview

Towards the end of the 1930s it was recognised from studies of the effect of the geomagnetic field on cosmic rays that the energy spectrum of the primary particles, not identified as being proton-dominated until 1941, extended to at least 10 GeV. The discovery of extensive air showers in 1938, however, radically changed this situation with the highest energy being pushed up by about 5 orders of magnitude, probably the single largest advance to our knowledge of energy scales ever made. It is now known that the energy spectrum extends to beyond $10^{20}$ eV, but it has taken over 60 years to consolidate this picture. In this chapter we trace the history of the discovery of extensive air showers, show how advances in experimental and theoretical techniques have led to improved understanding of them, and describe how some of the most recent work with contemporary instruments has provided important data on the energy spectrum, the mass composition and the arrival direction distribution of high-energy cosmic rays. These results are of astrophysical importance, but additionally some aspects of the shower phenomenon promise to give new insights on hadronic physics at energies beyond that reached by the LHC.

In Chap. 2, the measurement of the properties of cosmic rays $\lesssim 10^{14}$ eV per particle was discussed. The flux of particles falls so rapidly with energy ($\propto E^{-\gamma}$ with $\gamma \sim 2.7$) that around $10^{14}$ eV it becomes impractical to make measurements of high precision directly: the number of events falling on a detector of a size that can be accommodated on a balloon or a space-craft is simply too small. However, at this energy sufficiently many particles are produced in the atmosphere as secondaries to

K.-H. Kampert (✉)
Department of Physics, University Wuppertal, Wuppertal, Germany
e-mail: kampert@uni-wuppertal.de

A.A. Watson
School of Physics and Astronomy, University of Leeds, Leeds, UK
e-mail: a.a.watson@leeds.ac.uk

B. Falkenburg, W. Rhode (eds.), *From Ultra Rays to Astroparticles*,
DOI 10.1007/978-94-007-5422-5_5, © Springer Science+Business Media Dordrecht 2012

the incoming primary cosmic rays for some to reach mountain altitudes and, as the energy of the primary increases, even sea level. The transverse momentum acquired by secondary particles at production and the scattering which the shower electrons, in particular, undergo through interactions with the material of the atmosphere are such that the secondaries are spread over significant areas at the observational level. The phenomenon of the nearly simultaneous arrival of many particles over a large area is called an Extensive Air Shower (EAS): at $10^{15}$ eV around $10^6$ particles cover approximately $10^4$ m$^2$ while at $10^{20}$ eV some $10^{11}$ particles are spread over about 10 km$^2$. It was quickly recognised that the phenomenon of the air shower offered the possibility of answering four major questions.

1. *What particle physics can be learned from understanding air shower evolution?* A detailed understanding of how an air shower develops is crucial to obtaining an estimate of the primary energy and to learning anything about the mass spectrum of the primary particles. It is worth recalling that when the shower phenomenon was first observed that, in addition to the proton, neutron, electron and positron, only the muon was known, so that a realistic understanding of shower development had to wait until the discovery of the charged pion and its decay chain in 1947 and of the neutral pion in 1950. Indeed, much early thinking was based on the hypothesis that showers were initiated by electrons and/or photons. Once it was recognised that the initiating particle was almost always a proton or a nucleus, the first steps in understanding the nuclear cascade focused on such matters as whether a proton would lose all or only part of its energy in a nuclear collision and how many pions were radiated in such a collision. A combination of observations in air showers, made using Geiger counters and cloud chambers, of data from studies in nuclear emulsions and of early accelerator information was used to inform the debate. The issues of inelasticity (what fraction of the energy is lost by an incoming nucleon to pion production) and the multiplicity (the number of pions produced) are parameters which are still uncertain at most of the energies of interest.
2. *What can be inferred from the arrival direction distributions of the high-energy particles?*
From the earliest years of discovery of cosmic rays there have been searches for directional anisotropies. Hess himself, from a balloon flight made during a solar eclipse in April 1912, i.e. before his discovery flight in August of the same year, deduced that the Sun was not a major source (Hess, 1912). There are a few predictions of the level of anisotropy that might be expected. While there have always been speculations as to the sources, the fact that the primary particles are charged and therefore are deflected in the poorly known galactic and intergalactic magnetic fields makes it difficult to identify them. One firm prediction was made very early on by Compton and Getting (1935) that cosmic rays should show an anisotropy because of the motion of the earth within the galaxy. Eventually it was realised that this idea would be testable only with cosmic rays undeflected by the solar wind (discovered much later) so measuring the Compton–Getting effect became a target for air shower experiments. However, as the velocity of the earth is only about 200 km s$^{-1}$, the effect is $\sim$0.1 % and it has taken around

70 years for a convincing demonstration of its discovery. The search for point sources has been largely unsuccessful, but one of the motivations for searching for rarer and rarer particles of higher and higher energy has been the expectation that anisotropy would eventually be found.

3. *What is the energy spectrum of the primary cosmic rays?*
A power-law distribution of cosmic rays was first described by E. Fermi in 1949 (Fermi, 1949) but until 1966 there were no predictions as to the power-law index or to further structures in the energy spectrum. Observations in 1959 had indicated a steepening at around $3 \times 10^{15}$ eV (the "knee"), while in 1963 it was claimed from observations made with the first large shower array that the spectrum flattens just above $10^{18}$ eV. However, not only were there no predictions of these features, interpretation of them remains controversial. By contrast the discovery of the 2.7 K cosmic microwave background radiation in 1965 led, a year later, to the firm statement that if cosmic rays of energy above $\sim 4 \times 10^{19}$ eV exist, they can come only from nearby sources. It took about 40 years to establish that there is indeed a steepening in the cosmic ray spectrum at about this energy but whether this is a cosmological effect or a consequence of a limit to which sources can accelerate particles is unclear: $4 \times 10^{19}$ eV is within a factor of $\sim 5$ of the highest energy event ever recorded.
4. *What is the mass composition of the primary cosmic rays?*
One of the major tasks of the air shower physicist is to find the mass of the primary particles. This has proved extraordinarily difficult as even if the energy of the primary that produces an event is known, the uncertainties in the hadronic physics make it hard to separate protons from iron. Data from the LHC will surely help, but above $10^{17}$ eV one has reached a regime where the centre-of-mass energies in the collisions are above what is accessible to man-made machines. Indeed it may be that in the coming decades the highest-energy cosmic rays provide a test bed for theories of hadronic interactions, mirroring the fact that cosmic ray physics was the place where particle physics was born in the 1930s.

In what follows we have chosen to emphasise the progress made since the 1940s towards answering these four questions through an examination of the development of different techniques, both experimental and analytical, introduced in the last 70 years. While new techniques have enabled air showers to be studied more effectively, it is remarkable how the essentials of what one seeks to measure were recognised by the pioneers in the 1940s and 1950s. Increasingly sophisticated equipment, operated on increasingly larger scales has been developed, and had led to some answers to the key questions although many issues remain uncertain.

Galbraith (1958) and Cranshaw (1963) have written books in which details of early work, up to the end of the 1950s, are discussed in more detail than is possible below, while in Hillas's classic book on Cosmic Rays (Hillas, 1972) there is an excellent discussion of some of the earliest papers in a context which includes fundamental ideas of cosmic rays physics, including shower physics.

We now move on by reviewing the history of the discovery of the air shower phenomenon.

## 5.2 The Discovery of Extensive Air Showers

A technical development of crucial importance for the study of cosmic rays was the invention of the coincidence technique by Walther Bothe in the late 1920s (Bothe, 1929) for which he was awarded the Nobel Prize in 1954. Coupling the coincidence technique to the newly developed fast responding Geiger–Müller counters (Geiger and Müller, 1928) had already allowed verification that Compton scattering produces a recoil electron simultaneously with the scattered $\gamma$-ray. Bothe's coincidence circuit reached resolving times for singly charged particles of 1.4 ms but was limited to only two-fold coincidences. Only a few months later, the young Bruno Rossi described a coincidence circuit which was conceptually different from Bothe's, as it could accommodate many channels (Rossi, 1930) and it also pushed the resolving time down to 0.4 ms. This, together with the strong reduction of accidentals in triple coincidences, was an essential improvement for the detection of rare cosmic events. In the mid-1930s the coincidence method has also been used to trigger a cloud chamber inside a magnetic field. Instead of using the usual method of random expansion of the chamber, as had to be performed by Dimitry Skobeltzyn for his discovery of multiple production of fast $\beta$-particles in single interaction processes (Skobeltzyn, 1927, 1929), Blackett and Occhialini (1932) placed Geiger–Müller counters above and below a vertical cloud chamber, so that charged particles passing through the two counters would also pass through the chamber, triggering its expansion. This technique allowed the observation of apparently simultaneous production of numerous electrons and positrons (cf. Fig. 5.4). Blackett in his Nobel lecture of 1948 recalled that

> the development of the counter-controlled cloud chamber method, not only attained the original objective of achieving much economy in both time and film, but proved to have the unexpected advantage of greatly enhancing the number of associated rays photographed. (Blackett, 1948)

In retrospect, this experiment marked the birth of "rare event triggering", which became a key tool for progress in nuclear and particle physics experiments.

The development of the coincidence approach was crucial also for the discovery and study of extensive air showers. In 1933 Rossi made a key observation which was hard to accept for the scientific community and which, as Rossi recalled later (Rossi, 1985, page 71), even "raised doubts about the legitimacy of the coincidence method". In an experimental arrangement as shown in Fig. 5.1, Rossi observed that the coincidence rate between the three adjacent Geiger counters *increased* when an absorber was placed above the counters. Only when the absorber thickness reached a certain value did the coincidence rate fall, but much less than was expected even from $\beta$-rays, the most penetrating particles known at that time. All this was difficult to accept for the scientific community, and it became known as "*Rossi's transition curve*". Rossi, however, correctly concluded that soft secondary particles were produced by the cosmic particles entering the material. These secondary particles then suffer increasing absorption with increasing total thickness of the absorber (Rossi, 1933). It is interesting to note that the same basic observation was made a year later by Regener and Pfotzer (1935) when studying the vertical intensity of cosmic

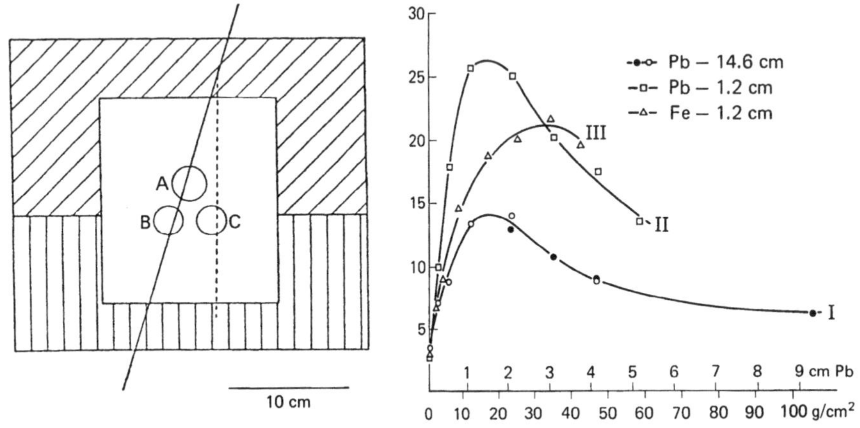


**Fig. 5.1** Rossi's transition curve: The experiment in which the abundant production of secondary radiation by cosmic rays was discovered. Coincidences between Geiger–Müller counters, arranged as shown on the *left*, are produced by groups of secondary particles generated by cosmic rays in the lead shield above the counters. The curves labelled *I–III* refers to Pb and Fe absorbers of different thicknesses placed above the counters (Rossi, 1933)

rays in the stratosphere up to a height of 28 km by recording the rate of three-fold coincidences. Flying and operating sensitive instruments in the stratosphere was a remarkable experimental achievement in itself which became possible because of Regener's long term experience in flying balloon-borne instruments for atmospheric studies and because of his tedious work in patching hundreds of tiny pinholes in the rubber balloons to prevent untimely bursting of the balloons in the upper atmosphere. All this of work paid off by observing an unexpected clear maximum in the coincidence rate at a pressure of 100 mm of mercury (about 14 km above sea level). This became known as the "Pfotzer Maximum". Regener correctly interpreted the maximum as being due to the multiplication of electrons – which he called "*Schauer*" – in the atmosphere such as had been suggested by Bhabha and Heitler (Regener and Ehmert, 1938; Bhabha and Heitler, 1937). However, neither Rossi nor Regener seem to have recognised that the same physical mechanism was behind their observations.

Schmeiser and Bothe, at the same time (Schmeiser and Bothe, 1938), pointed out that Rossi's transition curve implied the occurrence of showers in air – which they named "*Luftschauer*" – and showed that particles in air showers had separations up to 40 cm. Independently, Kolhörster et al. (1938) reported similar data on the rate at which coincidences between a pair of Geiger counters fell as a function of separation. The results of these pioneering measurements are shown in Fig. 5.2. It is clear, however, that Rossi had made the same discovery some years earlier. In 1934, he made observations in Eritrea (see Fig. 5.3) that suggested to him that there was a correlated arrival of particles at widely separated detectors. In his publication (Rossi, 1934) he gave the phenomenon the name "*sciami*". He was not able to follow up this work before he had to leave Italy and it seems to have been unknown to either Bothe or Kolhörster.

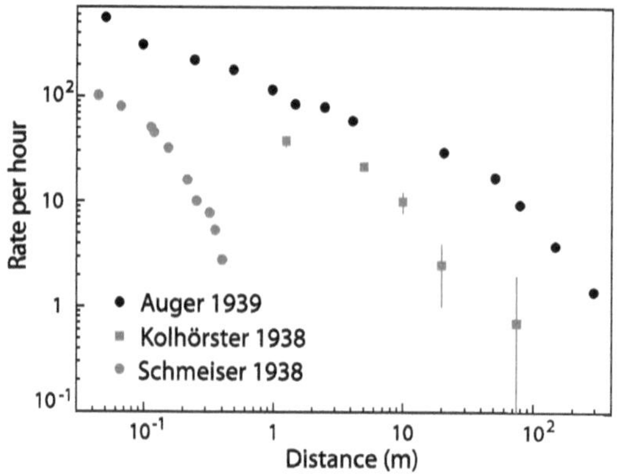


**Fig. 5.2** The discovery of extensive air showers (EAS): Decoherene curves measured with Geiger counters separated up to 300 m distance. Data of (Schmeiser and Bothe, 1938) and (Kolhörster et al., 1938) were measured at sea level with counters of 91 cm$^2$ and 430 cm$^2$ effective area, respectively, while data of (Auger et al., 1939a) were measured with counters of 200 cm$^2$ at the Jungfraujoch at 3 450 m

Despite the work of Rossi and the two German groups, credit for the discovery of extensive air showers has usually been given to Auger and his collaborators for what seems to have been a serendipitous observation (Auger et al., 1939a) depending strongly on the electronic developments by Roland Maze who improved the resolving time of coincidence circuits to 5 μs (Maze, 1938). Auger, Maze and Robley found that the chance rate between two counters separated by some distance greatly exceeded the chance rate expected from the resolving time of the new circuitry. For a while, the phenomenon was known as "Auger showers" (Auger, 1985, page 214). In their measurements performed at the Jungfraujoch in the Swiss Alps they were able to separate their detectors by up to 300 m. The decoherence curves are shown again in Fig. 5.2. Differences in the coincidence rates between the three groups of authors can be understood both by the different effective areas of the Geiger counters and by the different altitudes at which the measurements were performed. In view of the sequence of air shower observations, the important achievement of Auger and his group, which distinguishes their work from that of Rossi, Schmeiser and Bothe, and Kolhörster, appears to be not so much in separating their detectors by up to 300 m, but in estimating the primary energy to be around $10^{15}$ eV. This estimate was based on the number of particles in the showers, assuming that each particle carried, on average, the critical energy.[1] A factor of 10 was added to account for energy lost in the atmosphere. A similar conclusion came from using the work of Bhabha and Heitler, based on the ideas of quantum electrodynamics (QED). It is worth quoting the final remarks of Auger from his paper presented at the 1939 Symposium held in Chicago (Auger et al., 1939b):

[1]The critical energy is the energy at which energy losses by ionisation and bremsstrahlung are equal. The critical energy of electrons in air is appr. 79 MeV.

**Fig. 5.3** The 28 years old Bruno Rossi (*middle*) with an ascari soldier (*right*) and an Italian officer in Eritrea during his 1933 campaign to verify the existence of the East–West effect (Courtesy of the MIT Library)

> One of the consequences of the extension of the energy spectrum of cosmic rays up to $10^{15}$ eV is that it is actually impossible to imaging a single process able to give a particle such an energy. It seems much more likely that the charged particles which constitute the primary cosmic radiation acquire their energy along electric fields of very great extension.

The identification of mechanisms to accelerate particles to energies as great as $10^{20}$ eV as have now been observed remains a great challenge, though the mechanism suggested by Auger now seems unlikely as electric fields over great extensions are unavailable because of the conductivity of the interstellar plasma, unknown in 1939.

That Bothe, Kolhörster and Auger seem to have been unaware of Rossi's work perhaps reflects the fact that the research was done at a time when scientists wrote most commonly in their native languages: in addition, there was no preprint system such as operated in the post-war period and, of course, there was no arXiv.[2] Information about new results was sometimes exchanged by correspondence between senior scientists or during face-to-face meetings. The prominence given to Auger's work probably arises from his stay with Blackett in Manchester and from the fact that he was able to take advantage of his time in Chicago in the early 1940s relatively unhindered by war work. Presumably Bothe, who attended the Chicago meeting, and

[2]Electronic preprint archive: http://arxiv.org/.

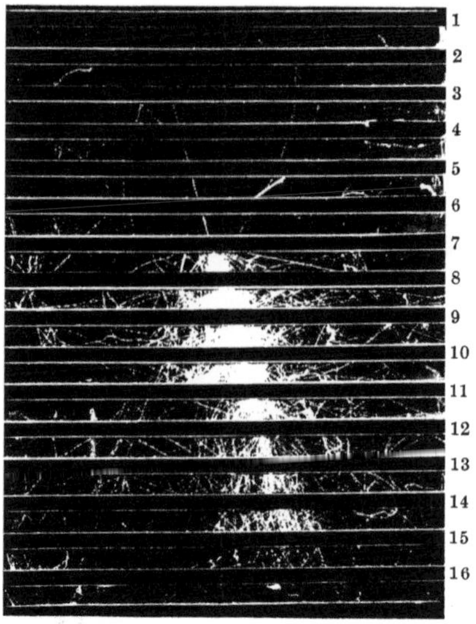


**Fig. 5.4** Image of a particle cascade, or shower, as seen in a cloud chamber at 3 027 m altitude. The primary particle is estimated to be a proton of about 10 GeV. The first interaction will most probably have been in one of the lead plates. Neutral pions feed the cascade which multiplies in the lead. Charged pions make similar interactions to protons, or decay into muons. The cross-sectional area of the cloud chamber is $0.5 \times 0.3$ m$^2$ and the lead absorbers have a thickness of 13 mm each (Fretter, 1949)

Kolhörster had little chance for cosmic ray work after 1939. Rossi left Manchester for Chicago and, during his brief stay there before joining the Manhattan project, his cosmic ray studies were focused largely on the problem of muon decay.

Only a few years after the discovery of extensive air showers, Skobeltzyn et al. (1947) at the Pamir mountains at an altitude of 3 860 m above sea level pushed measurements of coincidences out to distances of 1 000 m. To suppress random coincidences which would occur between single distant Geiger counters, they were the first to apply so-called double-coincidences, meaning that coincidences were first formed within trays of local Geiger counters, before a coincidence was formed between the distant trays.

## 5.3 Basic Ideas About Extensive Air Showers

Work by Auger and his colleagues using cloud chambers triggered by arrays of Geiger counter allowed features of air showers to be understood relatively quickly. By the late 1930s it was known that air showers contained hadronic particles, muons and electrons and major advances in understanding took place in the late 1940s and early 1950s after the existence of two charged and one neutral pion was established and it was recognised that muons were secondary to charged pions. The development of an air shower can be understood by studying Fig. 5.4 which we will reference on occasion. In the figure a cloud chamber picture of a shower created in lead plates by a cosmic ray proton of about 10 GeV is shown (Fretter, 1949). The features shown in this photograph, except for scale, are extremely similar to those present when a high-energy particle enters the earth's atmosphere and creates a shower.

Each lead plate (the dark bands running horizontally across the picture) is about two radiation lengths thick[3] and the cross-sectional area of the cloud chamber is $0.5 \times 0.3\ m^2$. The gas in the chamber was argon, effectively at atmospheric pressure, and thus most of the shower development happens within the lead plates. Little development of the cascade takes place in the gas, but the level of condensation gives a snapshot of how the particle number increases and decreases as the shower progresses through more and more lead. All of the important features of shower development, such as the rise and fall of the particle numbers, and the lateral spreading of the shower, are evident, as are some muons that penetrate more deeply into the chamber than most of the electrons. Had such a proton interacted near sea level in air, then the extent of the lateral spread of the shower would have been around 50 m.

The problem of identifying the nature and determining the energy of the particle that initiated this shower, if there were data available from only one layer of gas corresponding to the information available from a shower array at a single atmospheric depth, can be appreciated from Fig. 5.4. But until the 1980s, when a technique was developed that allowed the build-up of the air shower to be studied on an event-by-event basis, as is seen in the figure, this was the challenge faced by all air shower experimenters. Assumptions had to be made as to where the particle had its first interaction and what are the features of the hadronic interactions. Key parameters such as the cross sections for the interaction of protons (and heavier nuclei) with nuclei, pion–nucleus cross sections, the fraction of energy radiated as pions in each collision and the number of particles produced are needed. By contrast determination of the direction of the incoming primary is a relatively straightforward exercise.

The basic key processes of cascade multiplication occurring in EAS were laid out in 1934 by Bethe and Heitler based on QED (Bethe and Heitler, 1934) and were formulated in terms of pair-production and bremsstrahlung processes by Bhabha and Heitler (1937). Carlson and Oppenheimer (1937) finally completed the theory by accounting also for energy losses of electrons by ionisation and for practical calculations they pioneered the use of diffusion equations. Moreover, they demonstrated quantitative agreement of their calculations with the experimental results by Regener and Pfotzer (Pfotzer Maximum) (l.h.s. of Fig. 5.5), pointed out the importance of fluctuations of the shower maximum, and noted that a more penetrating burst like component, as suggested by Heisenberg (1936) based on measurements by Hoffmann[4] was needed to allow electrons to penetrate the atmosphere to a thickness of 30 radiation lengths (r.h.s. of Fig. 5.5). This paper presented the simple

[3]The radiation length is an appropriate scale length for describing high-energy electromagnetic cascades. It is both the mean distance over which a high-energy electron loses all but $1/e$ of its energy by bremsstrahlung, and 7/9 of the mean free path for pair production by a high-energy photon.

[4]In 1927 Hoffmann had discovered a phenomenon which became known as "Hoffmann bursts" (Hoffmannsche Stöße) (Hoffmann and Pforte, 1930). In measurements of ionisation currents in an ionisation chamber he found occasional discontinuities of strong currents which were interpreted as nuclear explosions.

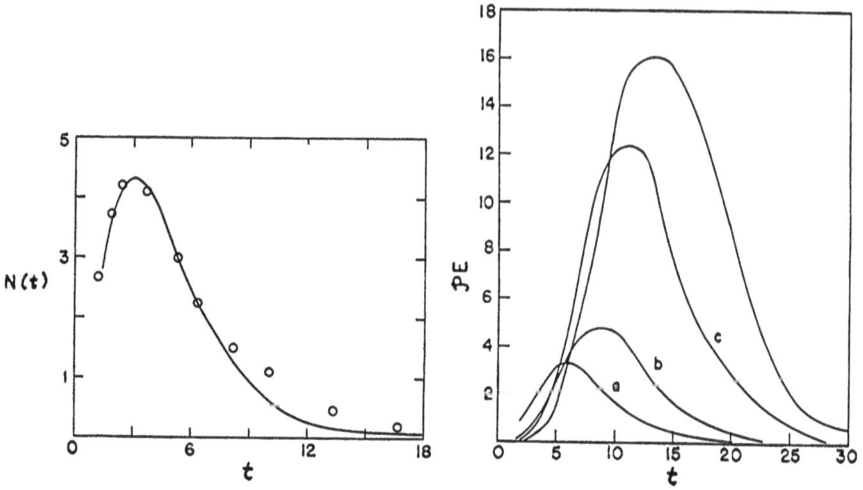


**Fig. 5.5** *Left*: Total number of electrons $N(t)$ against $t = X/X_0$ ($X_0$: radiation length), calculated by Carlson and Oppenheimer for 2.5 GeV electrons in air and compared to experimental results (*circles*) of Pfotzer (1936). *Right*: Estimated number of electrons with $E_e > 50$ MeV in Pb for $E_0 = 2.7, 20, 150$, and 1 110 GeV (Carlson and Oppenheimer, 1937)

concept of electromagnetic cascades, as is still found in any textbook and in introductory exercises about high-energy particle interactions in matter. Even though it does not capture accurately all details of electromagnetic showers, it accounts for its most important features: the total number of electrons, positrons, and photons at the shower maximum is simply proportional to the primary energy $E_0$ and the depth of maximum shower development is logarithmically proportional to $E_0$.

Nowadays, particle showers in the atmosphere are simulated on powerful computers by using sophisticated Monte Carlo codes, allowing many more details of interaction features to be added. Also, the cascade model by Bethe and Heitler and Carlson and Oppenheimer is applicable only to primary electrons or photons so that diffusion calculations have largely lost their importance. However, because of its advantage of being helpful to an understanding of the basic features of particle cascades, the concepts are still used and have been generalised to hadronic primaries, see e.g. (Matthews, 2005).

## 5.4 From Geiger Counters to Phototube Based Detectors

Electrostatic photomultipliers (PMTs) were invented in the late 1930s but were not available for cosmic ray studies until in the 1950s when studies of Cherenkov-light detection and work with liquid and plastic scintillators started. Nonetheless significant progress in the understanding of showers was made using arrays of Geiger counters at mountain altitudes and at sea level. The scale of early experiments, a few 10s of metres in diameter, meant that showers from primaries of $10^{14}$ to $10^{16}$ eV were the focus although there was always a drive to find the limiting energy that Nature reached.

In the 1950s a relatively large array of Geiger counters that eventually covered $\sim 0.6\ \text{km}^2$ was developed by Cranshaw and Galbraith (1954, 1957) at Culham near sea level, the site of UK Atomic Energy Establishment. In USSR, investigations of air showers were initiated by Skobeltzyn who encouraged George Zatsepin of the Lebedev Institute to develop a program in the last years of WW II. The first Russian activity was carried out in the Pamirs (3 860 m) and was the start of a major effort on shower work at mountain stations by Soviet scientists which continued for many decades, latterly at a well-serviced installation at Tien Shan (3 340 m) near Almata. The leaders of this work, in addition to Zatsepin, were N.A. Dobrotin, S.I. Nikolsky and S.A. Slavatinsky. There was also a major effort in Moscow, headed first by S.N. Vernov and later by G.B. Khristiansen. Until the start of construction of the Yakutsk array in the late 1960s, the Soviet program was largely focused on studying primary particles of less than $10^{17}$ eV.

The most important output from the early period of the Moscow work was the discovery of a feature in the size spectrum of showers[5] which became known as "knee" of the cosmic ray spectrum (Kulikov and Khristiansen, 1959) and it had considerable impact. It was verified with high precision relatively quickly by a number of groups (Fukui et al., 1960; Kameda et al., 1960; Allan et al., 1962; Kulikov et al., 1965). Estimating the energy of the knee from the track-integral method (see Sect. 5.8), Kulikov and Khristiansen had argued that the break may be caused by diffusion of cosmic rays out of the galaxy, so that cosmic rays at $E > 10^{16}$ eV may have an metagalactic origin. Thus, an astrophysical feature in the cosmic ray spectrum may have been discovered. This started a long running debate, picked up by Peters (1961) who proposed that what was being seen reflected a similar feature in the primary spectrum of cosmic rays induced either by a limitation of the acceleration processes or by a leakage of particles from the galaxy. There were competing claims that this feature was due to a characteristic of nuclear interactions with a dramatic change occurring near $10^{15}$ eV and the debate about astrophysics or particle physics origin, was not to be settled for a further 45 years until precise data from KASCADE became available (see below).

The increasing availability of PMTs led to some significant advances in the air shower technique including the use of Cherenkov radiation to study extensive air showers suggested by Blackett (1947), Galbraith and Jelley (1953), and Chudakov and colleagues (1960). These data were used independently by Greisen (1956) and Nikolsky (1962) to derive a relationship between the primary energy and the shower size which proved to be particularly important for early estimates of the primary energy.

[5]The "shower size spectrum" or just "size spectrum" is a common notion used for the distribution of the shower size, i.e. of the total number of particles that reached ground. The shower size, $N$, is obtained by fitting the lateral distribution $\rho(r)$ of shower particles at ground and evaluating the integral $N = 2\pi \int_0^\infty r\rho(r)\,dr$.

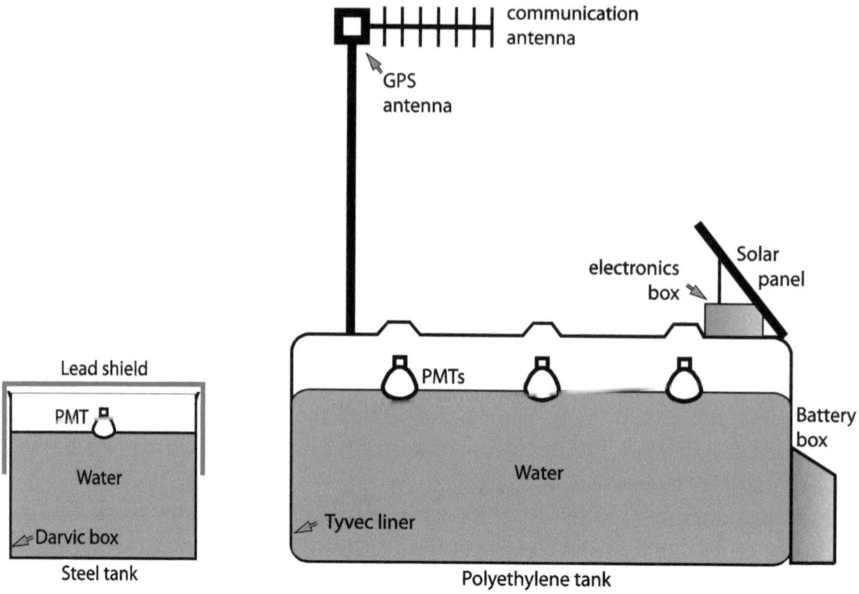


**Fig. 5.6** Scale comparison of the first water Cherenkov detectors used by Porter et al. (1958) of 1.44 $m^2$ read out by a single 5″ diameter PMT to those used by the Pierre Auger Observatory of 10 $m^2$ read out by three 9″ PMTs (Abraham et al., 2004)

Another key development made at Culham was of the water-Cherenkov detector. Credit for this work goes to N.A. Porter who, while a member of the team working with the Geiger counter array, became the first to succeed in preventing bacterial growth in unfiltered water long enough to realise a stable detector (Porter et al., 1958). One of several advantages of a water-Cherenkov detector is that it enables the energy flow in the shower to be measured. Porter's detector can be seen as the prototype of those that were used at Haverah Park (1967–1987) and at the Pierre Auger Observatory (from 2000). Indeed, there has been remarkably little advance over Porter's design in which the PMT looked downwards into the water, as becomes obvious from direct comparison of his design to that of the present Pierre Auger Observatory (Abraham et al., 2004) (cf. Fig. 5.6).

Another extremely important development arising from the availability of PMTs was made at MIT[6] under Rossi's leadership. He had realised that the short fluorescence-decay times that were found in the newly discovered liquid scintillators might make it feasible to construct large area detectors in which fast timing of the arrival of the particles of a shower would be possible. The scintillating material chosen was a solution of terpenyl in benzene held in 5 gallon ($\sim$20 litres) drums of 600 $cm^2$ cross section. Using three of these detectors, mounted in various configurations on the roof of the Physics Department at MIT, Bassi et al. (1953) showed

[6]Massachusetts Institute of Technology.

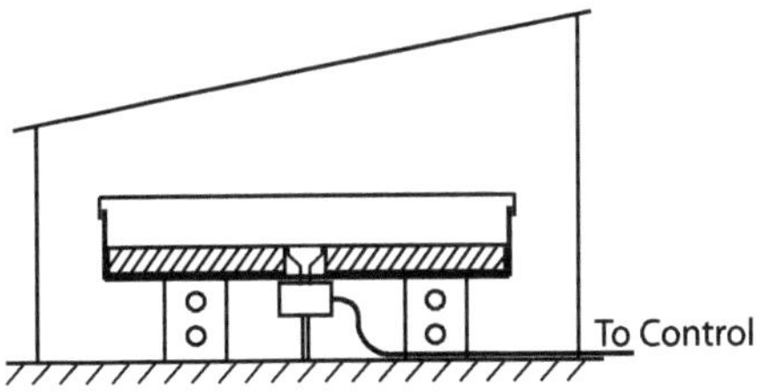


**Fig. 5.7** Schematic diagram of a scintillation counter used in the Agassiz shower array. The scintillator block was 105 cm in diameter and 10 cm thick. The *inside of the box* was painted white and the diffuse light reflected from the walls was collected by a Dumont 5″ diameter PMT (Reproduction from Clark et al., 1957)

that the particles in the disk of the shower were spread over a thickness of only a few metres and, by shielding one of them with up to 20 cm of lead, that the electrons in the shower lead the muons close to the shower axis. The discovery that the shower disk was relatively thin ($\sim$10 ns) opened up the possibility of measuring the direction of the primary particle. Assuming that the direction was perpendicular to a plane tangent to the surface defined by the leading particles in the shower, it was demonstrated that the direction of the shower could be found to within $\sim$2°. This was a major advance as hitherto the very crude collimating effect of the atmosphere had been used to define shower directions.

This pioneering work led to the construction of a larger array at a partially wooded site, the Agassiz Astronomical Station of the University of Harvard. Unfortunately the liquid scintillators were flammable and after a lightning-induced fire a method of making solid scintillator in large slabs with masses of $\sim$100 kg was developed (Clark et al., 1957). These could also be viewed by PMTs and a schematic diagram of one scintillation counter is shown in Fig. 5.7.

At the Agassiz site an array of 15 such detectors was operated between 1954 and 1957 with the layout shown in Fig. 5.8. Members of the group included George Clark, William Kraushaar, John Linsley, James Earl, Frank Scherb and Minoru Oda, who became a leading figure in air shower work in Japan. An excellent first-hand account of Rossi's work at MIT has been given by Clark (2006).

Cosmic-research began in Japan in the 1930s at RIKEN first under the guidance of Y Nishina and then under S Tomonaga. At the end of WW II, experimental work in nuclear physics in Japan was essentially terminated for some years following the destruction of the cyclotrons at RIKEN in Tokyo and those in Kyoto and Osaka. By contrast, cosmic ray work flourished: Tomanaga stimulated studies of extensive air showers at Mt Norikura (2 770 m above sea level) (Ito et al., 1997) and played a key role in establishing the Institute for Nuclear Studies (INS) in Tokyo. He was also instrumental in encouraging Nishimura and Kamata to develop three-dimensional analytical calculations of electromagnetic cascades, work which they began after reading the Rossi and Greisen article of 1941 (Rossi and Greisen, 1941) during daily visits to a US reading room in Tokyo. Japan has been one of the leading countries in cosmic ray physics ever since.

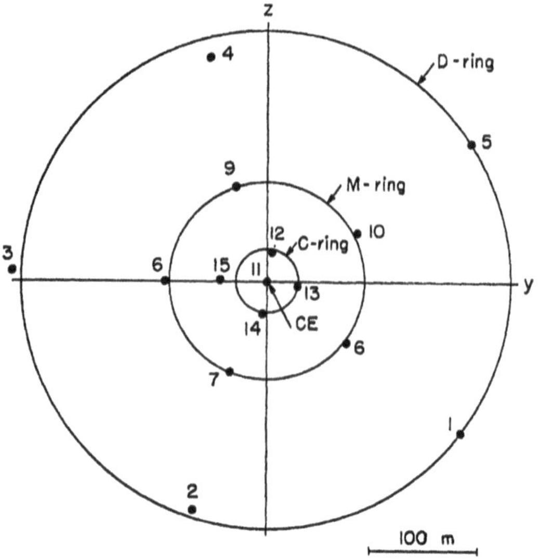


**Fig. 5.8** Schematic diagram of the MIT scintillation detector array. The four detectors in the *C-ring* were used only during a small part of the running time in order to extend the results to small shower sizes of $5 \times 10^5$ particles (Clark et al., 1957)

By far the most important insights at that time came from combined data from the four muon and 14 scintillator detectors operated at INS. Although the results have long since been surpassed, the group was the first to point out the key information that could be derived from a study of plots of muon versus electron number, $N_\mu$ vs. $N_e$ plots in modern language. One of the plots from the INS work is shown in Fig. 5.9 for nearly vertical events. This type of diagram, with improved statistics and smaller uncertainties in $N_\mu$ and $N_e$, when combined with detailed shower simulations, later proved to be a powerful tool for extracting information on primary mass. In addition, it was soon recognised, when Monte Carlo studies developed, that at fixed primary energy the fluctuations in electron number were greater than those for the muons. Accordingly the muon number came to be used as a proxy for shower energy.

The work at INS was the forerunner of other projects in air shower research. In addition to the BASJE project at Chacaltaya (cf. Sect. 5.7), the activities led to the Akeno and AGASA arrays in Japan and the Telescope Array in the USA.

## 5.5 The Impact of the MIT Group and of Rossi

The impact of the MIT group on the understanding of the extensive air showers was enormous. As well as seminal technical advances, they introduced the analysis techniques that are the basis of methods that have been used to deal with data from surface arrays ever since. The group was the first to develop routines to derive the direction of the shower from the fast-timing measurements and to find the position in the plane perpendicular to the shower where the signal size would be the largest, the shower core. The MIT group was also the first to report a measurement of the differential shower size spectrum above $N = 10^6$ and took the first steps towards determining the energy of the primary particles using models of shower development.

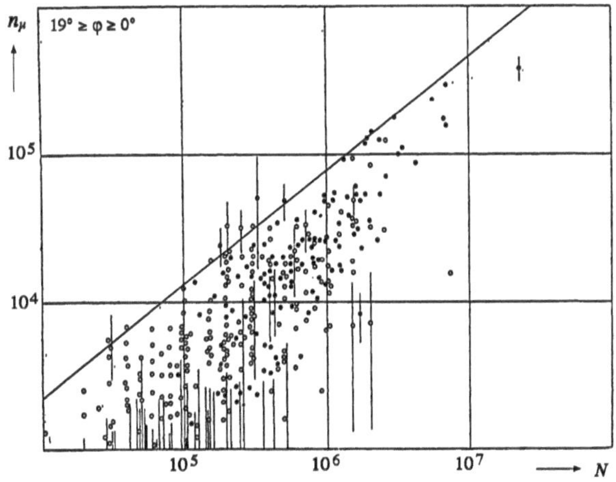


**Fig. 5.9** Reconstructed muon number $n_\mu$ vs. shower size $N$ for vertical showers as measured by Fukui et al. (1960)

The determination of the energy that has created a particular shower is not straightforward and it is instructive to appreciate the various approaches that have been adopted over the years. Although it was established relatively early that air showers contained nucleons, pions and muons in addition to an abundance of electrons and photons, the gross features of showers were found to be relatively well-described under the assumption that the primaries were electrons. It thus became the practice to infer the primary energy from a measurement of the total number of charged particles, $N$, – dominantly electrons and positrons – in a shower, relating this to the primary energy using theory provided by such as the Nishimura–Kamata equations that describe the lateral distribution of charged particles for showers produced by photons or electrons. The number of particles was straightforward to measure when the detectors were Geiger counters as they respond predominantly to charged particles. Also, for the study of the showers produced by primaries of energy less than $\sim 10^{17}$ eV, it was practical and economically feasible to build arrays in which the average separation of the detectors was less than the Molière radius,[7] about 75 m at sea level: roughly 50 % of the charged particles of a shower lie within this distance.

As greater understanding of showers developed, there were moves away from using the photon/electron approximation to estimate the primary energy from the number of charged particles measured in the shower. Also a difficulty in obtaining $N$ was recognised as scintillation counters were increasingly introduced during the 1950s. Because of the success of the approach with Geiger counters and the lack of other methods to find the energy on an event-by-event basis, considerable effort was initially expended in relating the scintillator measurements to what would have been the particle count had a Geiger counter been located at the same position as the scintillation counter. This adjustment to particle number was reasonable while the

[7]The Molière radius is the root mean square distance that an electron at the critical energy is scattered as it traverses one radiation length.

spacing between detectors remained small. For example, at the Agassiz array, measurements were made at distances much closer to the shower core than one Molière radius (see Fig. 5.8) and the scintillator response was converted to particle number using an array of Geiger counters operated for that purpose. As more understanding of shower structure developed, the importance of the thickness of the scintillators was recognised and it was also realised that the conversion from scintillator signal to number of charged particles depended on the distance of the scintillators from the shower core because the energy spectrum of electrons and photons was distance-dependent.

The MIT group also pointed out that to obtain an energy spectrum from the observed size spectrum required "a quantitative knowledge of the cascade processes initiated by primary particles of different energies". This problem of quantitative knowledge of the hadronic process is still an issue over 50 years on though there is a growing understanding of the key hadronic interactions, most recently from the LHC. S Olbert, one of the MIT group (Olbert, 1957) had solved the shower equations to relate the shower size at different atmospheric depths to the primary energy. Using two models of high-energy interactions then current (the Landau model (Belenki and Landau, 1956) and the Fermi model (Fermi, 1950, 1951)), making the assumptions of a collision mean free path for protons of 100 $\mathrm{g\,cm^{-2}}$ and complete inelasticity, Olbert obtained relations between $N$ and energy $E_0$.

A study of the muon content of showers was made using a hodoscoped system of Geiger counters shielded with lead. This work established that roughly 10 % of the particles in an air shower were muons (Clark et al., 1958).

A final report on the work of the MIT group was made in 1961 (Clark et al., 1961) where details of the largest event with a 'Geiger counter size' of $N = 2.6 \times 10^9$ were given. The array at Agassiz had been operated for about a year, from July 1956 to June 1957. In addition the group had run a small shower array at Kodaikanal in India to search for anisotropies in the arrival direction pattern at energies just above $10^{14}$ eV.

The work directed by Rossi subsequently led to the establishment of the shower array at Chacaltaya (with the Japanese group from INS as major partners). A particular motivation was to search for $\gamma$-rays by attempting to identify showers containing fewer muons than average. This attempt was unsuccessful but the first indirect deductions about the position at which the number of particles in showers of $\sim 10^{16}$ eV reach their maximum were obtained. Additionally, attempts were made to find the depth of shower maximum using the constant intensity cut method, a very important technical conception. Rossi also encouraged the work led by Linsley at Volcano Ranch in New Mexico to establish the first array with an area of over 1 $\mathrm{km^2}$, built to study the highest energy events.

## 5.6 Work of the Cornell Group and the Influence of Greisen

Greisen, a former student of Rossi, and who had also worked at Los Alamos, founded a group at the Cornell University in the 1950s. The first step was to build an array of radius 500 m with $15 \times 0.85$ m$^2$ scintillators with five near the centre on 3 to 80 m spacing and 5 each 150 and 500 m from the array centre. Like the MIT group, the Cornell team did not have a fast computer available to them initially and developed some ingenious analogue methods to find the direction and the shower core. This could only be adopted for a relatively small number of large events. Above $10^{13}$ eV, around $10^4$ events were recorded per day with several million accumulated in 1957 and 1958.

A measurement of the number spectrum above $N = 6 \times 10^6$ was made using an approach similar to, but independent of, that of the MIT group and the two measurements were found to be in good agreement. The largest event recorded with the Cornell array contained $N \simeq 4 \times 10^9$ particles.

Greisen and his group also studied muons in showers, extending what had been done at MIT and elsewhere and he derived useful formulae to describe the lateral distribution of muons above 1 GeV and also the energy spectrum of muons as a function of distance. Although the muon sample was only 559, and the shower analysis was not done on an event-by-event basis, the relations established have been found to fit a wide sample of modern work on the muon lateral distribution even for showers of greater energy. The parameterisations of both the electron and muon lateral distributions (LDF) presented in Greisen's seminal reviews (Greisen, 1956, 1960) described the data well over a large range of distances from the shower core and atmospheric depths. Greisen also noted that his parameterisation of the electromagnetic distribution was a close approximation to the analytical calculations for electromagnetic showers performed by Kamata and Nishimura (1958). Greisen's approximations to the Nishimura–Kamata functions become known as the Nishimura–Kamata–Greisen (NKG) function.

After work on the Cornell scintillator array had been completed, Greisen turned his attention to the development of the fluorescence technique. His two reviews remain important sources of insights. In particular, in the first of these reviews, Greisen developed a method to estimate the energy that a primary particle would need, on average, to produce a shower of a certain size.

## 5.7 Work in Bolivia

Of the several laboratories to be developed for the study of air showers, one of the most important, and certainly the highest, was constructed at Chacaltaya in Bolivia at 5 200 m and is still in operation. The mountain had already been used extensively for the exposure of nuclear emulsion plates in the 1940s. At Chacaltaya important steps were taken to infer the depth of shower maximum, to measure the energy spectrum and to study the mass of cosmic rays, including a search for photons.

As a first step to understanding the features of showers at high altitude 11 of the scintillators used in the Agassiz experiment were deployed in an array of 700 m diameter on the Altoplano at El Alto, near La Paz, Bolivia, at an altitude of 4 200 m in 1958. Showers of size $\sim 10^7$ were studied. It was found that, unlike those of a similar size at sea level, the steepness of the lateral distribution changed with zenith angle, being steeper for the more vertical showers. Furthermore, for $N \sim 3 \times 10^6$ the change in shower size with depth from 630 to ~800 g cm$^{-2}$ was small suggesting that these showers had their maxima close to 630 g cm$^{-2}$ (Hersil et al., 1961, 1962).

In 1958, following a proposal by Oda, the MIT, Tokyo and La Paz groups joined forces to establish the Bolivian Air Shower Joint Experiment (BASJE) at Mt Chacaltaya which started taking data in the early 1960s. The basic shower array comprised the 20 Agassiz-like scintillators deployed within a circle of 150 m diameter with five scintillators for fast timing, supplemented with a muon detector of 60 m$^2$ array. The muon detector was constructed from 160 tonnes of galena (the natural mineral form of lead sulfide) which was readily available locally. Modules of 4 m$^2$ commercial scintillator were developed by K Suga (INS) and were used together with a logarithmic time-to-height amplifier (Suga et al., 1961) to measure the muon flux in showers. The 60 m$^2$ of scintillator were placed below a concrete structure supporting the galena. The size of this muon detector exceed those built previously by about an order of magnitude and made practical a search for showers produced by primary gamma rays under the hypothesis that such showers would have low numbers of muons. Events with less than 10 % of the average number of muons were found but they were not clearly separated from the bulk of the data and did not show any anisotropy. In addition to the energy spectrum measurements and the photon search, innovative studies of the mass composition of cosmic rays were made. Further, Krieger and Bradt (1969) augmented the scintillator array with nine open PMTs to detect air-Cherenkov light and concluded that at $\sim 10^{16}$ eV the composition was much as it was at $10^{12}$ eV.

## 5.8 The First Surface Detector Arrays Covering More than 1 km$^2$

Many small arrays were built to study the cosmic rays in the region from $10^{14}$–$10^{17}$ eV at locations across the world with scientists in Australia, Germany, India, Italy, Japan, Poland, UK, the USA and the USSR making important contributions. The early measurements have been replicated with very superior statistics in the modern arrays built in Germany (KASCADE and KASCADE-Grande), in Italy (EAS-TOP) and in Tibet: this applies particularly to the energy region $10^{14}$ to $10^{16}$ eV which includes the region where the energy spectrum steepens. We shall discuss those briefly in Sect. 5.14.

By contrast the number of devices constructed with collecting areas of over 1 km$^2$ has been only 7, including the Pierre Auger Observatory, the Telescope Array and the Yakutsk array that are still operating, although with the latter reconfigured to study smaller showers. A Soviet proposal for a 1 000 km$^2$ array named EAS-1000, led by Khristiansen, was given formal approval and construction began (Khristiansen et al., 1989), but the project was hit by the political and economic problems that came with the glasnost and perestroika and was never realised. Data from the Telescope Array and the Pierre Auger Observatory currently dominate from the Northern and Southern Hemispheres, respectively. By contrast to the low-energy arrays, it is useful to discuss the pioneering large arrays in some detail first, as at each different features of technique and analysis were introduced which were important for later studies. The layout of these surface arrays can be found in the review by Nagano and Watson (2000): essentially all arrays are variations of the style developed at MIT shown in Fig. 5.8. While methods of data recording evolved, the analysis techniques were similar to those introduced at MIT.

The first of the giant shower arrays was constructed at *Volcano Ranch*, New Mexico (1 770 m) by members of the MIT group (Linsley et al., 1961). It consisted of 19 plastic scintillation counters of 3.3 $m^2$ area, each viewed with a 5″ PMT. The construction, maintenance and data analysis of Volcano Ranch was the almost single-handed effort of Linsley who made many contributions to the understanding of giant showers. Figure 5.10 shows him together with his colleague Livio Scarsi.

Data from this array yielded the first measurement of the energy spectrum of cosmic rays above $10^{18}$ eV, giving the earliest hint of a flattening of the spectrum in that region (Linsley, 1963a), a hint that took over 20 years to confirm convincingly. Linsley also made the first exploration of the arrival direction distribution of these exceptional events. The most energetic one was assigned an energy of $10^{20}$ eV (Linsley, 1963b), an energy that was subsequently revised to $1.4 \times 10^{20}$ eV (Linsley, 1980). This event, reported before the discovery of the 2.7 K cosmic microwave background radiation and the subsequent prediction of a steepening of the spectrum, remains one of the most energetic cosmic rays ever recorded.

Following the closure of the Culham array in 1958 it was decided, under the strong influence of Blackett, that work on extensive air showers should continue in the UK but be supported and developed within the university environment by a team drawn from several universities. This led to the construction of the *Haverah Park array* (1964–1987) under the leadership of J G Wilson until his retirement in 1976, with strong support in the initial stages from R M Tennant. Prototype studies were carried out at Silwood Park near London under H R Allan who led a small team to examine the potential of the Cherenkov detectors developed by Porter at Culham (Allan et al., 1962) and A W Wolfendale, who led an effort to evaluate the potential of neon flash tubes.

While the Silwood studies were underway, a site search identified land about 25 km from the University of Leeds (200 m) where an array covering 12 $km^2$ was established and which operated for 20 years from 1967 to study features of showers from $10^{15}$ to $10^{20}$ eV. The primary detectors were water-Cherenkov detectors of 2.25 $m^2 \times 1.2$ m with over 200 being deployed. In addition there was 10 $m^2$ of liquid scintillator shielded by lead to provide muon detectors with an energy threshold of 250 MeV, and a muon spectrometer.

**Fig. 5.10** Livio Scarsi (sitting) and John Linsley in 1960 at an age of 33 and 35 years, respectively (Linsley and Scarsi private archive©)

The determination of the energy of the primary presented particular problems as it was impossible to relate the observed signal to the number of charged particles, as had been done at Agassiz, in a reliable way. The corresponding modelling was done with exceptional skill by M Hillas, whose rare insight was supported by carefully analysed auxiliary experiments, and with early Monte Carlo calculations carried out on a mainframe (KDF9) computer with only 64 kByte of memory. His method of estimating the primary energy was used to argue that a shower of energy $>5 \times 10^{19}$ eV had been detected at Haverah Park soon after the Greisen–Zatsepin–Kuzmin prediction (see below) (Andrews et al., 1968) and has later been widely adopted in subsequent measurements with ground arrays, at AGASA, the Pierre Auger Observatory and the Telescope Array.

By far the most complex, and most northerly, of the early giant arrays, was the *Yakutsk Array* operated by the Institute of Cosmophysical Research and Aeronomy at Yakutsk, Siberia (105 m). It began taking data in 1970 and was developed to cover an area of 18 $km^2$ in 1974. The leaders were D.D. Krasilnikov and N.N. Efimov with the close involvement of Nikolsky and Khristiansen from Moscow. A detailed description of the array has been given by Afanasiev et al. (1993). A particularly important feature was the presence of 35 PMT systems of various areas to measure the air-Cherenkov radiation associated with the showers. These gave indirect information about the longitudinal development and provided a calorimetric approach to the energy estimates for the primary particles through the track-integral method (Greisen, 1956, 1960). Measurements relating to the energy spectrum, the mass composition and arrival direction distribution of cosmic rays above $10^{17}$ eV have been reported. In recent years the array has been contracted to study showers of lower energy and to make more detailed investigations of showers of higher energy.

The team from the University of Sydney who designed 'The Sydney University Giant Air Shower Recorder (*SUGAR*)' introduced a totally novel concept to the detection of extensive air showers by an array of ground detectors. Before this innovation, the practice had been to link the detectors with cables to some common point where coincident triggers between them could be made and the signals recorded, in the early days often using oscilloscopes. This method becomes impractical for areas much above 10 $km^2$ as it was rarely possible to have the relatively unrestricted land access enjoyed by Linsley at Volcano Ranch: the cost of cable, their susceptibility to damage and the problems of generating fast signals over many kilometres were further handicaps. The concept, due to Murray Winn, was first discussed in 1963 (McCusker and Winn, 1963). The Sydney group proposed the construction of an array of detectors that ran autonomously with the time at which a trigger above a certain number of particles was recorded being measured with respect to a timing signal transmitted across the area covered by the detectors. The concept was realised in the Pilliga State Forest near Narribri (250 m) where 47 stations were deployed over an area of $\sim$70 $km^2$. Most of the detectors were on a grid of 1 mile (1.6 km) with 9 on a smaller spacing to enable smaller showers to be studied. Time and amplitude data were recorded locally on magnetic tape and coincidences between different stations found off-line some time after the event. A difficulty was that the rate of triggers of a local station above a level that was low enough to be useful is very high and the rate could not be handled with technologies available at the time. The problem was solved by burying the detectors under 2 m of earth and placing them in pairs 50 m apart.

While the concept was brilliant it was somewhat ahead of its time in terms of the technology available. Calor gas had to be used to supply the power at each station and the reel-to-reel tape recorders proved difficult to operate in the dusty environment. The array was thus quite difficult to maintain and the problem of handling many magnetic tapes at a single computing site proved to be a challenge. The PMTs used were 7″ in diameter and suffered from after-pulsing which complicated the measurement of the signals as logarithmic time-to-height converters were used to find the amplitudes (Suga et al., 1961). Efforts were made to overcome this difficulty. There was also a serious problem in estimating the energy of events as only muons were detected and therefore there was total reliance on shower models with little ability to test which was the best to use because of a lack of different types of detector in the array. Attempts to overcome this with a fluorescence-light detector and with a small number of unshielded scintillators were unsuccessful. Energy spectra were reported in (Winn et al., 1986a). The measurement of the shower directions to a precision of a few degrees was a demonstration that the timing stamp method was effective and the most valuable data from the SUGAR array were undoubtedly from the measurements of directions, the first such measurement to be made from the Southern Hemisphere at energies above $10^{18}$ eV (Winn et al., 1986b). In later analyses of the SUGAR database, the Adelaide group reported the detection of a signal from the region of the Galactic Centre (Clay et al., 2000; Bellido et al., 2001).

The concept of autonomous detection was tested at Haverah Park in an early attempt to devise methods to construct an array of $\sim$1 000 $km^2$ but the method had its most effective realisation in the system that was designed for the surface detector array of the Pierre Auger Observatory and subsequently at the Telescope Array.

The largest shower array constructed before the advent of the Pierre Auger Observatory and the Telescope Array was the 'Akeno Giant Air Shower Array (*AGASA*)' which was built outside Tokyo at Akeno (900 m). The AGASA team was led by M Nagano and the array operated from 1990 until 2004. It consisted of 111 unshielded scintillator detectors each of 2.2 $m^2$ with an inter-detector spacing of $\sim$1 km. Muon detectors of various areas between 2.4 and 10 $m^2$ were installed at 27 of the 111 detectors. Each detector was serviced using a detector control unit that recorded the arrival time and size of every incident signal and logged monitoring information, the pulse height distribution, the voltage, counting rate and temperature in a manner that anticipated what is done at the Auger Observatory. An optical fibre network was used to send commands, clock pulses and timer frames from the central station to each module and to accept the trigger signals, shower data and monitoring data.

Some important claims were made about the energy spectrum and the arrival direction distributions at the highest energies. The energy spectrum was reported as extending beyond $10^{20}$ eV with the 11 events observed, showing no sign of any cut-off. The energies were estimated using model calculations and subsequent work, in which the energy spectrum has been found by the track-length integral method inferred from observations of fluorescence light, have shown that there were deficiencies in the model calculations used.

## 5.9 Use of the Monte Carlo Technique

The use of Monte Carlo techniques in the study of the cascade characteristics of air showers has grown enormously since they were first introduced in the early 1960s. The techniques developed have become indispensable for the interpretation of data, to model the performance of detectors and to understand the development of the cascade itself (Wilson, 1952). Wilson's work was carried out with what was essentially a roulette wheel but subsequent activities depended on the computing power available with particular ingenuity being shown in the earliest days to combat the limitations of the times.

Early calculations of the cascade development made use of phenomenological models of the hadronic interactions such as the CKP-model of Cocconi et al. (1962) developed to calculated particle fluxes from future accelerators. Other phenomenological models were developed and were used in interpretation of data from many experiments. A problem was recognised by Linsley in 1977 when he found that some of the Monte Carlo calculations produced results that were in violation of his elongation rate theorem (Linsley, 1977) in that the computation of the change of some shower parameters with energy was greater than was physically possible. This raised questions about the accuracy of some of the Monte Carlo codes. Accordingly Linsley and Hillas (1982) organised a discussion targeted at having interested groups use a common model within their codes to calculate the depth of shower maximum and how it varied with energy. This exercise was partially successful and the results from seven groups who contributed were reported and assessed. The problem of following all of the particles in a shower was first discussed by Hillas (1982): he introduced the concept of 'thinning' which has subsequently had very wide application. He pointed out that it was not necessary in some cases to follow every particle to get a good picture of a shower and reported that good results for muons were obtained efficiently by choosing a demarcation energy, $D$, set at $10^{-4}$ of the primary energy, and following all particles of energy $>D$ but only a fraction of particles of energy $E < D$. The technique was also used for electromagnetic cascades.

By the mid-1980s computing power had increased enormously and several major programs were developed. Hillas created the MOCCA program at this time, written in Pascal. Only a limited description of this code reached the literature but it was made available to the designers of the Auger Observatory for which purpose it was translated into FORTRAN in the early 1990s.

When work on the KASCADE project at Karlsruhe started by the end of the 1980s, it had been realised that most of the cosmic ray projects used their own specific tools which often became a source of errors. In case of disagreement between experiments it remained unknown whether the problem had been of purely experimental nature related to the apparatus or whether it had been due to differences in the EAS simulations applied. Thus, parallel to preparing for constructing KASCADE, an extremely important code was developed, with input by J Capdevielle and by P Grieder who were early pioneers of the Monte Carlo method. The CORSIKA code ('COsmic Ray SImulation for KAscade'), continuously maintained by a team at Karlsruhe with support from all over the world, has the merit of allowing different models of nuclear interactions to be included in an easy way and the authors made it widely available to the community.[8] Thus, over the years it had become a *de facto* standard in the field, similar to the GEANT simulation package in high-energy physics.

The important step made with CORSIKA is that, even though the EAS modelling may not be perfect, the very same modelling can be applied to all experiments in the field. As J Knapp, with D Heck one of the drivers behind CORSIKA, stated in his rapporteur talk at the ICRC in Durban (Knapp, 1997):

> Is the composition changing or not? The answer depends on the yardstick (i.e. the Monte Carlo program) used for comparison. Use the same yardstick to get consistent results, use a well-calibrated yardstick to get the correct result.

In addition to its application in shower modelling, the CORSIKA code has been used in many other investigations, ranging from mountain and pyramid tomography through muon measurements over neutrino searches to the possible link between cosmic rays and climate (see e.g. Usoskin and Kovaltsov, 2006).

---

[8] http://www-ik.fzk.de/corsika/.

## 5.10 The Impact of the Discovery of the Microwave Background Radiation

The primary purpose of the early km$^2$-scale EAS experiments was to study the energy spectrum and arrival directions of ultra-high-energy primary cosmic rays for the information which these data give about the origin of cosmic rays. It had been realised that cosmic ray particles beyond $10^{20}$ eV, which were believed to be atomic nuclei, would have a very great magnetic rigidity. Thus, the region in which such a particle originates must be large enough and possess a strong enough magnetic field so that $RB \gg (1/300) \cdot (E/Z)$, where $R$ is the radius of the region in cm, $B$ is the magnetic field in Gauss and $E$ is units of TeV. Also, anisotropies were expected to be seen. However, estimates of the particle flux were over-optimistic.

In May 1965 Penzias and Wilson reported their serendipitous observation of the cosmic microwave background radiation (CMB) (Penzias and Wilson, 1965). Only a few months later, Gould and Schréder (1966) pointed out that high-energy photons of a few $10^{14}$ eV traversing cosmic distances would suffer rapid energy losses due to electron-positron pair production by photon-photon collisions in the CMB. Thus, some earlier claims of high-energy muon-poor showers, supposed to be initiated by photons of extragalactic origin, were questioned by the authors and no "window" was open for extragalactic $\gamma$-ray astronomy until well above $10^{14}$ eV (Jelley, 1966). A few months later, Greisen (1966a) and independently Zatsepin and Kuz'min (1966) noted a related effect for proton primaries, in this case photo-pion production in the CMB being responsible for rapid attenuation of protons of energy beyond $4 \times 10^{19}$ eV. Figure 5.11 shows the key figure of Zatsepin and Kuz'min's paper including the data point from Linsley (1963b) which was hard to understand after this finding. The title of Greisen's paper "End to the Cosmic-Ray Spectrum?" expressed the situation perfectly and the effect became known as "GZK-effect". Its worth pointing out that both Greisen as well a Zatsepin and Kuz'min also noted that light and heavy nuclei would suffer rapid photo-disintegration above about the same energy threshold.

It is an interesting fact that the aforementioned large shower arrays that were developed in the UK, Siberia, and Australia which dominated the studies of cosmic rays above $10^{17}$ eV during in the 1970s and 1980s were all planned before this discovery which was to become one of the main motivations for their operation. By contrast, planning of the Fly's Eye detector, which detected fluorescence radiation, was begun in 1973 long after the interaction of the CMB and ultra-high-energy cosmic rays had been recognised and verification of the GZK-effect was one of the prime motivations for its construction. However, it turned out that none of these devices had a sufficiently large aperture to establish the existence of a steepening in the cosmic ray spectrum. In fact, the dispute between AGASA and Fly's Eye about the existence of a flux suppression at the highest energies became an important argument for the construction of Pierre Auger Observatory by the end of the 1990s.

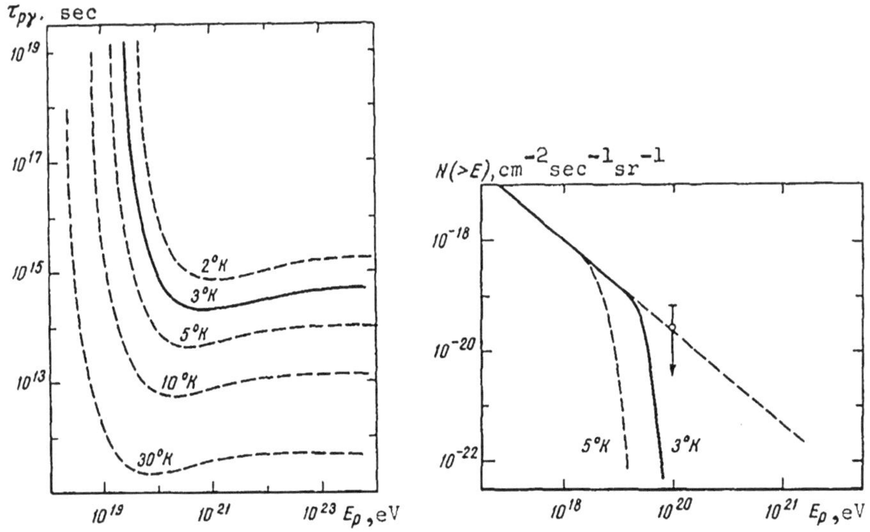


**Fig. 5.11** *Left*: Characteristic time for GZK-like collisions as a function of proton energy for different photon gas temperatures. *Right*: Expected suppression of the energy spectrum for a simplified source scenario (Zatsepin and Kuzmin, 1966)

## 5.11 The Development of the Fluorescence Technique

A powerful new technique for studying extensive air showers from the highest energy particles was developed during the 1960s. The approach depends on observing the faint fluorescence radiation which is emitted isotropically when the 2P and 1N band-spectra associated with molecular nitrogen are excited by ionising particles. It allows the atmosphere to act as a massive calorimeter and in principle gives the possibility of measuring the energy of cosmic rays without resorting to assumptions about hadronic physics. The key to large apertures is the isotropic emission, with application of the track-length concept used to give the shower energy.

It is not clear who first had the inspiration of using the excitation of $N_2$ for cosmic ray work and it may well be that the idea occurred to several people at about the same time. The concept of employing air-fluorescence to detect $X$-rays from nuclear explosions appears to have been discussed after the Manhattan project and it seems probable that Edward Teller was the first to have the idea of using air-fluorescence induced by $X$-rays produced in such explosions as a monitoring tool. This is the so-called "Teller light". The documents are still classified, but the application to NSF made by the Utah group in 1973 (J.W. Keuffel et al.) for construction of the Fly's Eye detector refers to the Teller light in the title of a classified paper. Greisen, who had been at the Trinity test, was perhaps aware of this activity and the idea may have been discussed informally in the US during the 1950s as a way of detecting the highest-energy cosmic rays.[9]

[9] S Colgate, private communication to AAW.

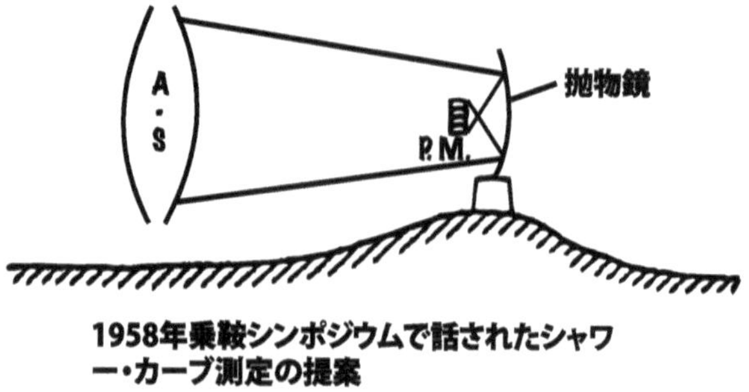


**Fig. 5.12** Concept of a PMT camera viewing the fluorescence light from an air shower collected with a mirror. The similarity of the layout shown here to the devices constructed by the Utah, Auger and TA groups is remarkable. Reproduction from Proceedings of Norikura Meeting in Summer 1957, INS Report 1958. The text translates as: *"parabolic mirror"* and *"A proposal for the shower curve measurement in Norikura symposium, 1958"* in *upper/lower lines*

The method was first discussed at an international forum in La Paz in 1962 where Suga outlined the idea and showed a spectrum of the emission in the ultra-violet part of the spectrum using $\alpha$-particle sources (Suga, 1962). The signal was expected to be small, even from showers produced by primary cosmic rays of $10^{20}$ eV, as the isotropic emission is only about 4 photons per metre of electron track in the wavelength range from 300 to 450 nm.

The fact that the light is emitted isotropically makes it feasible to observe showers 'side-on' from very great distances and thus it opens the possibility of monitoring large volumes of air. It is clear from a diagram taken from a Japanese publication of 1958 (Fig. 5.12) that discussions about using this method to detect high-energy cosmic rays must have taken place in Japan, under the guidance of Suga and Oda, for some years prior to Suga's report at La Paz.[10] During the discussions following Suga's presentation, Chudakov reported the results of measurements that he had made in 1955–1957 of the same phenomenon. He examined this effect as he was concerned that it might be a background problem in the detection of Cherenkov radiation, a technique that was being developed strongly in the Soviet Union in the 1950s, but he was slow to write up his observations (Belyaev and Chudakov, 1966). Chudakov also observed transition radiation in the same series of experiments.

The use of fluorescence radiation to detect air showers was already being studied in Greisen's group which included Bunner (1967, 1968) measuring the spectrum of the light produced by particles in air. Greisen did not mention this activity at La Paz but in an important review talk in 1965 (Greisen, 1966b) he pointed out many of the key issues and showed the band spectrum of the fluorescence light from 200 to 460 nm. This paper had a much wider distribution than did the report of Suga's talk.

The Japanese plans did not develop immediately. Goro Tanahashi from the INS group in Tokyo worked in Greisen's team at Cornell in the mid-1960s where efforts were being made to detect fluorescence radiation using a set of Fresnel lenses. On

[10] Tanahashi and Nagano, private communication.

his return to Japan Tanahashi played a major role in setting up a fluorescence detector at Mt Dodaira, with Fresnel lenses, and the successful detection of air showers by the fluorescence method was reported in 1969 (Hara et al., 1970). Greisen acknowledged this achievement generously[11] and recently Bruce Dawson has confirmed the INS conclusions using his experience from the Auger Observatory to re-examine the INS data (Dawson, 2011). The use of fluorescence light as a detection technique seems to have been thought of more or less simultaneously in three countries, but it is clear that the Japanese air shower physicists were the first to make convincing detections.

The work of Greisen's group at Cornell ended in 1972. Although unsuccessful his efforts had inspired many. Tanahashi attempted to introduce the fluorescence technique into the Sydney Air Shower array and Greisen's work was taken up in the USA by a team at the University of Utah, led first by Keuffel. Following the Japanese efforts, another convincing demonstration of the method was finally achieved through the operation of a small fluorescence detector in coincidence with the Volcano Ranch scintillator array (Bergeson et al., 1977). Fluorescence detectors could now be used as stand-alone devices.

Another lasting legacy of Greisen's work was the diagram made by Bunner for his 1964 master's thesis (Fig. 5.13). Here the essence of the reconstruction method is shown: the diagram has been reproduced many times but its source has rarely been acknowledged.

## 5.12 Development of Fly's Eye

Following the death of Keuffel, the Fly's Eye efforts in Utah were led by Haven Bergeson and then by George Cassiday and Gene Loh. Construction of a prototype detector (which later became known as Fly's Eye I) consisting of 67 camera units started near Dugway (Utah) in the early 1970s (Bergeson et al., 1975) and for cross-correlation and overall testing, three of those were taken to the Volcano Ranch array and positioned about 1.5 km from the ground array. With this set-up 44 showers were registered in 12 consecutive nights of operation. The events recorded ranged up to $2.5 \times 10^{18}$ eV and established the method of air-fluorescence detection, marking a major breakthrough in ultra high-energy cosmic ray methods.

Each of the 67 Fly's Eye I units used 1.5 m diameter aluminised spherical mirrors, associated Winston light collectors, PMTs and data acquisition electronics all of which were housed in corrugated steel pipes about 2.13 m long and 2.44 m in diameter giving the Fly's Eye detector a very specific look. In total, there were 880 PMTs at Fly's Eye I, each subtending a 5° by 5° pixel of the sky, which completely imaged the entire night sky. To improve shower reconstruction in the absence of a ground array at Dugway, the stereoscopic observations were pioneered by erecting Fly's Eye II at 3.4 km distance relative to Fly's Eye I. This smaller array of eight

[11] Letter from Greisen to Tanahashi, 29 Sept. 1969.

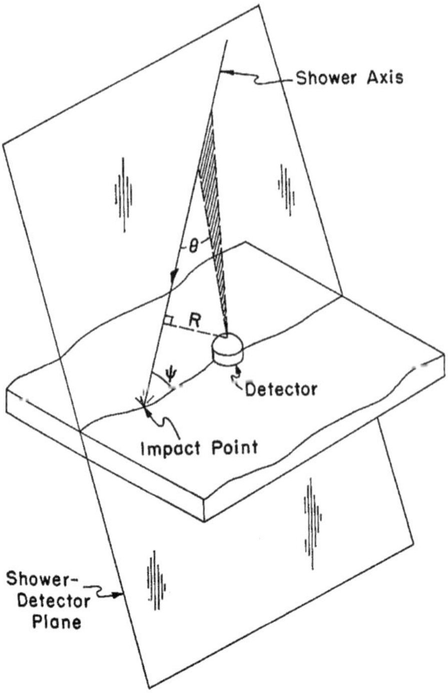


**Fig. 5.13** Perspective view of the shower geometry for fluorescence detector observations (Bunner, 1964)

(later extended to 36) identical units started operation in 1986. In monocular mode, Fly's Eye reached a collection area of about 1 000 km$^2$ (effectively about 100 km$^2$ if the ~10 % duty cycle of night time operation is taken into account).

A spectrum from a single eye was reported in 1975 along with a measurement of the mass composition above $10^{18}$ eV before work with two Eyes started. Full operation began in 1981. The science output culminated in the report of an event of $(3.2 \pm 0.9) \times 10^{20}$ eV (51 Joule) recorded in 1991 (Bird et al., 1995), still the highest energy ever claimed. The event fell only 12 km from the Fly's Eye I detector, allowing a good measurement of its profile and energy. However, it fell behind the Fly's Eye II detector, so it was not seen in stereo.

The aperture of this pioneering experiment was too small to measure the spectrum at $10^{20}$ eV, and hence to observe the GZK cut-off. However, the Fly's Eye and AGASA spectral measurements (see below) set the stage for work to come with the HiRes and the Pierre Auger Observatories.

## 5.13 The Cygnus X-3 Story and Its Impact

One of the consequences of the work on the cores of showers carried out at the Kiel array was the impact of an unexpected result that was never confirmed. In 1983 Samorski and Stamm (1983) reported a surprising observation suggesting that the

11 kpc distant $X$-ray binary system, Cygnus X-3, was a source of photons of above $2 \times 10^{15}$ eV. A signal of $4.4\sigma$ was found in the region around the object using data obtained between 1976 and 1980 based on 16.6 events above a background of $14.4 \pm 0.4$. Cygnus X-3 has a periodicity of 4.8 hours and 13 of the events in the on-source region were in one of the 10 phase bins into which the 4.8 hour period was divided. The Kiel conclusion appeared to be confirmed by results from a sub-array at Haverah Park (Lloyd-Evans et al., 1983), tuned to $\sim 10^{15}$ eV, and also by measurements made around the same time at lower energies using the air-Cherenkov technique. The claims stimulated great interest and, although now regarded as incorrect, gave a huge stimulus to activity in the fields of high-energy gamma ray astronomy and ultra-high-energy cosmic rays.

For the air shower field an important consequence was the interest that James W Cronin (University of Chicago) took in the subject. A Nobel Laureate for his work in particle physics, Cronin entered the cosmic ray field with vigour and led a team from the Universities of Chicago and Michigan to construct an air shower array, known as CASA-MIA, of $\sim 0.24$ km$^2$, to search specifically for signals from Cygnus X-3 (Borione et al., 1994). The array was on a different scale, in terms of numbers of detectors, from anything built previously with 1 024 scintillators of 1.5 m$^2$ laid out on a rectangular grid with 15 m spacing, above the muon detectors, each of 64 m$^2$, buried 3 m deep at 16 locations. As with the Chacaltaya array built 30 years earlier, the idea was that showers with small muon numbers were likely to be produced by gamma rays. The area of the muon detector was over 40 times that at Chacaltaya.

No signals were detected from Cygnus X-3 suggesting that the results from Kiel, Haverah Park and the TeV gamma ray observatories were spurious. However, what this enterprise showed was that it was possible to build much larger detectors than had been conceived previously and Cronin went on to be the leading player in the planning and implementation of the Pierre Auger Observatory. Another consequence of the Cygnus X-3 period was that other particle physicists, most notably Werner Hofmann and Eckart Lorenz, began work at La Palma to search for signals from Cygnus X-3 using a variety of novel methods, but they quickly moved into high-energy gamma ray astronomy.

## 5.14 Recent and Current Activities

The Cygnus X-3 observations revitalised experimental efforts for studying cosmic rays above $10^{14}$ eV and resulted in a new generation of devices with sophisticated instrumentation, including CASA-MIA, GRAPES, HEGRA, EAS-TOP, KASCADE, MAKET-ANI, Tibet-AS$\gamma$, and others.

In Italy a group led by Gianni Navarra in the mid 1980s started to install a multicomponent detector at the Campo Imperatore at 2 005 m a.s.l. on top of the underground Gran Sasso Laboratory, named *EAS-TOP*. It consisted in its final stage of an array of 35 modules of unshielded scintillators, 10 m$^2$ each, separated by 17 m near the centre, and by 80 m at the edges of the field, covering an area of about 0.1 $km^2$ for measurement of the shower size. A central 140 $m^2$ calorimeter of iron and lead, read out by 8 layers of positional sensitive plastic streamer tubes, allowed measurements of hadrons ($E_h \geq 30$ GeV) and muons ($E_\mu \geq 1$ GeV) in the shower core (Aglietta et al., 1989). Operation started in 1989 and a very important feature of EAS-TOP was the unique possibility of correlated measurements with the MACRO detector located underground in the Gran Sasso Laboratory, thereby combining shower information at ground with TeV energy muons measured underground.

EAS-TOP measured the cosmic ray mass composition across the knee, allowed tests of hadronic interaction models, measured the $p$-Air interaction cross section, and very importantly, it provided stringent tests of the cosmic ray anisotropy as a check of decreasing Galactic content of the cosmic rays. Before operation finally was terminated in 2000,[12] contacts were made to explore the possibility of shipping the scintillator stations to the KASCADE site in Karlsruhe to continue operation in an enlarged experiment there. A summary of the results from EAS-TOP has been given in Navarra (2006).

At the end of the 1980s, two institutes at the research centre in Karlsruhe, Germany (now KIT) led by G Schatz and B Zeitnitz joined efforts together with University groups from abroad, to construct *KASCADE* (KArlsruhe Shower Core and Array DEtector). Again, this endeavour was motivated largely by the surprising results from the Kiel array, so that $\gamma$-ray astronomy was on the agenda. However, concise measurements the cosmic ray composition and of hadronic interactions were realised to be of great need and the experiment was designed accordingly. Karlsruhe was chosen as the site mostly because of its direct proximity to all the infrastructure of the centre, needed to operate a most complex EAS experiment. It consisted of 252 array stations of $e/\gamma$- and $\mu$-detectors spread over $200 \times 200$ $m^2$ area, a highly complex 320 $m^2$ central detector, and a 130 $m^2$ area $\mu$-tracking detector, details of which are described in Antoni et al. (2003). The sampling fraction (fraction of counter area covered in the fiducial area of the EAS experiment) of 2.6 % and 3.3 % for the electromagnetic and muonic component, respectively, is the largest of all EAS experiments ever operated and was crucial for achieving good electron and muon number information. Figure 5.14 shows an example event measured with KASCADE.

Data taking started in 1996 and like the other projects already mentioned, KASCADE never found any significant diffuse or point-like $\gamma$-flux and only provided upper limits. Its main achievements, however, were tests of hadronic interaction models and most importantly measurements of the cosmic ray composition across the knee. The high experimental precision enabled a two-dimensional unfolding of the measured $N_e$ vs. $N_\mu$ distributions – 45 years after similar plots from the INS array (cf. Fig. 5.9) were analysed. The results convincingly demonstrated that the knee in the cosmic ray spectrum is caused by light particles and that the knee could be seen in five different mass groups with their position shifting to higher energies with increasing mass (Antoni et al., 2005), in good agreement with Peter's cycle (Peters,

[12] This was primarily for reasons of environmental protection arguments that applied to the Campo Imperatore area that was designated a National Park.

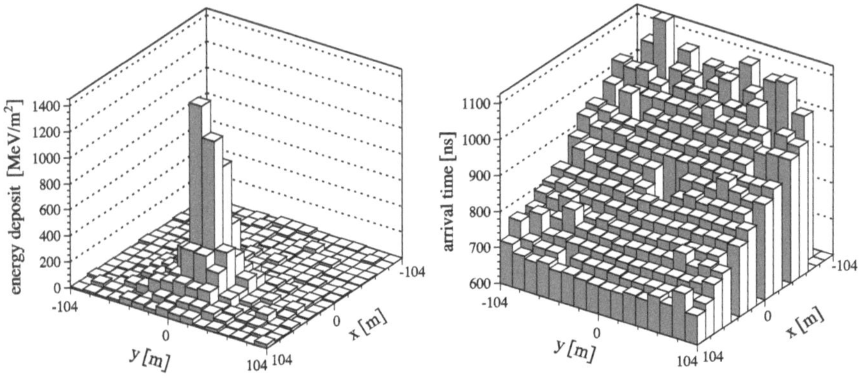


**Fig. 5.14** Example of an EAS registered by the $e/\gamma$ detectors of the KASCADE experiment in the energy range of the knee. *Left*: Energy deposits, *Right*: arrival times. The position of the shower core and the curvature of the shower front are well observed (Antoni et al., 2003)

1961). This achievement of combining high precision EAS data with sophisticated mathematical tools marked another milestone in cosmic ray physics.

Obviously this observation showed the need for improved data up to $10^{17}$ eV where the break of the iron-knee would be expected. The closure of EAS-TOP at about the same time triggered Navarra and Kampert to extend the KASCADE-Experiment with the scintillator stations of EAS-TOP to become KASCADE-Grande (2010). It covered an area of about 0.5 $km^2$, operated from 2003 to 2010, and recently (Apel et al., 2011) demonstrated a knee-like structure in the energy spectrum of the heavy component of cosmic rays at $E \simeq 8 \times 10^{16}$ eV. Does this mark the end of the galactic cosmic ray spectrum? In fact, the cosmic energy spectrum appears be much richer in its features than could be described by simple broken power laws, challenges to be addressed by future observations, such as the Tibet Array, IceTop as part of IceCube, GAMMA, GRAPES, and TUNKA.

At the highest energies, the second-generation air-fluorescence experiment, High-Resolution Fly's Eye (*HiRes*) became the successor of Fly's Eye. Proposed in the early 1990s it was completed in 1997 (HiRes I) and 1999 (HiRes II). It was also located at Dugway, Utah and also had two air-fluorescence detector sites, HiRes I and HiRes II spaced 12.6 km apart. This detector had smaller phototubes resulting in a pixel size of 1° by 1° in the sky. Amongst other improvements over the original Fly's Eye was an FADC data acquisition system at HiRes II which allowed a much more precise measurement of the longitudinal shower profile. The HiRes-I detector took data in monocular mode from 1997 to 2006, while HiRes II operated from 1999 to 2006. The last years of operation of HiRes suffered from an accident at the military site of Dugway which subsequently meant that a very small number of people could go to the site for shifts. Despite these operational problems, a rich spectrum of measurements of the cosmic ray composition, $p$-Air cross section, anisotropies, and the energy spectrum was reported. Most notably, clear signs of a cut-off in the energy spectrum, in good agreement with the GZK-effect was demonstrated. A comprehensive summary of the late Fly's Eye and early HiRes results can be found in Sokolsky and Thomson (2007).

The problem of the small number of events at the highest energies was recognised in the 1980s, even before the AGASA and HiRes detectors had completed construction, and a controversy about the existence or non-existence of a suppression of the cosmic ray flux at the GZK threshold of $5 \times 10^{19}$ eV became a major point of discussion. This led to the idea that 1 000 $km^2$ of instrumented area was needed if progress was to be made. Cronin argued that 1 000 $km^2$ was insufficiently ambitious and in 1991 he and Watson decided to try to form a collaboration to build two identical detectors of 3 000 $km^2$, each one in the Northern and Southern Hemisphere. Initially named Giant Air-shower Project (GAP) it later became the *Pierre Auger Observatory*, in honour of Pierre Auger's work on the discovery of extensive air showers. Argentina was selected as Southern site in a democratic vote at the UNESCO headquarter in Paris in November 1995 and construction of an engineering array finally began in 2001 near Malargüe, Argentina. Physics data taking started January 1st 2004 with about 150 water Cherenkov tanks and six fluorescence telescopes and construction of all 1 600 surface detector stations covering an area of 3 000 $km^2$ and 24 telescopes finished mid-2008. As of today, Auger has reached an exposure of nearly 25 000 $km^2$ sr yr, more than the sum achieved with all other experiments. The Northern part of the project could not be realised yet because of funding problems.

Highlights of results include the clear evidence for a suppression of the flux above $4 \times 10^{19}$ eV, observations of anisotropies in the arrival directions above $5.6 \times 10^{19}$ eV, suggesting a correlation to the nearby matter distribution, measurements of the primary mass favouring a change from a light to a heavier composition above $10^{19}$ eV, a measurement of the $p$-Air and $pp$ inelastic cross section at a centre-of-mass energy of $\sqrt{s} = 57$ TeV, almost 10 times higher in energy than recent LHC data, and the most stringent upper limits on EeV photon and neutrino fluxes, strongly disfavouring an exotic particle physics origin for the highest energy cosmic rays. The Auger Observatory will continue running for at least 5 more years and upgrade plans are being discussed.

When AGASA and HiRes were nearing the end of operation, a collaboration consisting of key members from AGASA and HiRes started to prepare for the construction of a large observatory, named *Telescope Array* (TA), in the Northern hemisphere. Like the Auger Observatory, TA combines a large area ground array, largely based on the AGASA design, with air-fluorescence telescopes based on the HiRes system. TA is located in the central western desert of Utah, near the city of Delta, about 250 km south west of Salt Lake City and covers with its 507 surface detector stations and 38 fluorescence telescopes a total area of about 730 $km^2$. Data taking started early 2008 and because of this, the total number of events recorded is still much less than from the Auger Observatory. Nevertheless, good agreement within the systematic uncertainties is seen for the energy spectrum. Analyses of composition and anisotropies still suffer strongly from limited statistics; thus final statements need to wait for more data.

## 5.15 Future

As discussed above, the situation at the upper end of the cosmic ray energy spectrum has changed considerably with the advent of new large scale observatories. No doubts now exist about the existence of a flux suppression above $\sim 5 \times 10^{19}$ eV. However, is this the observation of the GZK-effect which was predicted 45 years ago? From the experimental point of the view, the answer cannot be given, because the suppression could equally well be due to the limiting energy reached in nearby cosmic accelerators, just as discussed by Hillas in his seminal review (Hillas, 1984). In fact, the latter picture is supported by data from the Pierre Auger Observatory which suggest an increasingly heavier composition towards the end of the spectrum and seeing the suppression about 20 % lower in energy than expected for typical GZK scenarios. HiRes and TA, on the other hand find no significant change in their composition and their cut-off energy is in agreement with the GZK-expectation. Moreover, a directional correlation of ultra high-energy cosmic rays on a 3° scale is hard to imagine for heavy primaries. Could this indicate weaker extragalactic magnetic fields than thought, or could it point to deficiencies of hadronic interaction models at the highest energies? These models must be employed to infer the elemental composition from EAS data.

Obviously, nature does not seem ready to disclose the origin of the most energetic particles in the Universe yet. More work is needed and the main players in the field have intensified their co-operation sharing data and analysis strategies to better understand systematic uncertainties which, despite being small, appear to be quite relevant concerning conclusions to be drawn from the data. In parallel, experimental efforts are underway to increase the statistics more quickly and to further improve data quality. Most importantly muon detection capabilities, which are of key importance to understanding features of hadronic interactions at the highest energies, are being added.

Understanding the origin of ultra high-energy cosmic rays demands high quality data in the $10^{19}$ to $10^{20}$ eV energy range. While this is to be the major task of ground-based experiments during the next years, finding the long-searched point sources of cosmic rays simply requires much larger exposures. Plans for space-based experiments exist as well as for further efforts on the ground.

In 1979 Linsley developed the idea to observe giant air showers from space (Linsley, 1979). The advantages were obvious, as a fluorescence camera looking downwards from space could survey huge areas at ground simultaneously with only one atmospheric thickness between the light source and the sensor, the major challenge being the faint light because of the distance to the shower and the optical imaging required for geometrical reconstruction and $X_{\max}$ observations. Several projects of this type were proposed to space agencies in US, Europe, Japan and Russia with JEM-EUSO presently planned to be mounted at the International Space Station in 2018. The realisation of space-based projects involves some uncertainty, and it is clear that the energy and mass resolution for cosmic rays will be much worse than that achieved with ground-based observations. The prime goal is to collect event statistics at the highest energies to detect the long-awaited point sources of ultra high-energy cosmic rays.

Alternatives may also exist for ground-based experiments and they are presently being explored with great vigour. Most importantly, radio observations of EAS may allow measuring shower properties at moderate cost, thus allowing instrumentation of huge areas. The theoretical ideas were formulated in the 1960s (Askaryan, 1962) followed by successful experimental efforts to verify its existence. However, after a flurry of activities work on radio detection essentially ceased for about three decades mostly because of difficulties in monitoring the geo-electric field. With the advent of digital logic hardware, powerful low-cost computing, the ability to perform Monte Carlo simulations and above all the needs to considerably extend the aperture of ultra high-energy experiments, interest in radio observations revived and grew explosively a decade ago. D. Saltzberg and P. Gorham verified in beam measurements at SLAC the existence of the Askaryan effect in dense media (Saltzberg et al., 2001) and Falcke and Gorham in 2003 revived the possibilities of measuring ultra-high-energy cosmic rays and neutrinos with radio techniques (Falcke and Gorham, 2003), and Falcke and Kampert developed the idea of the LOPES experiment to test the potential of the radio technique with state of the art electronics at the KASCADE air shower experiment which convincingly demonstrated the detection and imaging of atmospheric radio flashes from cosmic ray air showers and the geo-synchrotron effect as the underlying mechanism (Falcke et al., 2005). Similar observations were made independently at the CODALEMA experiment in France. With the Netherlands entering the field, scientists from all three countries have joined forces to construct the Auger Engineering Radio Array (AERA) at the Pierre Auger Observatory. The goal of the still ongoing activities here is to verify the practicability of the radio technique for a giant future observatory and to explore the performance for energy and mass measurements. Moreover, radio observations by balloon-borne experiments offer the possibility of surveying huge areas with only a few antennas. In this case, the reflected radio beam off the surface is being detected. The feasibility of such a concept has been demonstrated very recently by the ANITA experiment flown over Antarctic ice (Hoover et al., 2010).

Very recently, again triggered by Gorham et al. (2008), the possible detection of microwave radiation from extensive air showers became another hot topic of ongoing experimental activities. The continuum radiation in the microwave range is expected to be caused by free-electron collisions with neutrals in the tenuous plasma left after the passage of the shower. Again the process seems to be confirmed by accelerator experiments, but the proof of emission from EAS remains to be demonstrated. Other than the radio emission, the microwave emission should occur isotropically which would make it an extremely powerful experimental tool, if confirmed.

The possibility of detecting very large air showers by reflecting a radar beam from the ionisation column that they create in air is another example of an idea of several decades ago being revived at present at the Telescope Array and KASCADE site.

## 5.16 Concluding Remarks

In this year, 2012, the centennial of the discovery of cosmic rays will be celebrated all around the globe. The enormous progress that has been made during this period is directly linked to the invention of new experimental tools and instrumentation and could not have been made without the ideas and skills of some ingenious pioneers. Almost no nuclear and particle physics experiment could be done without making use of the coincidence technique and also triggering on rare events, as another example, has been pioneered in cosmic ray data. The subsequent discoveries of new particles made by cosmic ray observations, including the positron, muons, pions, kaons, hyperons, and likely also charmed particles was discussed in Chap. 2.

The cosmic energy spectrum has been measured in great detail over more than 32 decades in flux, making this observable unique in Nature. The spectrum initially thought to follow a pure power-law distribution has exhibited more and more structure, starting with the discovery of the "knee" at about $4 \times 10^{15}$ eV by Khristiansen's group at Moscow State University in 1959, followed by the observation of the "ankle", first hinted at by Linsley (1963a) at Haverah Park, Akeno, and Fly's Eye in 1991 and the suppression at the GZK threshold in 2008 by HiRes and Auger. Very recently, a second knee caused by the heavy cosmic ray component has been reported by KASCADE-Grande and its not unlikely that further departures from a simple power-law distribution will be exhibited providing important clues about the origin of cosmic rays. Also, great detail about the primary mass could be extracted from the data with remarkable changes seen in the composition coinciding with the position of the structures in the energy spectrum (Kampert and Unger, 2012). The sky in cosmic rays is surprisingly isotropic up to the highest energies and is challenging our understanding of both cosmic ray propagation within the galactic and intergalactic environments and about their sources. Only at the highest energies are departures from isotropy seen (Abraham et al., 2007), but data suffer still from statistics.

Particles at the upper end of the spectrum have such breath-taking energies, a hundred million times above that provided by the LHC accelerator, that questions about how cosmic accelerators can boost particles to these energies, and about what the nature of the particles themselves is, are still open. The mystery of cosmic rays is nowadays tackled – and is perhaps going to be solved – by an interplay of sophisticated detectors for high-energy $\gamma$-rays, charged cosmic rays and neutrinos. Moreover, plans for the next generation experiments are being worked out and it is now realised that the true high-energy frontier in Nature provides unique opportunities to test particle and fundamental physics, such as of space time, at its extreme. Further surprises from future cosmic ray observations are almost guaranteed.

**Acknowledgements** We gratefully acknowledge stimulating discussions and generous support given by Luisa Bonolis, Antonella Castellina, Bruce Dawson, Piera Ghia, Antoine Letessier-Selvon, Maria Concetta Maccarone, Marco Segala, Mike Walter, and many other colleagues for helping us to access to some of the original key papers distributed in various archives around the world. KHK also acknowledges financial support by the German Ministry for Research and Education (BMBF) and by the Helmholtz Alliance for Astroparticle Physics and AAW acknowledges the UK Science and Technology Council and the Leverhalme Foundation.

## References

Abraham, J., et al.: Properties and performance of the prototype instrument for the Pierre Auger Observatory. Nucl. Instrum. Methods **A523**, 50 (2004)

Abraham, J., et al. (Pierre Auger Collaboration): Correlations of the highest energy cosmic rays with nearby extragalactic objects. Science **318**, 938 (2007)

Afanasiev, B.N., et al.: Recent rezults from Yakutsk experiment. In: Nagano, M. (ed.) Proc. Tokyo Workshop on the Techniques for the Study of the Extremely High Energy Cosmic Rays, p. 35 (1993)

Aglietta, M., et al. (EAS-TOP Collaboration): The EAS-TOP array at $E_0 = 10^{14}$–$10^{16}$ eV: stability and resolutions. Nucl. Instrum. Methods A **277**, 23 (1989)

Allan, H.R., et al.: The distribution of energy in extensive air showers and the shower size spectrum. Proc. Phys. Soc. **79**, 1170 (1962)

Andrews, D., et al.: Evidence for the existence of cosmic ray particles with $E > 5 \times 10^{19}$ eV. Nature **219**, 343 (1968)

Antoni, T., et al. (KASCADE Collaboration): The cosmic-ray experiment KASCADE. Nucl. Instrum. Methods **A513**, 490 (2003)

Antoni, T., et al. (KASCADE Collaboration): KASCADE measurements of energy spectra for elemental groups of cosmic rays: results and open problems. Astropart. Phys. **24**, 1–25 (2005)

Apel, W.D., et al. (KASCADE-Grande Collaboration): The KASCADE-Grande experiment. Nucl. Instrum. Methods **A620**, 202–216 (2010)

Apel, W.D., et al. (KASCADE-Grande Collaboration): Kneelike structure in the spectrum of the heavy component of cosmic rays. Phys. Rev. Lett. **107**, 171104 (2011)

Askaryan, G.A.: Excess negative charge of an electron-photon shower and its coherent radio emission. JETP **14**, 441–443 (1962)

Auger, P.: In: Sekido, Y., Elliot, H. (eds.) Early History of Cosmic Ray Studies. Reidel, Dordrecht (1985)

Auger, P., Maze, R., Robley: Extension et pouvoir pénétrant des grandes gerbes de rayons cosmiques. Comptes Rendus **208**, 1641 (1939a)

Auger, P., et al.: Extensive cosmic ray showers. Rev. Mod. Phys. **11**, 288 (1939b)

Bassi, P., Clark, G., Rossi, B.: Distribution of arrival times of air shower particles. Phys. Rev. A **92**, 441 (1953)

Belenki, S.Z., Landau, L.: Hydrodynamic theory of multiple production of particles. Suppl. Nuovo Cim. **3**(1), 15 (1956)

Bellido, J.A., et al.: Southern hemisphere observations of a $10^{18}$ eV Cosmic Ray Source near the direction of the Galactic centre. Astropart. Phys. **15**, 167 (2001)

Belyaev, V.A., Chudakov, A.E.: Ionization glow of air and its possible use for air shower detection. Bull. USSR Acad. Sci. Phys. Ser. **30**(10), 1700 (1966)

Bergeson, H.E., Boone, J.C., Cassiday, G.L.: In: Proc. 14th ICRC, Munich, vol. 8, p. 3059 (1975)

Bergeson, H.E., et al.: Measurement of light emission from remote cosmic-ray air showers. Phys. Rev. Lett. **39**, 847 (1977)

Bethe, H., Heitler, W.: On the stopping of fast particles and on the creation of positive electrons. Proc. R. Soc. A **146**, 83 (1934)

Bhabha, H., Heitler, W.: The passage of fast electrons and the theory of cosmic showers. Proc. R. Soc. A **159**, 432–458 (1937)

Bird, H.E., et al.: Detection of a cosmic ray with measured energy well beyond the expected spectral cutoff due to cosmic microwave radiation. Astrophys. J. **441**, 144–150 (1995)

Blackett, P.M.S.: In: Proceedings of the International Conference on the Emission Spectra of the Night Sky and Aurorae, pp. 34–35. Physical Society, London (1947)

Blackett, P.M.S.: Cloud chamber researches in nuclear physics and cosmic radiation. Nobel Lecture (13 December, 1948)

Blackett, P.M.S., Occhialini, G.: Photography of penetrating corpuscular radiation. Nature **130**, 363 (1932)

Borione, A., et al.: A large air shower array to search for astrophysical sources emitting $\gamma$-rays with energies $> 10^{14}$ eV. Nucl. Instrum. Methods **346**, 329 (1994)

Bothe, W.: Zur Vereinfachung von Koinzidenzzählungen. Z. Phys. **59**, 1–5 (1929)

Bunner, A.N.: The atmosphere as a cosmic ray scintillator. Master Thesis, Cornell University (1964)

Bunner, A.N.: The atmosphere as a cosmic ray scintillator. PhD Thesis, Cornell University (1967)

Bunner, A.N., Greisen, K., Landecker, P.B.: An imaging system for EAS optical emission. Can. J. Phys. **46**, S266 (1968)

Carlson, J.F., Oppenheimer, J.R.: On multiplicative showers. Phys. Rev. **51**, 220 (1937)

Chudakov, A.E., et al.: In: Proc. 6th ICRC, Moscow, vol. II, p. 50 (1960)

Clark, W.G.: The scientific legacy of Bruno Rossi. Università degli Padova, 7 (2006)

Clark, G.W., et al.: An experiment in air showers produced by high-energy cosmic rays. Nature **180**, 353 (1957)

Clark, G., et al.: The M.I.T. air shower program Suppl. Nuovo Cim. **8**, 623 (1958)

Clark, G.W., et al.: Cosmic-ray air showers at sea level. Phys. Rev. **122**(2), 637 (1961)

Clay, R.W., et al.: Cosmic rays from the galactic center. Astropart. Phys. **12**, 249 (2000)

Cocconi, G., Koester, L.J., Perkins, D.H.: Calculation of particle fluxes. Lawrence Berkeley Laboratory Report LBL 10022, pp. 167–192 (1962)

Compton, A.H., Getting, I.A.: An apparent effect of galactic rotation on the intensity of cosmic rays. Phys. Rev. **47**, 817 (1935)

Cranshaw, T.E.: Cosmic Rays. Clarendon Press, Oxford (1963)

Cranshaw, T.E., Galbraith, W.: Philos. Mag. **45**, 1109 (1954)

Cranshaw, T.E., Galbraith, W.: Philos. Mag. **2**, 797 (1957)

Dawson, B.: Comment on a Japanese detection of fluorescence light from a cosmic ray shower in 1969 (2011). arXiv:1112.5686

Falcke, H., Gorham, P.W.: Detecting radio emission from cosmic ray air showers and neutrinos with a digital radio telescope. Astropart. Phys. **19**, 477 (2003)

Falcke, H., et al. (LOPES Collaboration): Detection and imaging of atmospheric radio flashes from cosmic ray air showers. Nature **435**, 313 (2005)

Fermi, E.: On the origin of the cosmic radiation. Phys. Rev. **75**, 1169 (1949)

Fermi, E.: High energy nuclear events. Prog. Theor. Phys. **5**(4), 570 (1950)

Fermi, E.: Angular distribution of the pions produced in high energy nuclear collisions. Phys. Rev. **81**(5), 683 (1951)

Fretter, W.B.: In: Proceedings of Echo Lake Cosmic Ray Symposium (1949)

Fukui, S., et al.: A study on the structure of the extensive air shower. Prog. Theor. Phys. Suppl. **16**, 1–53 (1960)

Galbraith, W.: Extensive Air Showers. Academic Press, San Diego (1958)

Galbraith, W., Jelley, J.V.: Light pulses from the night sky associated with cosmic rays. Nature **171**, 349 (1953)

Geiger, H., Müller, W.: Elektronenzählrohr zur Messung schwächster Aktivitäten. Naturwissenschaften **31**, 617–618 (1928)

Gorham, P.W., et al.: Observations of microwave continuum emission from air shower plasmas. Phys. Rev. D **78**, 032007 (2008)

Gould, R.J., Schréder, G.: Opacity of the universe to high-energy photons. Phys. Rev. Lett. **16**, 252 (1966)

Greisen, K.: Prog. Cosm. Ray Phys. **3**, 1–141 (1956)

Greisen, K.: Cosmic ray showers. Annu. Rev. Nucl. Part. Sci. **10**, 63 (1960)

Greisen, K.: End to the cosmic ray spectrum? Phys. Rev. Lett. **16**, 748 (1966a)

Greisen, K.: In: Proc. 9th ICRC, London, vol. 2, p. 609 (1966b)

Hara, T., et al.: Detection of the atmospheric scintillation light from air showers. Acta Phys. Acad. Sci. Hung. **29**, 369 (1970)

Heisenberg, W.: Zur Theorie der Schauer in der Höhenstrahlung. Z. Phys. **101**, 533 (1936)

Hersil, J., et al.: Observations of extensive air showers near the maximum of their longitudinal development. Phys. Rev. Lett. **6**(1), 22 (1961)

Hersil, J., et al.: Extensive air showers at 4200 m. J. Phys. Soc. Jpn. **17**, 243 (1962)
Hess, V.F.: Über die Beobachtungen der durchdringenden Strahlung bei sieben Freiballonflügen. Phys. Z. **8**, 1084 (1912)
Hillas, A.M.: Cosmic Rays. Pergamon Press, Elmsford (1972)
Hillas, A.M.: Two interesting techniques for Monte Carlo simulation of very high-energy hadron cascades. In: Linsley, J., Hillas, A.M. (eds.) Proc. of the Paris Workshop on Cascade Simulations, p. 193 (1982).
Hillas, A.M.: The origin of ultra-high-energy cosmic rays. Annu. Rev. Astron. Astrophys. **22**, 425–444 (1984)
Hoffmann, G., Pforte, W.S.: Zur Struktur der Ultrastrahlung. Phys. Z. **31**, 347 (1930)
Hoover, S., et al.: Observation of ultrahigh-energy cosmic rays with the ANITA Balloon-Borne radio interferometer. Phys. Rev. Lett. **105**, 151101 (2010)
Ito, N., et al.: In: Proc. 25th ICRC, Durban, vol. 4, p. 117 (1997)
Jelley, J.V.: High-energy $\gamma$-ray absorption in space by a 3.5 K microwave field. Phys. Rev. Lett **16**, 479 (1966)
Kamata, K., Nishimura, J.: The lateral and the angular structure functions of electron showers. Prog. Theor. Phys. Suppl. **6**, 93 (1958)
Kameda, T., Toyoda, Y., Maeda, T.: J. Phys. Soc. Jpn. **15**, 1565 (1960)
Kampert, K.-H., Unger, M.: Measurements of the cosmic ray composition with air shower experiments. Astropart. Phys. **35**, 660 (2012)
Khristiansen, G.B., et al.: The EAS-1000 array. Ann. N.Y. Acad. Sci. **571**, 640 (1989)
Knapp, J.: In: Proc. 25th ICRC, Durban, vol. 8, p. 83 (1997)
Kolhörster, W., Matthes, I., Weber, E.: Gekoppelte Höhenstrahlen. Naturwissenschaften **26**, 576 (1938)
Krieger, A.S., Bradt, H.V.: Cherenkov light in extensive air showers and the chemical composition of primary cosmic rays at $10^{16}$ eV. Phys. Rev. **185**, 1629 (1969)
Kulikov, G.V., Khristiansen, G.B.: On the size spectrum of extensive air showers. JETP **35**, 441 (1959)
Kulikov, K., et al.: In: Proc. 9th ICRC, London (1965)
Linsley, J.: In: Proceedings of the 8th International Cosmic Ray Conference, Jaipur, vol. 4, p. 77 (1963a)
Linsley, J.: Evidence for a primary cosmic-ray particle with energy $10^{20}$ eV. Phys. Rev. Lett. **10**(4), 146 (1963b)
Linsley, J.: In: Proc. 15th ICRC, Plovdiv, vol. 12, p. 89 (1977)
Linsley, J.: Study of $10^{20}$ eV cosmic rays by observing air showers from a platform in space, response to *Call for Projects and Ideas in High Energy Astrophysics for the 1980's*, Astronomy Survey Committee (Field Committee) (1979)
Linsley, J.: In: Wada, M. (ed.) Catalogue of Highest Energy Cosmic Rays. World Data Center of Cosmic Rays, Institute of Physical and Chemical Research, Itabashi, Tokyo (1980)
Linsley, J., Hillas, A.M.: In: Proc. of the Paris Workshop on Cascade Simulations. Texas Center for the Advancement of Science and Technology, Texas (1982)
Linsley, J., Scarsi, L., Rossi, B.: Extremely energetic cosmic-ray event. Phys. Rev. Lett. **6**, 485 (1961)
Lloyd-Evans, J., et al.: Observations of $\gamma$-rays $> 10^{15}$ eV from cygnus X-3. Nature **305**, 784 (1983)
Matthews, J.: A Heitler model of extensive air showers. Astropart. Phys. **22**, 387 (2005)
Maze, R.: Étude d'un appareil à grand pouvoir de résolution pour rayons cosmiques. J. Phys. Radium **9**(4), 162–168 (1938)
McCusker, C.B.A., Winn, M.M.: A new method of recording large cosmic-ray air showers. Nuovo Cimento **28**, 175 (1963)
Nagano, M., Watson, A.A.: Observations and implications of the ultrahigh-energy cosmic rays. Rev. Mod. Phys. **72**, 689 (2000)
Navarra, G.: Cosmic ray composition and hadronic interactions in the knee region. Nucl. Phys. B, Proc. Suppl. **151**(1), 79–82 (2006)

Nikolsky, S.I.: In: Proceedings of 5th Interamerican Seminar on Cosmic Rays, vol. 2. Universidad Mayor de San Andreas, La Paz, Bolivia (1962)
Olbert, S.: Theory of high-energy N-component cascades. Ann. Phys. **1**, 247–269 (1957)
Penzias, A.A., Wilson, R.W.: A measurement of excess antenna temperature at 4080 Mc/s. Astrophys. J. **142**, 419 (1965)
Peters, B.: Primary cosmic radiation and extensive air showers. Nuovo Cimento **22**, 800 (1961)
Pfotzer, G.: Dreifachkoinzidenzen der Ultrastrahlung aus vertikaler Richtung in der Stratosphere. Z. Phys. **102**, 41 (1936)
Porter, N.A., et al.: Philos. Mag. **3**, 826 (1958)
Regener, E., Ehmert, A.: Über die Schauer der kosmischen Ultrastrahlung in der Stratosphäre. Z. Phys. **111**, 501 (1938)
Regener, E., Pfotzer, G.: Vertical intensity of cosmic rays by treefold coincidences in the stratosphere. Nature **136**, 718 (1935)
Rossi, B.: Method of registering multiple simultaneous impulses of several Geiger's counters. Nature **125**, 636 (1930)
Rossi, B.: Über die Eigenschaften der durchdringenden Korpuskularstrahlung im Meeresniveau. Z. Phys. **82**, 151 (1933)
Rossi, B.: Misure sulla distribuzione angolare di intensita della radiazione penetrante all' Asmara. Suppl. Ric. Sci. **1**, 579 (1934)
Rossi, B.: In: Sekido, Y., Elliot, H. (eds.) Early History of Cosmic Ray Studies. Reidel, Dordrecht (1985)
Rossi, B., Greisen, K.: Cosmic-ray theory. Rev. Mod. Phys. **13**, 240 (1941)
Saltzberg, D., et al.: Observation of the Askaryan effect: coherent microwave Cherenkov emission. Phys. Rev. Lett. **86**, 2802 (2001)
Samorski, M., Stamm, W.: Detection of $2 \times 10^{15}$ to $2 \times 10^{16}$ eV $\gamma$-rays from Cygnus X-3. Astrophys. J. **268**, L17 (1983)
Schmeiser, K., Bothe, W.: Die harten Ultrastrahlschauer. Ann. Phys. **424**, 161 (1938)
Skobeltzyn, D.V.: Die Intensitätsverteilung in dem Spektrum der $\gamma$-Strahlen von RaC. Z. Phys. **43**, 354 (1927)
Skobeltzyn, D.V.: Über eine neue Art sehr schneller $\beta$-Strahlen. Z. Phys. **54**, 686 (1929)
Skobeltzyn, D.V., Zatsepin, G.T., Miller, V.V.: The lateral extension of auger showers. Phys. Rev. **71**, 315 (1947)
Sokolsky, P., Thomson, G.B.: Highest energy cosmic rays and results from the HiRes experiment. J. Phys. G **34**, R401 (2007)
Suga, K.: In: Proceedings of 5th Interamerican Seminar on Cosmic Rays, vol. 2. Universidad Mayor de San Andreas, La Paz, Bolivia (1962)
Suga, K., Clark, G.W., Escobar, I.: Scintillation detector of 4 $m^2$ area and transistorized amplifier with logarithmic response. Rev. Sci. Instrum. **32**, 1187 (1961)
Usoskin, I.G., Kovaltsov, G.A.: Cosmic ray induced ionization in the atmosphere: full modeling and practical applications. J. Geophys. Res. **111**, D21 (2006)
Wilson, R.R.: Monte Carlo study of shower production. Phys. Rev. **86**(3), 261 (1952)
Winn, M.M., et al.: The cosmic-ray energy spectrum above $10^{17}$ eV. J. Phys. G **12**, 653 (1986a)
Winn, M.M., et al.: The arrival directions of cosmic rays above $10^{17}$ eV. J. Phys. G **12**, 675 (1986b)
Zatsepin, G.T., Kuzmin, V.A.: Upper limit of the spectrum of cosmic rays. JETP Lett. **4**, 78 (1966)

# Chapter 6
# Very-High Energy Gamma-Ray Astronomy: A 23-Year Success Story in Astroparticle Physics

**Eckart Lorenz and Robert Wagner**

## 6.1 Introduction

Since early times, astronomy has been an important part of human culture. Astronomical observations started way back in the Stone Age. Galileo opened the window of modern astronomy in 1609 AD, by making astronomical observations based on using optical telescopes. By means of a simple optical telescope, he could observe the four largest moons of Jupiter for the first time. The same year Kepler published the fundamental laws of planetary movements in the *Astronomica Nova*. During the following 350 years, astronomers were exploring the Universe in the wavelength range of visible light, successively investigating more and more of the so-called thermal Universe, which comprises all emission coming from thermal emission processes. In the year 1912, the Austrian physicist Victor Hess showed that some type of high-energy radiation is constantly bombarding the Earth from outer space (Hess, 1912). These so-called cosmic rays (CR), later identified mostly as charged particles, were a clear evidence of the existence of high-energy processes in our Universe exceeding energies that could possibly be created in thermal emission processes. A fundamental problem of CRs (below some $10^{18}$ eV) is that these charged particles do not allow their trajectories to be traced back to any astrophysical object, as they are deflected by (unknown) intergalactic magnetic fields and thus lose any directional information: the sources of the CRs cannot be identified. Even today, after 100 years of CR studies, many questions about the sources of CRs remain unsolved.

Shortly before and after the Second World War, new windows in energy bands below and above visible wavelengths of the electromagnetic spectrum were successfully opened, by observations in radio waves, infrared and ultraviolet light, X-rays,

E. Lorenz · R. Wagner (✉)
Max-Planck-Institut für Physik, Föhringer Ring 6, 80805 München, Germany
e-mail: robert.wagner@mpp.mpg.de

R. Wagner
Excellence Cluster "Origin and Structure of the Universe", Boltzmannstraße 2, 85748 Garching b. München, Germany

B. Falkenburg, W. Rhode (eds.), *From Ultra Rays to Astroparticles*,
DOI 10.1007/978-94-007-5422-5_6, 

and, eventually, in gamma rays. At around 1980, it was possible to observe cosmic radiation in the entire range of the electromagnetic spectrum, from $10^{-6}$ eV up to $10^9$ eV. By such observations it could be shown that, besides the thermal Universe (dominated by stellar production of photons), high-energy reactions are an essential part of what can be observed from the Universe.

In 1989, the window of very-high energy (VHE) gamma-ray astronomy was opened by the detection of TeV gamma rays from the Crab nebula by the Whipple collaboration (Weekes et al., 1989). This seminal detection started a very productive research field in an energy domain mostly accessible only by ground-based instruments. In 2012, we are celebrating the 100th year of cosmic-ray studies. This article will give an overview of the development of VHE gamma-ray astronomy. The richness of the results achieved over the years necessitated a selection of experiments, discussed here, that reflect the steady progress in VHE gamma-ray astronomy. Obviously, this selection is somewhat personal, and emphasis is put on such experiments that made initial breakthroughs in new detection methods and new results, while less emphasis is put on later experiments, using very similar techniques, which may, however, be of same scientific productivity. Also, experiments that were optimized for energies above 100 TeV are mostly skipped, because up to now no sources have been discovered in that energy domain.

VHE gamma-ray astronomy is part of high-energy cosmic-ray astrophysics. Many experiments of the past aimed both at the search for VHE gamma-ray emitting sources, as well as at solving fundamental questions concerning the nature of cosmic rays. Here, we concentrate on the discussion of gamma-ray studies and refer to Chap. 5 for details on CR studies.

### *6.1.1 VHE Gamma Rays, Messengers of the Relativistic Universe*

Cosmic rays result from and thus transmit information about distant high-energy processes in our Universe. Besides their energy (and particle type), the most important information they carry is the location of the astrophysical object of their origin. However, nearly all CRs are charged and therefore suffer deflection from their original trajectories by the weak magnetic fields ($\ll$1 Gauss) in our Galaxy and, if originating from somewhere in the extragalactic space, also by very weak extragalactic magnetic fields, which are known to exist. Their direction and size is, however, unknown. CRs up to about few times $10^{19}$ eV are nearly completely randomized in direction and cannot be associated with any astrophysical object. Even if the magnetic fields were known, it would currently be impossible to extrapolate observed charged CRs back to their point of origin because the uncertainty in determining their energy would result in a much too large correlated area. Therefore, only neutral particles are currently suited to serve as messenger particles. The two particle types that ideally fall into this category are photons – gamma ($\gamma$) quanta – and the neutrinos. All other neutral particles are too short-lived. The neutron with just below 15-minute lifetime in its rest frame would, even at the highest energies of $\approx 10^{19}$ eV, on average just travel over a distance from the center of our Galaxy to the Earth. Neutrinos, being weakly interacting particles, are very difficult to detect and huge volumes of dense material are required to observe a minuscule fraction of them impinging on the earth. A review of neutrino astronomy and its historical development is given in Chap 9. VHE $\gamma$ rays are therefore currently the best-suited messengers of the relativistic Universe. The challenge to explain $\gamma$-ray production is that, experimentally, currently two fundamentally different production processes (or a combination of these!) can be at work, namely, leptonic or hadronic processes. Neutrinos, however, can be created only in hadronic processes, therefore one could solve this ambiguity. The main production processes of $\gamma$ rays are as follows.

Inverse Compton scattering: VHE electrons upscatter low-energy photons over a broad energy range above the initial one,

$$e + \gamma_{\text{low energy}} \longrightarrow e_{\text{low energy}} + \gamma_{\text{VHE}}.$$

Normally, there are plenty of low-energy photons in the environment of stars due to thermal emission or due to synchrotron emission by the high energy electrons in the normally present magnetic fields. In the lower energy range, the dominant production process of gamma rays from leptons is via synchrotron radiation processes, where electrons lose a fraction of their energy by synchrotron radiation when passing through local magnetic fields.

Another production process is by hadronic interactions. Accelerated protons or heavier nucleons interact hadronically with other protons or nucleons in stellar environments or cosmic gas clouds. Dominantly, charged and neutral pions are produced. Charged pions decay in a two-step process into electrons and two neutrinos while neutral pions decay with >99 % probability into two gamma quanta; schematically:

$$p + \text{nucleus} \rightarrow p' \cdots + \pi^{\pm} + \pi^{0} + \cdots \quad \text{and}$$
$$\pi^{0} \rightarrow 2\gamma; \qquad \pi \rightarrow \mu\nu_{\mu}; \qquad \mu \rightarrow e\nu_{\mu}\nu_{e}.$$

Heavier secondary mesons, much rarer, normally decay in a variety of lighter ones and eventually mostly into $\pi^{\pm}$ and $\pi^{0}$ and/or $\gamma$. It is impossible to distinguish from observing gamma rays only whether they originate from either a leptonic or a hadronic parent particle, while the observation of neutrinos would be an unambiguous proof that these messengers come from hadronic interactions. Nevertheless, by analyzing gamma-ray spectra the dominant parent particle process can sometimes be deduced.

### *6.1.2 The Long Road to the Discovery of the First VHE-Emitting Gamma-Ray Source*

The main driving force for VHE gamma-ray astronomy was initially the search for the sources of the charged cosmic rays, while now, after the discovery of many sources, the interest has shifted to general astrophysics questions. In earlier times the searches were hampered by a few fundamental questions:

1. How large is the fraction of cosmic $\gamma$ rays of the total CR flux?
2. What exactly happens when cosmic particles hit the atmosphere?
3. What are the secondary decay products when VHE $\gamma$ rays hit the atmosphere?
4. How can VHE $\gamma$ rays be distinguished from the charged VHE CRs?
5. How transparent is the Universe for $\gamma$ rays of a certain energy, respectively, how far can one look with $\gamma$ rays of a certain energy into the Universe?

It took many years with the detection techniques available in those times to solve these problems step by step – largely due to inadequate instruments, slowly developing theories about particle interaction, slowly oncoming additional information from accelerator experiments and the lack of powerful computers.

The exact flux fraction of $\gamma$-rays of the total CR flux as a function of energy is still unknown today. Shortly after the discovery of CRs, Kolhörster speculated that CRs originated from cosmic $\gamma$-rays, but the first experiments were too simple to prove or disprove this assumption (Kolhörster, 1913). In 1925, R. Millikan, who introduced the name cosmic rays, was convinced that CRs originally were all $\gamma$-rays (Millikan and Cameron, 1928). In 1930, Millikan and Compton disagreed about the origin of CRs with Millikan pursuing their photonic origin, while Compton was convinced that CRs were originally primary, positively charged particles. This was later proven to be correct when it was possible to observe ionizing particles at the top of the atmosphere or with satellite-borne detectors. In the 1930s and 1940s it was still believed that a significant part of CRs were $\gamma$ rays, while in the early 1980s it was mostly thought that about 1 % of the CRs were $\gamma$ rays. Nowadays this question is still not completely resolved and much smaller flux ratios are assumed. One supposes that at most $10^{-4}$ of all particles coming from the Galactic plane are $\gamma$ rays while at most only $10^{-5}$ of the particles from outside the galactic plane are $\gamma$ rays. It took 27 years after Hess' first discovery of CRs until Pierre Auger discovered extended air showers initiated by CRs when hitting the atmosphere (Auger et al., 1939). Furthermore, it took many decades to understand the basics of the showering process; still today only approximate models describe some subtle effects.

## 6.2 Attempts Between 1960 to Late 1980s to Find the Sources of CRs

### *6.2.1 A Short Excursion: The Basic Detection Techniques*

The three decades from 1960 to the end of the 1980s saw very little progress in discovering one or more sources of VHE gamma rays. Experiments were in a vicious circle: Poor experiments gave very doubtful results and the funding agencies were not willing to finance large installations. Many physicists, that started their career in cosmic-ray physics, turned to high energy physics (HEP) experiments at accelerators; this field was and still is in an extremely productive phase. In contrast to HEP, very-high energy cosmic-ray experimentalists basically developed no

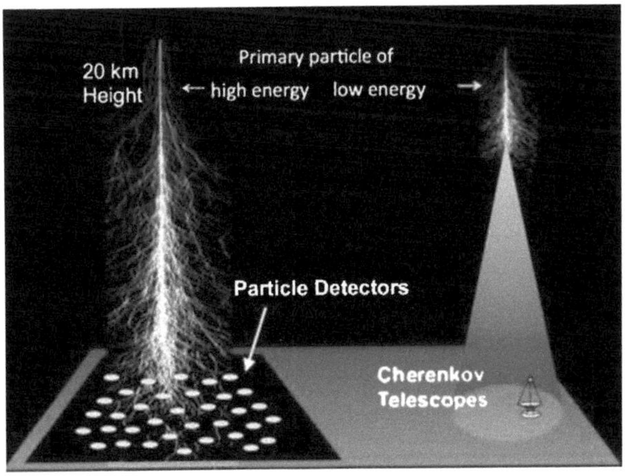


**Fig. 6.1** Principle of the two commonly used detector techniques for observing cosmic VHE particles. **a** An extended air shower array. Primary particles hitting the Earth's atmosphere initiate an extended air shower. Shower tail particles, which penetrate down to ground level, are detected by an array of particle detectors. **b** Cherenkov light detection of air showers that do not need to penetrate down to ground. Cherenkov light generated by the shower particles can be observed by one or more so-called imaging atmospheric Cherenkov telescopes, comprising a large mirror focusing the light onto a matrix of high-sensitivity photosensors in the focal plane. Both detector principles are used for the observation of charged cosmic-ray showers as well as gamma-ray induced showers. Courtesy C. Spiering

new techniques; very little progress in understanding the fine structure of shower developments was made because of lack if sophisticated experimental instruments, insufficient computing power and limited theory in high energy hadronic interaction. Two basic detector concepts[1] were used (Fig. 6.1): detectors that measure particles of the shower tail hitting the ground, so-called extended air shower arrays (EAS) or, as particle physicists call them: "tail-catcher detectors", and Cherenkov telescopes for observing showers that are essentially stopping high up in the atmosphere. Both methods make use of the atmosphere as calorimeter, in combination of either a tracking detector or a light sensor as a calorimetric measuring device.

High-energy cosmic rays (mostly protons and heavier nucleons, rarely gamma rays, electrons and positrons) enter the Earth's atmosphere and generate a cascade of secondary particles, forming an extended air shower. Initially, in this shower process the number of secondary particles is rapidly increasing. During this multiplication process the energy of the primary is partitioned onto the secondaries until the energy of the secondary particles becomes so low such that the multiplication process stops. Due to energy loss of the charged particles by ionization, the shower eventually dies out. Depending on the primary energy and nature of the incident particle,

[1] Other detection principles like the fluorescence detectors make use oft the very weak fluorescence light of an air shower; radio detectors detect radio waves emitted by the shower. Both detection principles are currently unsuited for VHE gamma-ray astronomy, because of an extremely high threshold.

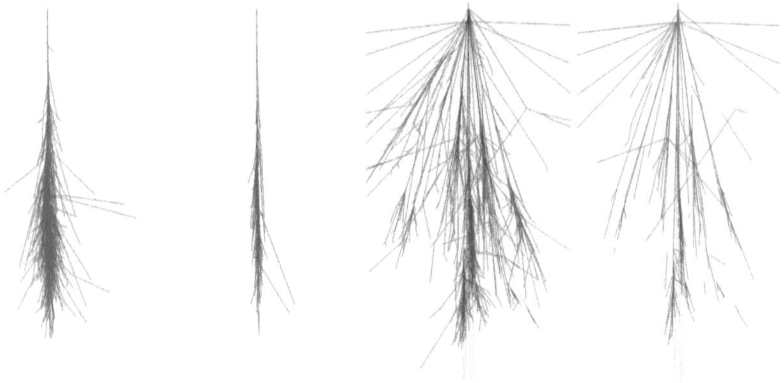

**Fig. 6.2** Simulations of air showers. **a** secondaries of a 50 GeV $\gamma$-ray primary particle. **b** Same, but only those secondaries that produce Cherenkov light are plotted. **c** Secondaries of a 200 GeV proton primary particle. **d** Same, but only those secondaries that produce Cherenkov light are plotted. In all figures, the particle type of the secondaries is encoded in their track color: *red* = electrons, positrons, gammas; *green* = muons; *blue* = hadrons. Figures courtesy Dario Hrupec (Institut Ruder Bošković, Zagreb), produced using code done by Fabian Schmidt (Leeds University), using CORSIKA (Color figure online)

the shower might stop at high altitudes or reach ground. Showers originating from hadrons ("hadronic showers") and electromagnetic showers, initiated by gamma rays or electrons (positrons) can be discriminated by their development: Fig. 6.2 shows examples of a gamma-ray induced shower and a proton-induced hadronic shower. If the charged secondary particles are moving faster than the speed of light in the atmosphere, they emit Cherenkov light within a small angle, which depends also on the (altitude-dependent) atmospheric density and particle energy. A hadronic shower starts normally with many secondary pions and a few heavier mesons. Due to the fact that about one third of the secondary particle in each interaction are $\pi^0$ particles, the electromagnetic component of hadronic showers becomes more and more enriched due to the decay $\pi^0 \rightarrow 2\gamma$. Rarely, charged pions decay into muons, which can penetrate deeply into ground. Gamma-ray induced cascades are much narrower in transverse extension. The dominant multiplication processes in electromagnetic showers is electron/positron bremsstrahlung, producing gamma rays and $e^+e^-$ pair production from gamma-ray conversion. The vertical atmosphere corresponds to 27 radiation lengths and 11 hadronic absorption lengths. Due to transverse momentum in hadronic interactions, multiple scattering and Earth magnetic field deflections the showers are widened, facilitating their detection. In case of the Cherenkov detector principle, the small emission angle of the Cherenkov light still illuminates a large area at ground, of typically 200–220 meters in diameter. A telescope anywhere in this area can detect an electromagnetic shower, provided the Cherenkov light intensity is high enough. Further details of the showering process can be found in Weekes (2003) or in numerous publications about calorimetry in high energy physics experiments.

The air-shower array detectors used in most cases are derivatives of the initial Geiger tube counters and nearly all followed the low active-density array concept. Most advanced detectors used large scintillation counters viewed by photomultipliers read out by simple electronics. These array detectors sampled the shower tail and measured the arrival signals in each hit counter, thus allowing to determine the energy and direction of the shower. The active area fraction of the array area covered by detectors was normally below 1 % resulting in rather large uncertainties in energy determination and modest angular resolution. A big problem was the precise angular calibration of the detectors, as no reference source was available. Special variants of the air shower array detectors were tracking charged particles passing the instruments. It was hoped to determine the incident particle direction from the measurement of a few angular measurements of the secondary tracks in the shower tail. These measurements, however, provided only a very poor directional determination because most of the secondary particles at the shower tail were of low momentum and multiple scattering was large. The air-shower arrays had basically a 24 h up-time and thus allowed the monitoring of a large fraction of the sky, i.e., they were in principle well suited for searching the sources of CRs. Depending on the altitude of the installation the threshold was very high. At sea level, one achieved a threshold of around $10^{14}$ eV for showers with vertical incidence. For large zenith angle showers, the energy threshold scales with a strong dependence of the zenith angle $\theta$ of around $\cos^{-7}\theta$. The main deficiency of air shower array detectors is their weak gamma/hadron separation power, poor energy and angular resolution at, and still quite far above their energy detection threshold. Muons might be used as discriminators. Gamma-ray induced showers contain, however, only very few muons (originating from rare photo-production processes), while hadronic showers contain quite a few muons mostly going down to ground level. Muons are normally identified by their passage of substantial amounts of matter. Therefore muon detectors had to be installed a few meters underground, thus making them an expensive component of the detector and, consequently, only few muon detectors could normally complement a small fraction of the arrays. The general procedure of searching for the sources by means of cosmic $\gamma$ rays was to look for locally increased rates in the sky maps, because hadronic events would be isotropically distributed and should form a smooth background. By means of the muon detectors it was hoped to suppress the hadronic background further.

The alternative techniques to the air shower arrays were detectors based on the observation of Cherenkov light from air showers. In 1934, Pavel Cherenkov discovered that charged particles emit some prompt radiation in transparent media when moving faster than the speed of light in those media (Cherenkov, 1934). Later, Ilia Frank and Igor Tamm developed the theory for this radiation, dubbed after its discoverer Cherenkov radiation. All three were awarded the Nobel Prize in 1958. In 1947, the British physicist P.M.S. Blackett predicted that relativistic cosmic particles passing the atmosphere should produce Cherenkov light and even contribute to a small fraction ($\approx 10^{-4}$) of the night sky background light (Blackett, 1948). In 1953, B. Galbraith and J.V. Jelley built a simple detector and proved that air showers generate Cherenkov light, which could be detected as a fast light flash during

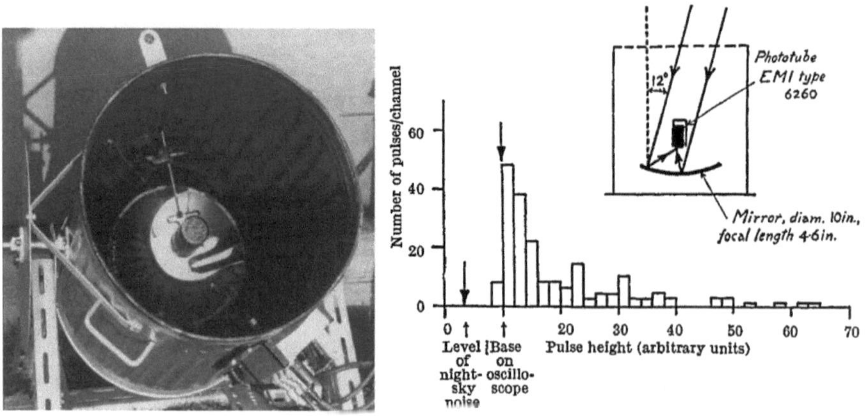


**Fig. 6.3** **a** The first design of an air Cherenkov counter in a garbage can used by B. Galbraith and J.V. Jelley in 1953 (Galbraith and Jelley, 1953). Photograph courtesy T.C. Weekes. **b** Setup and results of the observations of Galbraith and Jelley (figure taken from the original article)

clear dark nights (Galbraith and Jelley, 1953). With a threshold of around four times the night sky noise level they observed signals with a rate of about one event per two to three minutes. This was, by the way, the first demonstration that Cherenkov light was generated also in gases. Later, they could demonstrate that these signals were actually caused by air showers due to coincidences with the nearby Harwell air shower array. The first detectors consisted of a very simple arrangement, i.e., a search-light mirror viewed by a photomultiplier, as shown in Fig. 6.3. The first setup was installed in a garbage can for shielding from stray light. In the following years the technique was refined by using larger mirrors, replacing the single photomultiplier tube (PMT) by a few arranged in the focal plane and even a few of these simple telescopes in coincidences. As in optical astronomy, the air Cherenkov telescopes had to track the source under observation. Nevertheless, all these many pioneering efforts were not rewarded by any important discovery. The so-called air Cherenkov telescopes had some important advantages compared to the air shower arrays. The telescopes collected light from the entire development of the particle shower and one could, in principle, measure the energy of the initial particle with much higher precision and with a typically two orders of magnitude lower threshold compared to "tail catcher" detectors. The main disadvantages were that one could only observe with a very limited field of view of a few degrees. Thus one could study only a single object at a time and observations could only be carried out during clear, moonless nights. Similarly to the air shower arrays, the first-generation Cherenkov telescopes could not discriminate between hadronic and electromagnetic showers. Therefore observers tried to identify sources by just a change in the counting rate when pointing their telescope(s) to the sources and later for the same time slightly off the source. As the gamma-ray flux was very low compared to the CR flux, such a method was prone to secondary effects generating rate changes, for example fluctuations due to atmospheric transmission and the night sky light background from stars.

Because Cherenkov detectors measured the light coming from the entire shower, they had, besides their better energy resolution, also a better angular resolution. Basically, the combination of the atmosphere and the detector forms a fully active calorimeter with some imaging quality due to the directional distribution of the Cherenkov light. Figure 6.2 shows simulations of the shower development of typical $\gamma$-ray and proton-induced air showers, particularly illustrating those secondaries that produce Cherenkov light.

One should be aware that only a fraction of less than $10^{-4}$ of the total shower energy is converted into photons, and quite a few of these photons get lost before hitting the ground. Losses are due to absorption by ozone molecules below around 300 nm, Rayleigh scattering (normally well predictable) and Mie scattering due to fine dust or thin clouds or haze in the atmosphere. In the early times of Cherenkov detectors, losses due to Mie scattering were quite unknown even until the 1980s; these losses could not be fully explained because no adequate instruments for measuring them were used. Even around 1990, the predictions of the transmission of the atmosphere for Cherenkov light varied by up to a factor four. Adding to these uncertainties the systematic errors of the instruments, in particular the photon detection efficiency (PDE) of the photomultipliers, provided observers with measurements, which were hardly consistent. Also, as previously mentioned, the first-generation Cherenkov detectors did not allow one to discriminate between electromagnetic and hadronic showers. The early Cherenkov telescopes plainly did not have the necessary sensitivity to even observe the strongest sources, and often excesses of three standard deviations ($\sigma$) in the rate difference between *On* and *Off* source observation were claimed as a discovery.

HEP has made considerable progress in particle studies in laboratories in the years since 1960. This success could be traced to advances in accelerator developments as well as for the replacement of optical readout techniques for bubble chambers or optical spark chambers by a continuous development of more powerful electronic devices, intense use of computers and the formation of large collaborations. At the same time, CR physics progressed very little. Detectors were small and completely inadequate for the necessary collection of complete shower information and the important discrimination between $\gamma$-ray and hadronic showers was very much hampered by a poor knowledge of the shower development, i.e., by the lack of adequate VHE measurements of especially the high energy hadronic interaction. Modest progress in technology – completely different compared to the progress in HEP experiments – was achieved because of a lack of resources. Often leftover material from dismantled HEP experiments was used, thus reflecting the state of the art electronics of the 1950s and 1960s. Also, the use of computers was very restricted.

In HEP experiments, discoveries were often made as soon as an excess of at least 3-$\sigma$ above background was observed. When cuts based on poor knowledge of shower developments were applied to CR data to find sources, failures were guaranteed because the used selection procedures did not deliver unbiased samples. Thus, detections often were reported when a subset of cuts provided a 3-$\sigma$ excess, and this was then interpreted as a signal. Particularly the air shower arrays, although simple to operate, suffered from their high and rapidly changing threshold with the zenith angle. The results of that time were often highly controversial and often disagreed at the level of spectral analyses. These mostly and often contradictory 3-$\sigma$ observations of claimed sources contributed very much to the low reputation of cosmic-ray physics. Only a few physicists who did not change their focus to HEP in the 1950s and 1960s continued this research. Even the Cherenkov technique, which looked quite promising, was not delivering. In retrospective, the lack in finding the sources of the CRs is quite obvious. The reasons were:

- A very low gamma-ray flux compared to the charged CR flux
- Interaction of $\gamma$-rays with cosmic low-energy photons suppressed the detection of ultra-high energy $\gamma$-ray sources
- Shower tails contain very poor information on primary particles
- Poor energy resolution just above threshold
- Poor angular resolution
- Poor understanding of shower processes, in particular hadronic showers, as no precise accelerator measurements existed
- (The most severe problem) Poor $\gamma$/hadron separation power from data recorded with too simple detectors

In summary, the reasons for failure were the use of detectors of insufficient sensitivity, the lack of information from precision VHE experiments at accelerators, the lack of understanding the details of the dominant hadronic shower development and the atmospheric response. Nevertheless, one of the reasons for the activities in the field not completely fading away was due to a controversial high-significance result from an array detector set up by the University of Kiel.

### 6.2.2 Cyg X-3 in 1983: A Controversial Large Signal Pushes the Searches for Gamma-Ray Sources

At the University of Kiel a small but very active group pursued cosmic-ray research. In the mid 1970s, the group improved their cosmic-ray experiment by extending the existing scintillator detector array and measuring more parameters of air showers. They added quite a few scintillation counters up to a distance of 100 m from the previous core detector arrangement. Also, they improved the measurement of the different shower components in the shower tail, such as a measurement of the electron, the muon and the hadron parameters of individual showers. Figure 6.4 shows the layout of the inner part of their array. The array comprised the following detectors:

- 27 unshielded scintillation counters for measuring the shower size and core position.

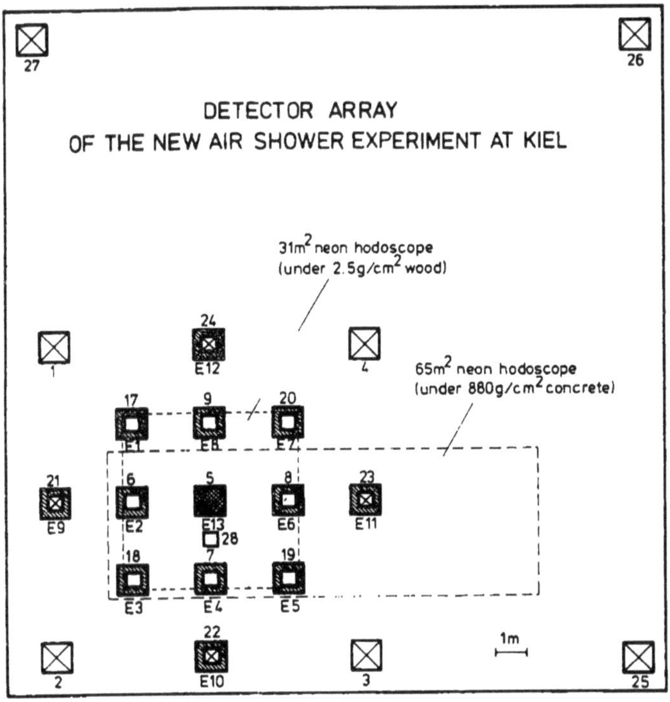


**Fig. 6.4** Layout of the inner part of the Kiel detector (figure taken from Bagge et al., 1977): Central part of the EAS detector array. Squares represent scintillation counters of 1(0.25) m$^2$ area each. Shaded areas indicate a 1.25 × 1.25 m shielding of 2 cm lead plus 0.5 cm of iron on top of the scintillator. Detectors with additional fast timing photomultipliers are labeled with diagonal crosses

- 11 scintillation counters connected to 22 fast timing circuits for determining the arrival directions by means of time of flight measurements.
- A 31 m$^2$ neon hodoscope with 176,000 flash tubes for measuring the electron core structure. This hodoscope was protected by wood mounted on the ceiling with 2.5 g cm$^{-2}$ density. The flash tubes were recorded on film by four cameras in case of a trigger from a few scintillation counters.
- 13 scintillation counters shielded by 2 cm of lead and 0.5 cm of iron for measuring the energy flow of the electromagnetic shower component.
- A 65 m$^2$ neon hodoscope of 367,500 flash tubes under a layer of 880 g cm$^{-2}$ concrete. These flash tubes registered shower muons and the hadronic component. Three cameras recorded the flash tube pattern when an event trigger occurred.

Operation of this array, one of the most powerful detectors of that time, started in 1976. As the installation was at sea level the threshold of the detector was quite high, around 1–2 × 10$^{15}$ eV. In 1983, the group published the results from four years of data recording, concentrating on the search of possible gamma-ray emission from Cygnus X-3 (Samorski and Stamm, 1983). Cygnus X-3 is one of the strongest X-ray emitting binary star systems and was for a long time a prime search

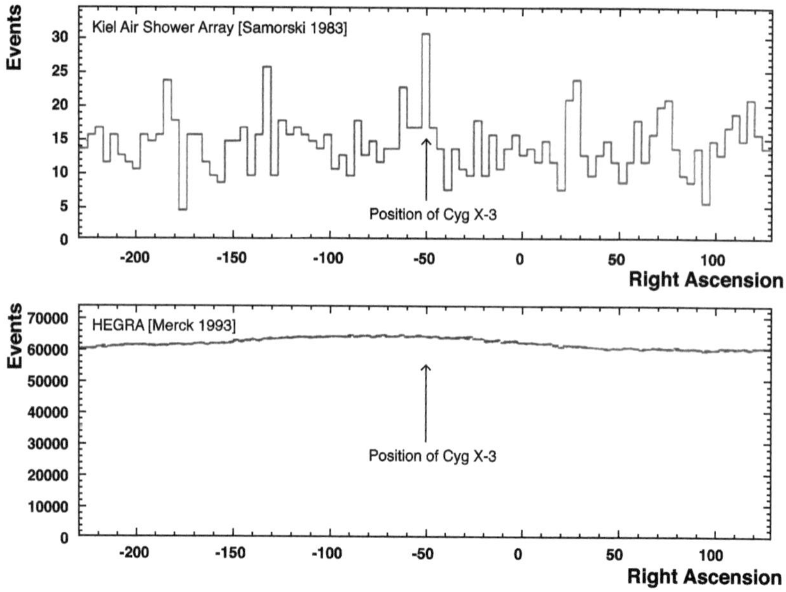


**Fig. 6.5** Measurement of the cosmic-ray flux from the direction of Cygnus X-3 (right ascension band at 40.9° ± 1.5° declination) and the surrounding sky region as measured in 1983 by Samorski and Stamm (1983) (*upper panel*) and 8 years later by HEGRA (*lower panel*). Figures taken from Merck (1993)

candidate for the emission of TeV gamma rays. Quite a few experiments claimed to have seen gamma rays with roughly a 3-$\sigma$ excess. In their 1983 publication, the Kiel physicists claimed a 4.4-$\sigma$ excess at the position of Cygnus X-3 from 3,838 hours of observation time and a sensitive area of 2,800 $m^2$, i.e., at a declination angle of 40.9° ± 1.5°, and 307.8 ± 2.0° in right ascension. Figure 6.5 shows the published results. What enhanced the belief in the result was the finding of a strong peak in the phase diagram of a 4.8 h periodicity, derived from X-ray data ten orders of magnitude lower in energy (Fig. 6.6). As early as 1973, the SAS-2 satellite had reported gamma-radiation within a narrow phase interval of the 4.8 h-phase (Parsignault et al., 1976). Common belief was that Cygnus X-3 comprises a binary system generating gamma rays and that the eclipsing of the compact star by its companion was most likely causing this periodic signal. This result created quite some interest and intense discussion not only in the CR community, but also in the HEP community, and quite some groups started to observe specifically Cygnus X-3. Basically, this result triggered a revival in interest in the search for the sources of CRs. In the wake of the Kiel experiment quite a few other experiments confirmed the results, mostly claiming also to observe a 4.8-h periodicity signal. Details go beyond the scope of this article and we mention only a few references (which include also some general discussion): Lloyd-Evans et al. (1983); Marshak et al. (1985); Watson (1985). Later, some additional results surfaced, which could have reduced the excitement. It turned out that the excess was

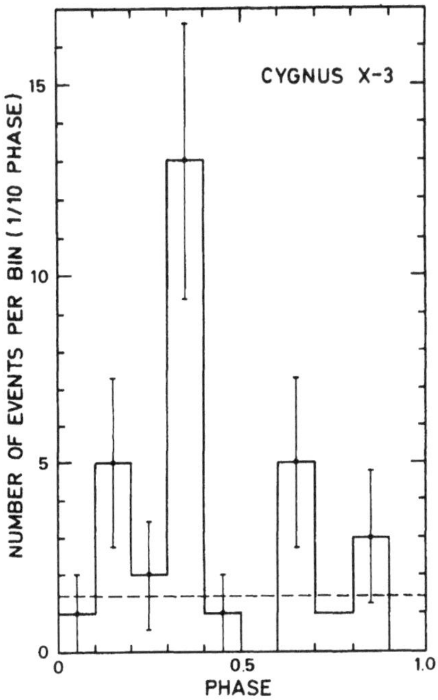


**Fig. 6.6** Phase diagram of Cygnus X-3, indicating a 4.8-h periodicity (Samorski and Stamm, 1983)

about 1.5° off the position of Cygnus X-3 (Samorski, private communication), but this was considered consistent with the systematic uncertainty of the measurement of the shower arrival direction by means of time of flight measurements. Also, not published in the 1983 article, the muon hodoscope results showed that nearly all showers in the Cygnus X-3 bin had a very similar muon flux as that of hadronic showers, i.e., also the excess showers were consistent with hadronic showers. In the absence of reliable gamma experiments at accelerators it was speculated that electromagnetic showers above $10^{15}$ eV had a strong hadronic component, explaining the presence of a strong muon component. Cygnus X-3 is about 12 kiloparsecs away from the earth and $>10^{15}$ eV photons from this distance should already be strongly attenuated by interaction with the cosmic microwave background, see Sect. 6.3. Again, CR physicists speculated that in the absence of trustworthy accelerator experiments PeV gamma rays behaved quite differently compared to low-energy gamma rays. The Kiel results again triggered quite a few 3-$\sigma$ observations as well a similar number of contradicting results, and a flood of exotic theoretical predictions for an energy range inaccessible to HEP accelerator experiments. It is interesting to note that the Kiel physicists assumed that the gamma-ray flux to be about 1.5 % of the total VHE CR flux (Samorski and Stamm, 1983). Eight years later, the HEGRA (high-energy gamma-ray array) experiment, started by the Kiel group on the Canary island of La Palma at a height of 2,200 meters above sea level, with much higher precision and higher data statistics could not confirm any signal from Cygnus X-3 (Merck et al., 1991), see the lower panel of Fig. 6.5. The CASA-MIA experiment, at that time the EAS experiment with the highest sensitiv-

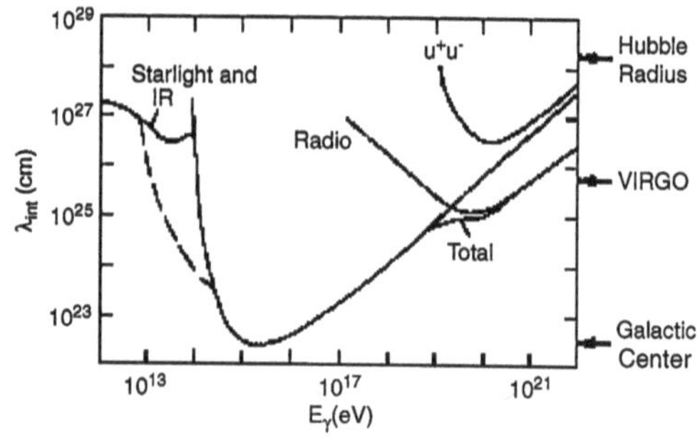


**Fig. 6.7** Absorption length of VHE gamma rays in the Universe due to interaction with the low-energy photon fields. The dominant absorption around $10^{15}$ eV is caused by interactions with photons of the 2.7-K CMB. From Chi et al. (1992). Figure taken from Lorenz (2006)

ity (array size $500 \times 500$ m$^2$; median energy of 100 TeV) could not find any signal from Cygnus X-3 (Borione et al., 1997). As it cannot be excluded that a signal from Cygnus X-3 is variable, the Kiel result might not be contradicting later negative observations.

## 6.3 How Far Can We "See" with VHE Gamma Rays?

An important issue, which was not very much considered by the CR physicists of the 1960s to 1980s, was the question of how far VHE gamma rays would, on average, propagate through the Universe before being lost by absorption or scattering. In 1965, Arno Penzias and Robert Wilson discovered the 2.7-K cosmic microwave background radiation (CMB, Penzias and Wilson, 1965). Later, it was found that besides the dominant CMB the Universe is also "filled" with a wide spectrum of low-energy photons, although the 2.7-K CMB photons with about 420 photons per cm$^3$ were by far the most dominant ones. These low-energy photon fields are basically a calorimetric measure of all past and present radiating cosmic objects. VHE gamma rays have to pass this soup of low-energy photons and might occasionally interact with them forming an $e^+e^-$ pair and because of this are lost to the observer. This process is described by quantum electrodynamics (QED). The cross section peaks close to the double electron mass in the center of mass of the two photons. Depending on the density of low-energy photons, the propagation length of the VHE gamma rays is more or less limited. As early as around 1966, R.J. Gould and G. Schréder (1966) made a first prediction of the opacity of the Universe. At around $10^{15}$ eV the absorption length is just around 10 kiloparsec, i.e., approximately the distance of the center of our Galaxy to us. Figure 6.7 shows the prediction of the absorption length from a later work of X. Chi, J. Wdowczyk and A.W. Wolfendale (1992) for the gamma-ray absorption length due to the different contributions of the density of low-energy cosmic photon fields. The density of infrared (IR) photons is more or less a crude estimate. This plot indicates that detections of PeV gamma rays from observing extragalactic sources are impossible, and it is quite obvious that air shower arrays with their high threshold had close to no chance of significantly contributing to the search for gamma-ray emitting sources of CRs, except if these were very close (<10 kpc) to the

Earth. Also detectors needed to be placed at high altitudes of a few thousand meters.

## 6.4 The Whipple Collaboration Opens the Window of VHE Gamma-Ray Astronomy in 1989

It took over 35 years until the air Cherenkov technique was rewarded with the first discovery of a VHE $\gamma$-ray emitting source since the initial observation of Cherenkov light from air showers by J.V. Jelley. The first-generation Cherenkov telescopes were in general using relatively small mirrors and very simple readouts in the form of a single PMT. In 1968, a large 10-m telescope was completed at the Fred Lawrence Whipple Observatory in Arizona, USA (Fazio et al., 1968). Figure 6.9 shows a photograph of the 10-m Whipple telescope at Mount Hopkins. Again, during the first phase only a single PMT was used as a "camera" and thus, $\gamma$/hadron discrimination was impossible. Therefore, no source could be detected although the light-collecting mirror was sufficiently large. Then, under the leadership of Trevor Weekes, both the instrument and the analysis methods were developed further to increase the sensitivity, and a method for the crucial $\gamma$/hadron separation was implemented, enabling the search for sources of much lower $\gamma$-ray fluxes than in other experiments. In 1989, the Whipple collaboration published the first convincing observation of gamma-ray emission from the Crab nebula (Weekes et al., 1989). It was basically a culmination of 10 to 20 years of hard experimental work with many steps of improvements. While quite a number of discoveries in particle physics are just surprise results, like for example the discovery of the $\psi$ particle at the SPEAR storage ring at SLAC (Augustin et al., 1974), the opening of the new window in VHE $\gamma$ astronomy was a long and tediously prepared search for the first VHE $\gamma$ source over many years.

The collaboration concentrated on a source that turned out to be the strongest steady state galactic source. Already in 1958, Philip Morrison (1958) and, independently in 1959, Guiseppe Cocconi (1959) had put forward strong arguments for observing VHE gamma rays from the Crab nebula and made predictions for high $\gamma$-ray fluxes. Ever since that time the Crab nebula was a target of VHE $\gamma$-astronomy, but the Whipple collaboration spent a remarkably long observation time of 80 h spread over three years.

They used a telescope of a large light collection area and for the first time a camera allowing an efficient $\gamma$/hadron separation of the data. The use of an "imaging camera" was at first proposed by T.C. Weekes and K.E. Turver (1977), but it took another 10 years until the first useful imaging camera was built. This camera with only 37 PMTs covered a field of view (FOV) of 3.5 degrees diameter. It allowed the recording of coarse pictures of air showers and making a simple discrimination of electromagnetic and hadronic showers. This rudimentary camera was nevertheless the start of the design of consecutively improved cameras with finer and finer pixel sampling while the FOV of 3.5° is quite a standard even of today's telescopes.

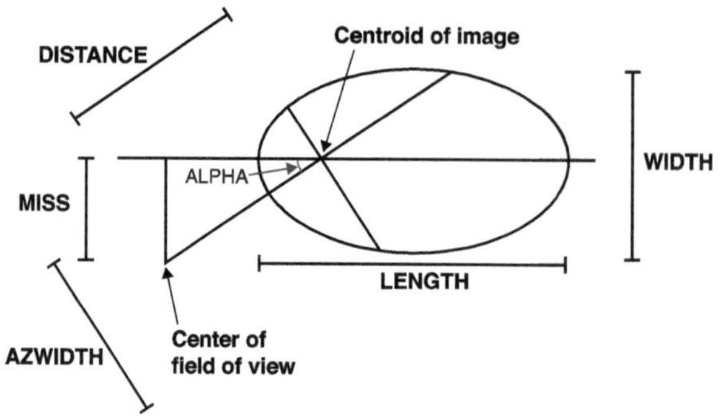


**Fig. 6.8** The image parameterization employed by Weekes et al. (1989): The shower image is characterized by the width and length of the shower ellipse along with some parameters describing the position and angle of the shower in the camera plane – showers originating from the source should point back to the source position, i.e., have a small MISS value. Weekes et al. (1989) showed that there are distinct differences in all parameters given between gamma- and hadron-initiated showers. Today, the original MISS parameter has been superseded by the ALPHA parameter, describing the angle between the weighted center of the shower ellipse and the camera center, and, later still, by the $\theta^2$ parameter, allowing for analyses without assumptions about the source position

The third and most important achievement was the introduction of a refined $\gamma$/hadron separation method based on the calculation of image moments. This analysis developed by the Whipple collaboration in the mid-1980s was based on the combination of both a measurement of the shower image orientation, originally proposed by T.C. Weekes in 1981 (Weekes, 1983) and an analysis to evaluate the difference in images between gamma-ray showers and hadron showers, originally proposed by A.A. Stepanian, V.P. Fomin and B.M. Vladimirsky (1983). The shower image should align with the position of the source in the camera (Fig. 6.8) and images of gamma and hadron showers should distinctly differ in shape with gamma-showers being rather slim and concentrated, while hadron showers are much wider and more irregular. Of course, shower fluctuations could sometimes make discrimination difficult and limit discrimination power. The originally rather simple moment analysis, commonly known as Hillas parameterization analysis (Hillas, 1985) became the basic concept for $\gamma$/hadron separation in future Cherenkov telescope experiments. It is still in use in most of today's experiments with some refinements based on additional information retrieved from better cameras with finer resolution and better shower timing data.

While the classical analysis method gave just a 1-$\sigma$ excess, the $\gamma$/hadron analysis bases on the Hillas moments allowed the hadronic background to be reduced by 98 % (with a loss of around 50 % of $\gamma$ events). Eventually, an excess at a 9-$\sigma$ level was found. This observation was confirmed in the following years by a number of other Cherenkov telescope experiments and opened the window for VHE $\gamma$ astronomy. Also, many other experiments followed the concept of the second-generation Cherenkov telescope with a pixelized camera and confirmed the VHE $\gamma$ emission of the Crab nebula (Table 6.1).

**Fig. 6.9** Photo of the Whipple 10-m telescope at Mount Hopkins. Courtesy Brian Humensky

A very important, but hardly noticed, byproduct of the detection of gamma rays from the Crab nebula was the first trustworthy measurement of the $\gamma$ flux of $\approx$0.2 % of the CR flux in a FOV of 2 degrees around the Crab nebula position and above about 0.7 TeV. This low value explains why past experiments had no chance of finding a real signal due to their low $\gamma$/hadron separation power.

## 6.5 Experiments of the Decade 1990–2000

### *6.5.1 A Small Sensation: Whipple Finds an Extragalactic Source Five Billion Light Years Away and Opens the Window for AGN Studies*

Not long after the discovery of VHE $\gamma$-emission from the Crab nebula and the search for some other galactic sources, the Whipple collaboration started a search for $\gamma$-emission from extragalactic sources. Candidates were AGNs of the blazar type, that had been detected in X-rays and low-energy gamma rays in satellite observations. Amongst the five candidate AGNs they selected for their study, only the weakest low-energy $\gamma$-emitter, the AGN Markarian (Mkn) 421, showed a strong VHE signal of about 30 % of the Crab nebula flux (Punch et al., 1992). If converted naively to the intrinsic brightness of the source nearly 5 billion light years away, Mkn 421 must emit over $10^6$ times more VHE gamma rays than the Crab nebula. This observation

**Table 6.1** Some Imaging Cherenkov telescopes in the 1990s, similar to the Whipple telescope, which later confirmed the Crab nebula VHE gamma-ray emission and detected also some other gamma-ray sources

| Telescope | #Cameras/Pixels | Collaboration | Ref. |
|---|---|---|---|
| Crimean GT48 | 2×37 pixels | Crimean Astronomical Observatory | 1,2 |
| Yerevan | 37 pixels | Yerevan | 3 |
| Ala-Too | 144 pixels | Lebedev | 4 |
| Cangaroo I | 2 Telescopes | Japan/Australia | 5 |
| HEGRA | $37 + 5 \times 271$ pixels | HEGRA collaboration | 6 |
| Granite (Whipple+11-m Tel.) | $109 + 37$ pixels | extended Whipple collaboration | 7 |
| Narrabri | 24 pixels | Durham | 8 |
| Telescope array prototype | $8 \times 256$ pixels | TA coll. | 9 |
| CAT | 600 pixels | CAT collaboration | 10 |
| ASGAT | $7 \times 7$ pixels | ASGAT collaboration | 11 |
| (not a telescope array with genuine imaging quality) | | | |

References – 1: Vladimirsky et al. (1989), 2: Fomin et al. (1991), 3: Aharonian et al. (1989), 4: Nikolsky and Sinitsyna (1989), 5: Kifune (1992), 6: Aharonian et al. (1991), 7: Akerlof et al. (1991), 8: Bowden et al. (1991), 9: Aiso et al. (1997), 10: Barrau et al. (1998), 11: Goret et al. (1991)

opened the window of extragalactic $\gamma$-search. Later, quite a few AGNs were detected and now nearly the same number compared to galactic sources are observed. Nearly all of them are blazars, i.e. galaxies with an accreting super-massive black hole in the center, a large accretion disc, and two jets orthogonal to the accretion disc (sometimes only one is seen, presumably due to beaming effects). Most current models assume that $\gamma$-rays are produced in the jets. In case one jet points towards the earth, they are called blazars. Many gamma-detected blazars show rapidly varying $\gamma$-activity, which is called "flaring". Intensity variations by a factor ten or more are observed, in extreme occasions up to a factor of $\approx$50 with respect to the lowest gamma-ray fluxes seen from the respective blazars. It is likely that most blazars have not yet been detected because they are currently in a "dormant" state. Also, the sensitivity of current Cherenkov telescopes might only allow one to see the strongest flaring sources, as up to now nearly all observed blazars have a super-massive black hole of at least $10^8$ solar masses (Wagner, 2008).

### 6.5.2 A Persistently Flaring Blazar: Mkn 501 Flares for Over Six Months

Soon after the discovery of Mkn 421, the Whipple collaboration discovered another blazar, Mkn 501 (Quinn et al., 1996) at a redshift of $z = 0.034$, at nearly the same

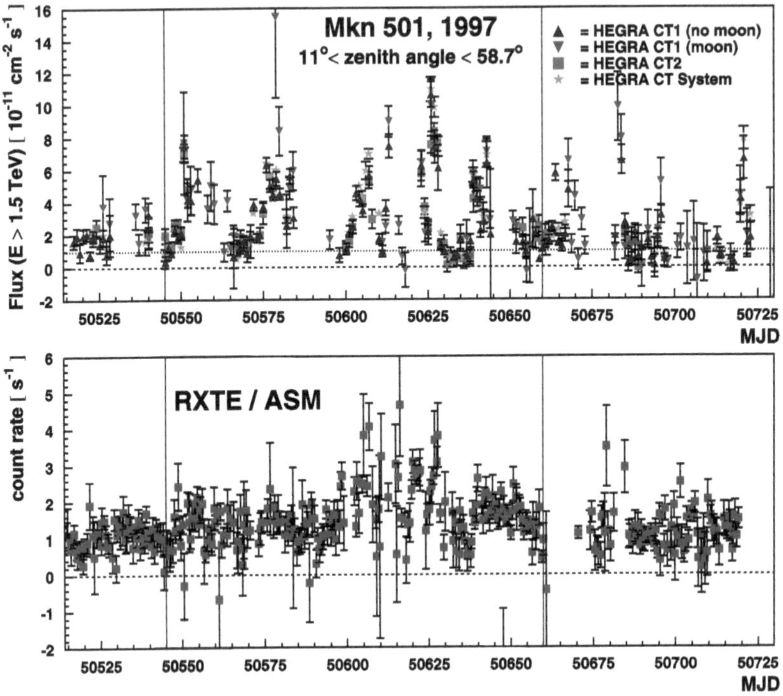


**Fig. 6.10** The light curve of Mkn 501 in summer 1997. Flux variations in the range of 20 were observed in the VHE domain. The flaring activity extended over the entire observation period of 6.7 months. Due to a new observation method introduced by HEGRA it was possible to observe such a strong source also during partial moonlight. Data from D. Kranich's Ph.D. thesis (Kranich, 2002). The TeV data show much larger fluctuations than the X-ray data recorded by RXTE (Remillard and Levine, 1997)

distance as Mkn 421 and of very similar performance. The VHE $\gamma$-emission of Mkn 501 was soon afterwards confirmed by the HEGRA collaboration (Bradbury et al., 1997).

In 1997, Mkn 501 showed a series of extremely large outbursts extending in time over the entire observation period in 1997 and, up to now, never seen from any other AGN. The flare intensities reached peak values exceeding the low state by up to approximately a factor 20. The flaring activity was observed by HEGRA stereoscopic system (Aharonian et al., 1999), TACTIC (Joshi et al., 2000), and the Whipple telescope (Quinn et al., 1999). At that time, the HEGRA collaboration introduced a new method for observing strong sources also during partial moonlight, thus HEGRA was able to collect a nearly continuous light curve over nights during nearly 6 months. Figure 6.10 shows this flux measurement above 1.5 TeV from the HEGRA collaboration during the observation period in 1997. The data are compared with the X-ray data from the RXTE satellite between 2 keV $< E <$ 10 keV (Remillard and Levine, 1997). Figure 6.10 highlights the enormous variation in the

**Fig. 6.11** Photograph of the Crimean multi-telescope setup. Each barrel contains one Cherenkov telescope. Three telescopes form a unit. Each of those units can be positioned along a railway system and view the air showers under slightly different angles. This arrangement allowed both coincidence measurements and simple multi-telescope observations. The system was used from 1960 until 1963. Figure courtesy T.C. Weekes

highest energy domain while at lower energies also a change in the X-ray flux was observed but with a much smaller and smoother flux variation.

### *6.5.3 Stereo Observations improve the Sensitivity of Cherenkov Telescopes*

Soon after the first Cherenkov telescopes were used to look for the sources of cosmic rays one tried to improve the sensitivity by means of the stereo technique, i.e. by viewing the showers from spaced telescopes. Chudakov and coworkers at the Catsiveli site in Crimea were the first to attempt designing a multi-telescope stereo system (Chudakov et al., 1963), which also facilitated simple stereo observations. They used 12 detectors each comprised of a large mirror and only one photomultiplier per telescope. Units of three detectors each were installed on a simple mount, which could be separated on rails. Figure 6.11 shows a photo of their arrangement. With normally only 20 m separation and a single large diameter photomultiplier/telescope, the stereo quality was rather poor and more a coincidence measurement for reducing accidental triggers. Some time later, J. Grindlay (Grindlay et al., 1975) tried another stereo approach (Fig. 6.12) with only two similar telescopes mounted on a circular rail system allowing a separation of up to 180 m. Later, some other similar attempts were made, but again none of them, however, led to a high-significance source detection. The lack of any discovery can be traced back to the missing $\gamma$/hadron separation power. After the breakthrough discovery of

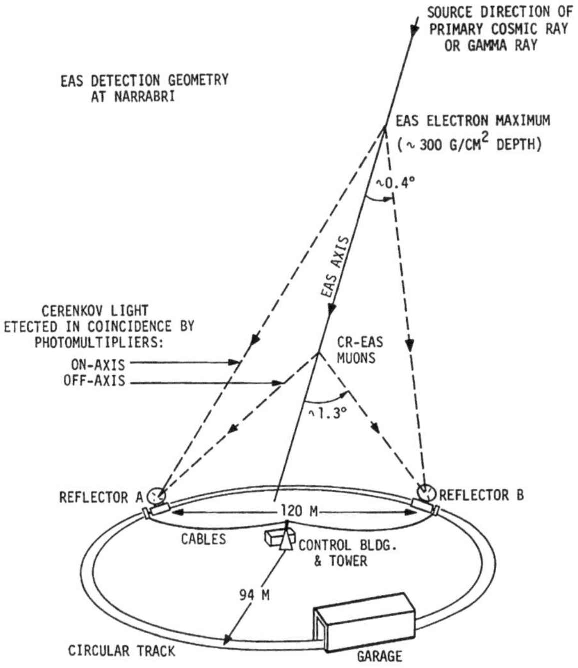


**Fig. 6.12** Another stereo telescope configuration by J. Grindlay and coworkers (Grindlay et al., 1975) used between 1972 and 1976 at Mount Hopkins and Narrabri. The two single-PMT equipped telescopes run on rails and can be operated at different distances

the Whipple collaboration using a pixelized camera, part of the extended Whipple collaboration converted an 11-m solar telescope, originally located in New Mexico, into a 37-pixel camera Cherenkov telescope, dubbed Granite, and genuine stereo observations were pursued. Unfortunately, the sensitivity of the stereo system was worse than the Whipple telescope alone. Reasons were mirrors of poorer optical quality and a tendency of icing due to radiation cooling caused by low heat conductivity of the foam backing of the mirror. Additionally, the spacing between the two telescopes of $\approx$120 m did not yield enough events simultaneously detected in both telescopes and thus was far from optimal. The first successfully operating stereo system with significantly improved sensitivity was build by the HEGRA collaboration.

### *6.5.4 The First High-Sensitivity Stereo Imaging Cherenkov Telescope System as Part of the HEGRA Observatory*

After the publication of a 4.4-$\sigma$ excess from the direction of Cygnus X-3, the Kiel physicists in 1987 started to build an improved scintillation counter array, the HEGRA experiment, at the Roque de los Muchachos (2,200 m asl) observatory on the Canary island of La Palma.

Already in the early 1990s, the Kiel institute leader, the late Otto Claus Allkofer, had discussed with Felix Aharonian from the Armenian group in Yerevan about the

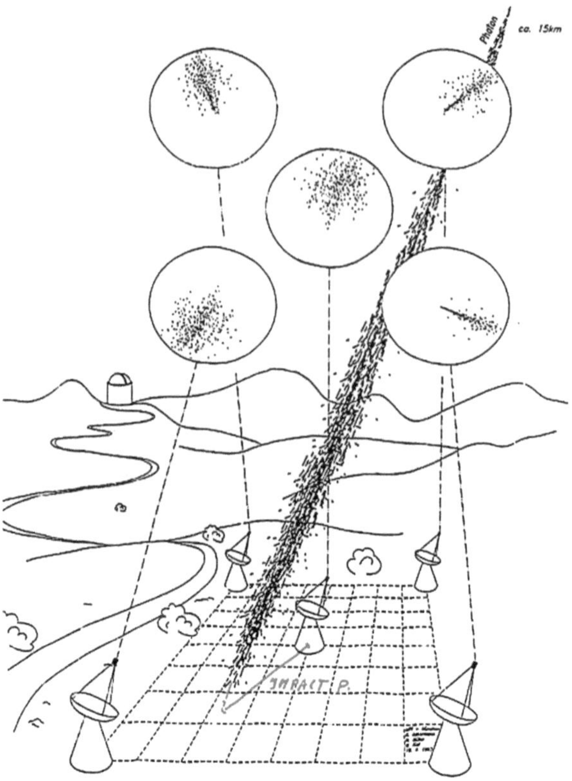


**Fig. 6.13** A sketch of the principle of stereo observations. In most cases one or two telescopes record a shower image of excellent quality from different angles. Altitude not to scale

possibility of adding five Cherenkov telescopes because of the excellent optical conditions at the La Palma site. The Armenian group had already built a small imaging Cherenkov telescope on Mount Aragats and had plans for a stereo system.

Eventually, a prototype Cherenkov telescope and five telescopes, operating in a stereo system, were built. The system was very successful with an increase in sensitivity of about a factor 10 compared to a single telescope of the same size. The reasons were manifold and are shown in a sketch in Fig. 6.13. With a stereo system, showers are observed from different directions. This can improve the $\gamma$/hadron separation by means of viewing the shower in part under optimal condition and by suppressing the so-called head-tail ambiguity of single telescopes. In single telescope pictures recorded by a classical gated analog-to-digital converter (ADC) readout, there is an ambiguity about the shower direction pointing either towards or away from the potential source location. In stereo systems one can cut the background by a factor two by solving this ambiguity. Stereo observations also provide a much better shower energy determination and a better angular resolution allowing the study of extended sources. The HEGRA stereo system was the first one that used regularly a readout with flash ADCs, now common in all Cherenkov stereo systems.

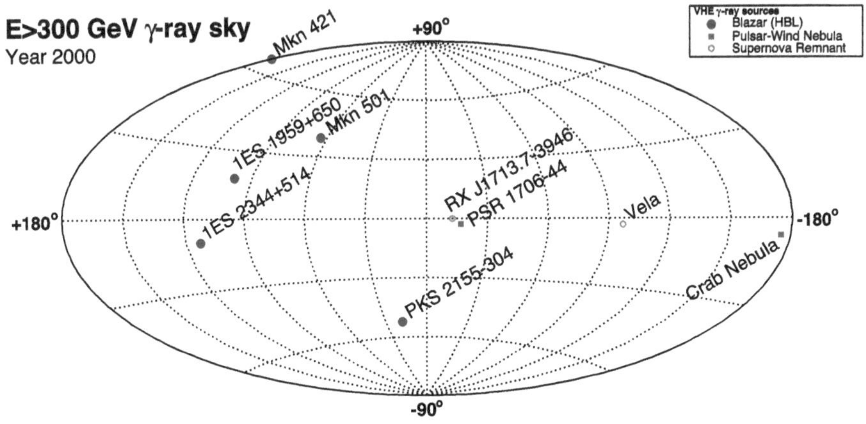


**Fig. 6.14** The VHE ($E > 300$ GeV) sky map at the year 2000

In the last decade of the last century a few other stereo systems were built (Table 6.1), but none reached the sensitivity of the HEGRA experiment. Nowadays stereo telescope systems are the main tool in VHE $\gamma$-astronomy.

## 6.6 Progress in the First Decade of the New Millennium

The progress in discovering new VHE gamma-ray emitting sources after the discovery of the Crab nebula was initially rather slow. Figure 6.14 shows the VHE sky map in the year 2000. Only eight more sources were discovered, all of them by "imaging" Cherenkov telescopes, which became the "workhorse" for the searches. These second-generation Cherenkov telescopes were simply not sensitive enough to observe sources that emit VHE gamma rays below 10 % of the Crab nebula flux. Nevertheless, confidence in the observation techniques and analysis methods developed. For nearly every group observing on the northern half of the Earth the Crab nebula was the test bench. The number of extragalactic sources found was equal to that of galactic ones detected. All extragalactic sources were blazars, while two galactic sources were pulsar-wind nebulae and two supernova remnants (SNR). The community followed a suggestion of Trevor Weekes that observed sources were accepted as discoveries only if their significance exceeded 5 $\sigma$ and all sources on the sky map were at least confirmed by one other experiment.

### *6.6.1 The Large Third-Generation Imaging Cherenkov Telescopes*

As in any emerging area of scientific research, the financing of large detectors is the issue of hard negotiations. On the whole, the majority of the astrophysicists

**Table 6.2** Table of the third-generation observatories with large mirror telescopes. The overview lists location and altitude of the observatories, the diameter and number ("#") of the individual telescopes, and the start dates of operations

| Name | Location | Diameter | # | Altitude | Start |
|---|---|---|---|---|---|
| Cangaroo III | Australia | 10 m | 4 | 160 m asl | 1999 (4 telescopes in 2003) |
| H.E.S.S. | Namibia | 12 m | 4 | 1,800 m asl | 2002 (4 telescopes in 2003) |
| MAGIC | La Palma | 17 m | 2 | 2,225 m asl | 2004 (2 telescopes in 2009) |
| VERITAS | Arizona | 12 m | 4 | 1,270 m asl | 2006 (2 telescopes in 2006, 4 in 2008) |

and astronomers were still not convinced that the new field would really contribute to the fundamental understanding of the relativistic Universe and the meager results of the past times did not justify the diversion of funding from other areas. Nevertheless, the results from mainly the last decade of the last century made it obvious that new, better telescopes would lead to a breakthrough in the field. Also, the stereo-observation technique was generally accepted as the approach that would reach sensitivities around 1 % of the Crab nebula flux within 50 h observation time for achieving a 5-$\sigma$ excess signal. Eventually, four large projects materialized: Cangaroo III, H.E.S.S., MAGIC and VERITAS.

The plans for these third-generation improved telescopes started to evolve around the year 1994 onwards. The construction of the first Cangaroo III telescope started already in 1997, the main activities of H.E.S.S. basically around 2000, MAGIC in 2002, and VERITAS in 2003. Table 6.2 lists some essential information about the third-generation observatories.

### *6.6.2 Cangaroo III*

Cangaroo III (Collaboration of Australia and Nippon for Gamma-Ray Observation in the Outback) was built by a Japanese-Australian collaboration at low altitude near Woomera in Australia, at 31°06′ S, 136°47′ E and 160 m above sea level (Enomoto et al., 2006). The telescopes had 57 $m^2$ mirror area each, i.e., about half that of one of the H.E.S.S. and VERITAS telescopes. The Cangaroo III telescopes used plastic mirrors with rather modest focusing quality and significant aging (both for reflectivity and focusing), while the PMTs with a square cathode had a relatively low photon detection efficiency (PDE). The telescopes were located at a low altitude of $\approx$160 m above sea level, where usually significant Mie scattering from fine dust leads to significant light losses. In summary, the Cangaroo III telescopes were not very competitive with H.E.S.S., MAGIC and VERITAS. Eventually, the activities in Australia were stopped in 2011.

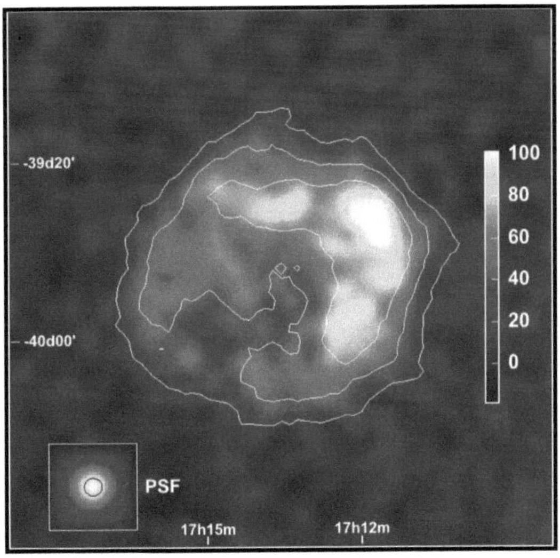


**Fig. 6.15** Scan of the extended source RX J1713.7-3946, overlaid with a radio scan (*black lines*) from the satellite-borne $\gamma$-detector ASCA in the 1–3 keV energy range. The *insert* in the *left lower corner* shows the resolution of a point-like source (Aharonian et al., 2006a)

### *6.6.3 H.E.S.S.*

H.E.S.S. (High Energy Stereoscopic System) was built by a large international collaboration in the years 2000–2003 in Namibia at 23°16′ S, 16°30′ E, at 1,800 m above sea level (Hofmann, 2001). H.E.S.S. comprises four 12-m diameter imaging Cherenkov telescopes with a 110-m$^2$ mirror and a multi-pixel camera of 960 PMTs each. The observatory is suited for the study of gamma-ray sources in the energy range between 100 GeV and 100 TeV. The stereoscopic system has a sensitivity of 0.7 % of the Crab nebula flux within 25 hours of observation time when pointing to zenith. Like Cangaroo III, H.E.S.S. is located in the Southern hemisphere and is particularly suited for the observation of sources in the central region of the galactic plane. H.E.S.S. is currently the most successful observatory, as it has discovered more than half of all known VHE sources. Due to their large diameter cameras of 5° FOV, H.E.S.S. has studied quite a number of extended sources. For example, a scan of the supernova remnant RX J1713.7-3946 in the Galactic plane (discovered in X-rays by *ROSAT*, Pfeffermann and Aschenbach, 1996) highlights the detection power for extended sources and is shown in Fig. 6.15 (Aharonian et al., 2006a). In 2012/2013, H.E.S.S. will be extended by a central fifth telescope with a 28-m diameter reflector and an energy threshold of 30–40 GeV.

### *6.6.4 MAGIC*

The MAGIC collaboration pursued another path in the development. They designed an ultra-large Cherenkov telescope with a 17-m diameter mirror (Baixeras et al., 2003) on La Palma (28.8° N, 17.8° W, 2,225 m above sea level). A second one, which was constructed later. The telescope is based on numerous novel concepts, such as a low-weight carbon-fiber reinforced plastic space frame, supporting the diamond-turned, low-weight, sandwich aluminum mirrors. To counteract small deformations during tracking, the matrix of small mirror elements, approximating a parabolic mirror profile, was corrected by an active mirror control system. The total moving part of the telescope has a weight of only $\approx$70 tons and could be repositioned to any point on the sky within 20 seconds in order to observe at least part of gamma-ray bursts (GRB). A second telescope was built only after the new items of the first one proved to work. The first telescope started to take data in 2004, and stereo observations with both telescopes commenced in 2009. The first telescope has a threshold of 60 GeV and initially a sensitivity of $\approx$1.5 % of the Crab nebula flux while the stereo system has a threshold of 50 GeV and a sensitivity of 0.8 % of the Crab nebula flux.

### *6.6.5 VERITAS*

The fourth of the third-generation imaging Cherenkov telescopes is the VERITAS telescope complex (Holder et al., 2006). VERITAS stands for Very Energetic Radiation Imaging Telescope Array System (for gamma-ray astronomy). VERITAS comprises four 12-m telescopes and is located in Arizona (31.75° N; 110.95° W, 1,268 m asl). The four telescopes at the base camp of the Mount Hopkins telescope site are quite similar to the H.E.S.S. telescopes in mirror size, but the cameras have a smaller FOV. As for H.E.S.S., the threshold is $\approx$ 100 GeV; the sensitivity is also better than 1 % of the Crab nebula flux. The first telescope started operation in late 2005, while the full system saw first light in 2007. The VERITAS telescopes have already undergone major upgrade, in which cameras with new electronics, photomultipliers with increased quantum efficiency, and a new trigger were installed.

### *6.6.6 Milagro*

Milagro was the first really successful tail-catcher detector. Progress in understanding the shower development at its tail and using a detector with 100 % active area around the shower core axis finally produced the first convincing detection of some VHE gamma-ray sources. This detector, dubbed Milagro (Sinnis, 2009), made use of a large water pond of 80 $\times$ 50 m with a depth of 8 m. The detector was located near Los Alamos at an altitude of 2,630 m above sea level. 175 small water tanks surrounded the water pond to collect information about the radial shower extension. The charged shower tail particles generated Cherenkov light when passing the water. Electrons from $\gamma$-showers stop normally in the first 2 meters while hadronic showers contain some particles that penetrate deeply into the water pond. The water pond was subdivided into two layers of 2.8 $\times$ 2.8 m cells. Each cell was viewed by

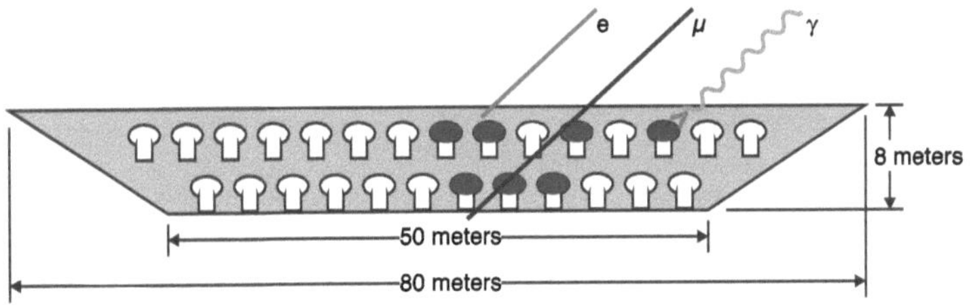


**Fig. 6.16** Schematic cross section of the water pond of the Milagro detector. Depending on the type of incident particle, PMTs in the *upper* and *lower region* of the pond would detect light, as illustrated. The gamma/hadron separation of Milagro was based on these different penetrating powers

one large PMT. The top layer of 450 PMTs was under 1.4 meters of water and the bottom layer of 273 PMTs was under 6 m of water, as illustrated in Fig. 6.16.

Milagro had some considerable $\gamma$/hadron separation power. Air showers induced by hadrons contain a penetrating component (muons and hadrons that penetrate deeply into the reservoir). This component resulted in a compact bright region in the bottom layer of PMTs. A cut based on the distribution of light in the bottom layer removed 92 % of the background cosmic rays while retaining 50 % of the gamma-ray events. The detector was suited for the observation of showers above 2 TeV (from showers coming close from the zenith) and had an up-time of 24 h. At 45° zenith angle the threshold was 20 TeV. The collaboration operated the detector from 2002 to 2006.

Milagro with its rather high threshold was best suited for the search for galactic sources in the outer part of the galactic plane. During a survey of the galactic plane (Abdo et al., 2007) three new, in part quite extended sources were discovered and a few already known sources confirmed. Milagro stopped operation in 2007. Another successful air-shower array is Tibet AS operated by a Japanese collaboration (Huang et al., 2009). This detector at 4,300 m asl comprises a large number of scintillation counters but still has only a fractional sampling of the surface and has therefore a threshold of 3 TeV. Air-shower detectors have a 24 h up-time and should in principle be well suited for the detection of gamma-ray bursts, but their currently high threshold has prevented any detection up to now.

### *6.6.7 A Bonanza of Galactic Sources: H.E.S.S. Scans the Galactic Plane*

Shortly after completion of the four H.E.S.S. telescopes, the collaboration started scanning the inner part of the galactic disk with a sensitivity of 2 % of the Crab nebula flux above 200 GeV. In order to achieve a nearly uniform sensitivity across the galactic disk, the four telescopes were slightly re-adjusted to cover a strip of ±3° latitude relative to the Galactic plane. The scan extended from −30° to +30° in

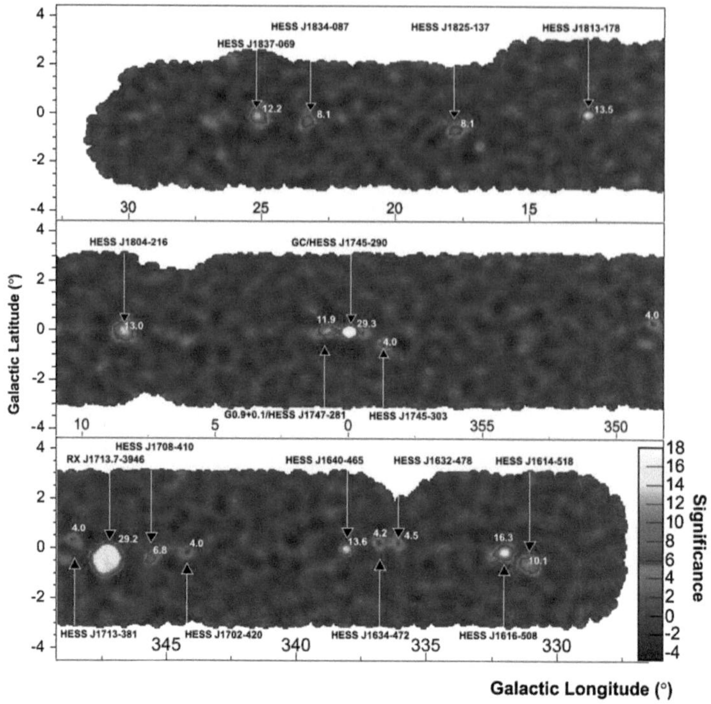


**Fig. 6.17** The H.E.S.S. scan of the inner region of the Galactic plane with 13 newly discovered sources (Aharonian et al., 2006a)

longitude, covered by 500 pointings in a total of 230 hours. In total, 14 new sources were discovered (Fig. 6.17), about half of them unidentified sources and the other half in part pulsar-wind nebulae (candidates for the sources of CRs) and SNR with $\geq$4-$\sigma$ significance after all trials (Aharonian et al., 2006a). Later, a partial rescan with higher sensitivity, respectively with an improved analysis method, increased the number of detected sources to over 30. Also, a few binary objects were found to be gamma-ray emitters. This scan made H.E.S.S. the most successful observatory for the detection of galactic sources. Quite a few sources could not be classified. The richness of sources found in the galactic plane tells us that one could expect a significantly larger number with the next generation higher sensitivity telescopes.

## *6.6.8 H.E.S.S. and MAGIC Discover the First Binaries*

About one third of all stars are arranged in binary systems. Already during the Cygnus X-3 studies by the Kiel and other groups, the mostly accredited model for the VHE gamma-ray production was assumed to be a binary system with a periodicity of 4.8 hours. In the 1980s, binaries were considered as *the sources* of cosmic gamma rays. Later, after quite a few VHE gamma-ray sources were discovered and none of them could be explained as binary systems, the question after the discovery of the Crab nebula was raised at nearly every International Cosmic Ray Conference before 2005: Where are the binaries? Eventually, both H.E.S.S. and MAGIC detected binaries in the Galactic plane. H.E.S.S. published the first discovery of a VHE binary, PSR B1259-63 (Aharonian et al., 2005a) and LS 5039 on the Southern sky (Aharonian et al., 2005b). Soon afterwards MAGIC discovered the first binary on the Northern sky, LS I+61 303 (Albert et al., 2006). Figure 6.18 shows the light curves of the three binaries. The composition of the binaries is not evident; Fig. 6.19 shows the two preferred models.

### *6.6.9 MAGIC Discovers the First VHE Pulsar*

Pulsars are one of the most interesting stellar objects. In the high-energy domain, satellite-borne gamma-ray detectors detected a few gamma-ray pulsars. The EGRET detector on board the *Compton* gamma-ray satellite confirmed the observation of seven high-significance pulsars in the MeV region while recently the follow-up satellite *Fermi* added many more pulsars (6 % of all newly discovered stellar objects were pulsars Nolan et al., 2012) and measured the spectra of the brightest ones up to $30/40$ GeV. Ever since the discovery of VHE gamma-ray emission from the Crab nebula, groups have searched for pulsed emission from pulsars in the VHE domain, but up to 2009 without success. In 2009, the MAGIC collaboration developed a new low threshold trigger, which could record data down to 26 GeV, i.e., with considerable overlap with *Fermi*-LAT data. Although *Fermi*-LAT had predicted a cutoff in the pulsed gamma-ray emission at 12 GeV for the Crab pulsar, MAGIC discovered pulsed emission from 26 GeV upwards to nearly 100 GeV, cf. Fig. 6.20 (Aliu et al., 2008). Later, the VERITAS collaboration found pulsed gamma-ray emission from the Crab pulsar using data from 100 to 400 GeV (Aliu et al., 2011). Some months later the MAGIC collaboration confirmed these results (Aleksić et al., 2012b). These two measurements had opened the window of VHE pulsed gamma-ray studies.

### *6.6.10 What to Expect in the Next Decade: The Next Generation Detectors for VHE Gamma-Astronomy*

The first decade of the 21st century saw considerable progress in VHE gamma-ray astronomy. The third-generation Cherenkov telescopes achieved a sensitivity of $\approx 1$ % of the Crab nebula flux and currently about one new source per month is discovered. Nevertheless, one sees a gradual shift from "source hunting" to the study of the underlying physics and to fundamental physics issues. The recent successes have triggered ideas for quite a few new detectors with another large step in sensitivity increase and which should be realized in the coming years. There follows a very short overview of the new ideas.

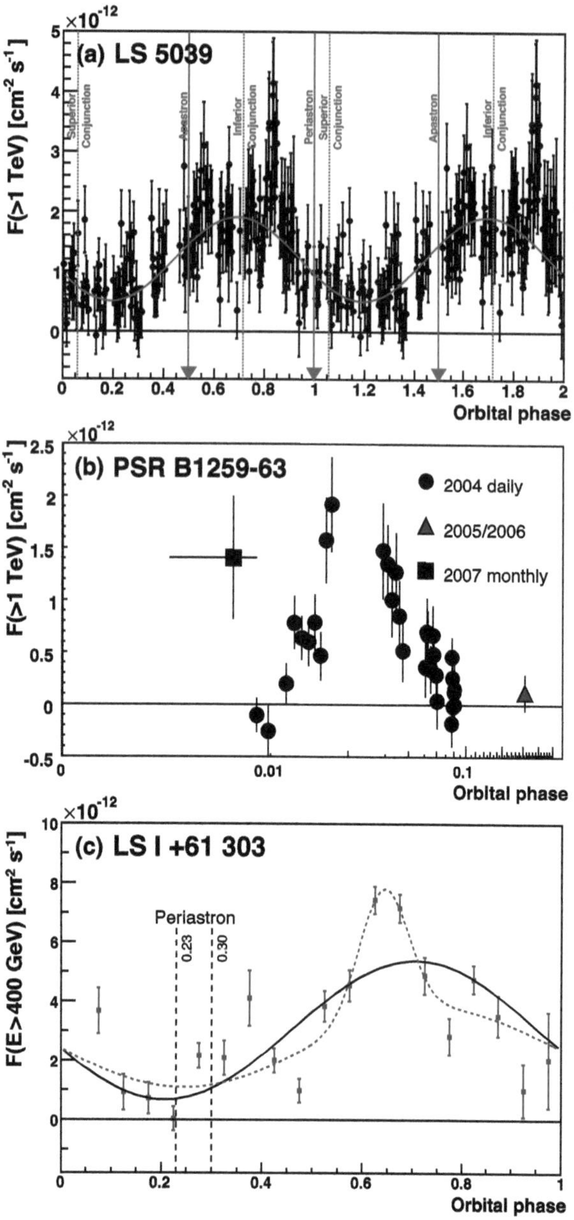


**Fig. 6.18** Light curves of the binaries **a** LS 5039, periodicity 4 d (Aharonian et al., 2006b), **b** PSR B1259-63, periodicity 3.9 y (Aharonian et al., 2009d) **c** LS I+61 303, periodicity 26 d (periodicity 26 d)

## *6.6.11 Towards a Large VHE Gamma-Ray Observatory: The CTA Project*

Around 2007 it became evident that a further large increase in sensitivity could not be achieved by improving single telescopes but by considerably increasing the

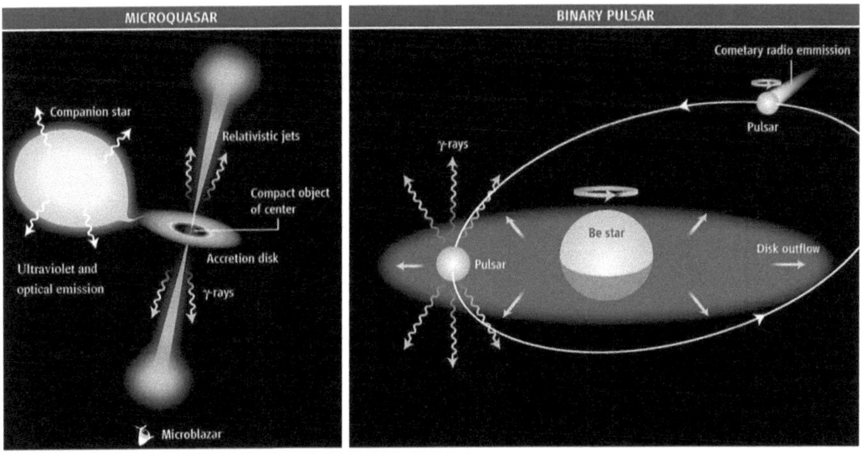


**Fig. 6.19** The two preferred models of binary system emitting gamma rays. **a** the so-called microquasar model with a small black hole accreting mass from the companion star. Gamma rays are produced in the jets. **b** a binary system proposed by Felix Mirabel (2006). A pulsar circulates around a Be star

number of telescopes in an array configuration. The idea for CTA (Cherenkov Telescope Array) was born. Building a detector covering the energy range of 20 GeV to 100 TeV (Actis et al., 2011) requires a large number of three different sizes of telescopes (23 m, 12 m, and 3 to 5 m diameter, respectively) in order to achieve a sensitivity 10 times higher compared to H.E.S.S. (see Fig. 6.21 for the predicted sensitivity). The sites have not yet been selected. For covering the entire sky, it will be necessary to select one site in the Southern hemisphere and one in the Northern hemisphere. The energy range of CTA South will be extended to about 100 TeV for the study of galactic sources while CTA North will need the two larger size telescopes types, because multi-TeV gamma rays from higher redshift extragalactic sources are suppressed by the interaction with the low-energy photon fields (see Sect. 6.3) and consequently no longer detectable. The initially European project is now enlarged to a worldwide collaboration approaching 900 members. CTA will start observations around 2015–2017. In their initial phase, the telescopes will be relatively conservative copies of current third-generation telescopes.

### *6.6.12 Other Projects: AGIS, MACE, HAWC, LHAASO*

Four other projects have passed the level of first ideas and are currently under detailed evaluation or in a first phase of construction. AGIS (Vandenbroucke, 2010) and MACE (Koul et al., 2005) are Cherenkov telescopes, while HAWC (Salazar, 2009) is an extended air-shower (EAS) array at high altitude for achieving a low

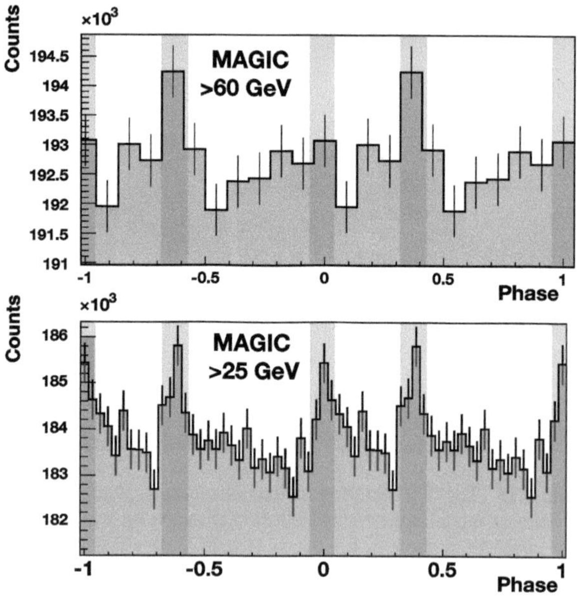


**Fig. 6.20** First detected pulsed VHE gamma-ray emission of the Crab pulsar as measured by MAGIC (Aliu et al., 2008). A signal of 6.3-$\sigma$ significance, $8{,}300 \pm 1{,}300$ pulsed events over a background of 6,106 events, has been detected. Figure courtesy the MAGIC collaboration

threshold. LHAASO (Cao et al., 2011) is a facility that combines various air shower detector elements and Cherenkov telescopes.

## 6.7 A Short Summary of Physics: What Have We Learned from VHE Gamma-Ray Sources?

The biggest success of ground-based gamma-ray astronomy, besides promoting VHE gamma-ray astronomy from the astronomy of a single source in 1989 by increasing the number of detected VHE gamma-ray sources to nearly 150 in 2012 is the diversity of source classes that could be established in this energy range.

A significant increase in the number of gamma-ray sources was due to the systematic scan of the Galactic plane performed by the H.E.S.S. collaboration from 2003 onwards. At almost the same pace, the extragalactic VHE gamma-ray sky became populated by VHE gamma-ray sources, dominantly blazars, due to systematic searches by the three VHE gamma-ray instruments H.E.S.S., MAGIC and VERITAS.

### *6.7.1 Supernova Remnants*

A final stage of stellar evolution is reached when a star runs out of the fuel necessary for the fusion reactions that counteract the gravitational pressure. If the star

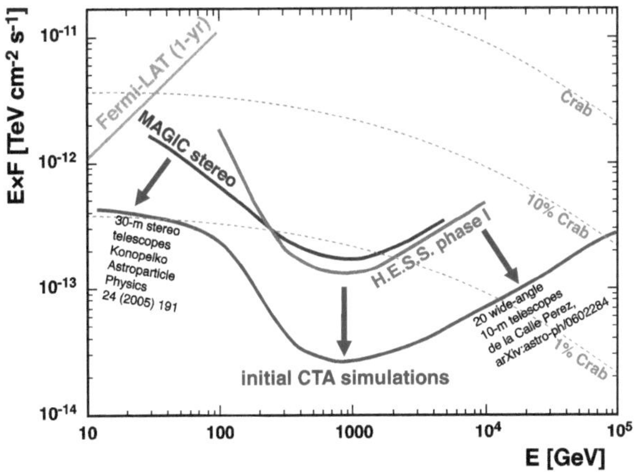


**Fig. 6.21** Predicted sensitivity of CTA (in comparison to H.E.S.S., MAGIC, and *Fermi*-LAT) and the Crab nebula flux

is heavy enough, the collapse of the stellar core is followed by the ejection of the outer shells of the stellar material. Depending on the mass of the remaining object, a neutron star or a black hole is formed; the ejected material may interact with interstellar material. This expanding structure is called a supernova remnant. For a long time, supernova remnants have been suspected to be the sources of charged cosmic rays up to energies of at least $10^{15}$ eV. SNR generally are extended objects, and any VHE gamma-ray emission observed traces either, in case of hadronic origin, regions in which cosmic rays interact with target material, or, in case of leptonic origin, target electrons that exist in SNR. Showcase examples for detected and spatially resolved SNRs in gamma rays so far are the four objects RX J1713.7-3946 (Aharonian et al., 2006a), RX J0852.0-4622 (Lemoine-Gourmard et al., 2007), RCW 86 (Aharonian et al., 2009b), and SN 1006 (Acero et al., 2010). Generally, the VHE emission seems to resemble the X-ray morphology in these SNR, favoring a leptonic origin of the VHE emission, and particularly SN 1006 and RX J1713.7-3946 are most certainly dominated by leptonic acceleration. On the other hand, an association of the gamma-ray emission with the presence of a molecular cloud (traced by CO density), which may serve as target material for hadronic gamma-ray production. Such an association is given in IC 443 (Albert et al., 2007), whereas in Tycho's supernova remnant, a combination of Fermi-LAT (GeV) and VERITAS spectra (Acciari et al., 2011) rule out leptonic acceleration models. The energy spectra from SNR are particularly hard, with a cutoff that sets in at about 20 TeV, indicating that the primary particles responsible for the gamma-ray emission must have had energies of some hundred TeV.

## *6.7.2 Pulsars and Pulsar-Wind Nebulae*

If a rotating neutron star remains in the system, it is referred to as a plerion or pulsar-wind nebula (PWN). The Crab nebula is a showcase example of a PWN. In such systems high energy electrons originating from the pulsar power the gamma-ray emission. PWN are the most commonly found type of galactic gamma-ray sources. Nonetheless, not only the nebula itself may emit gamma radiation: As recently discovered (Aliu et al., 2008), the pulsar in the center of the Crab nebula emits pulsed VHE gamma radiation.

About one third of the sources found in scans of the galactic plane could not yet be associated with counterpart objects. For these, spectral and temporal properties of the TeV emission, and spatial co-location with known emission at other wavelengths are being investigated to learn about their nature.

### *6.7.3 Compact Objects and Binary Systems*

The source of high-energy particles in binary systems is the accretion of matter on one of the companions. Such systems provide vastly different conditions than the previously discussed objects, like high magnetic fields, high radiation densities, and high-energy photon fields. Due to this, particle acceleration and cooling timescales are short (typically in the order of the orbital periods of the systems). Compact objects (stellar-mass black holes or neutron stars) may also exhibit relativistic jet outflows. Such objects are then called microquasars in analogy to quasar-type active galactic nuclei. Well-known binary system TeV gamma-ray sources are PSR B1259-63 (Aharonian et al., 2005a), LS 5039 (Aharonian et al., 2005b), LS I +61 303 (Albert et al., 2006), and HESS J0632+057 (Maier et al., 2011).

### *6.7.4 Stellar Clusters and Stellar Winds*

Strong stellar winds, as they typically exist in star-forming regions and stellar clusters, may accelerate particles and lead to VHE gamma-ray production. Stellar winds seem natural candidate regions for VHE gamma-ray production as they also drive particle acceleration in binary systems and outflows in pulsar systems. Recently, TeV gamma-ray emission has been discovered in the young star system Westerlund 2 (Aharonian et al., 2007; Abramowski et al., 2011a), and indications have been found in the Cyg OB2 star association.

### *6.7.5 Unidentified Sources*

The galactic plane scan revealed a substantial number of sources with no evident counterpart at any other wavelength – about 20 such "dark accelerators" are now known. Some objects could later on be identified as PWN or SNR by catalog searches, by the revision of the likeliness of an association to a known object (e.g., HESS J1303-631/PSR J1301-6305) or by targeted follow-up observations (e.g., HESS J1813-178, Helfand et al., 2007). However, for quite a few unidentified sources, such methods have failed to reveal their nature (Aharonian et al., 2008). Particularly a lack of X-ray emission may hint at a hadronic origin of the gamma-ray emission. Detailed studies of the (temporal, spectral, and morphological) features of these TeV-only emitters may help to identify the particle acceleration process at work and may also help answering the question whether these objects represent a source class of their own. However, as particle acceleration that leads to gamma-ray production generally requires certain rather characteristic parameters of the accelerator (like magnetic field strength, extension, densities), it may be difficult to establish a new class of TeV emitters.

In a certain sense, also the gamma-ray source at the center of our Galaxy is an unidentified TeV source (Kosack et al., 2004; Aharonian et al., 2004). Here, the difficulties come from source confusion, as the Galactic center region is a very busy one: Besides star-forming regions (Sgr B1, Sgr B2, Sgr D), the most prominent source towards the Galactic center is Sgr A, within which Sgr A* has been identified as possibly being a super-massive black hole. In addition, also a dark-matter annihilation signal could be expected from the center of our Galaxy. The gamma-ray energy spectrum determined from the Galactic center source is rather hard, favoring a PWN origin, and disfavoring a dark-matter origin. Dedicated searches for a dark-matter signal are reported, e.g., in (Abramowski et al., 2011c).

### *6.7.6 Extragalactic Gamma-Ray Sources: Active Galactic Nuclei*

The second VHE gamma-ray source to be detected in 1992 was the active galactic nucleus Mkn 421. This source, like most of the well over 20 AGNs discovered as of today, is a blazar, which is a subclass of AGN with relativistically beamed emission towards the observer. Blazars have been detected at a redshift range of $z = 0.031$ (Mkn 421, Punch et al., 1992) up to $z = 0.536$ (3C 279, Albert et al., 2008a) so far. Active galactic nuclei are powered by accretion of matter by super-massive black holes with some billion solar masses and show high variability down to timescales of minutes and below, indicating complex particle acceleration and cooling processes working within the jet acceleration regions. The most remarkable flaring activity so far has been observed in PKS 2155-304 (Acero et al., 2012) with flux intensities exceeding by an order of magnitude the otherwise mostly "dormant" emission (Aharonian et al., 2009c; Abramowski et al., 2010) and flux variations on timescales of minutes.

The TeV AGNs were for a long time dominated by so-called high-peaked BL Lac objects (Fig. 6.22), which are AGNs with a peak of their synchrotron emission in the X-ray range of the energy spectrum. In leptonic acceleration models the TeV emission is then interpreted as photons scattered off the same electron population that created the X-ray emission. Lately, some "low-peaked" BL Lac objects (with the synchrotron peak in the optical regime; e.g. BL Lac itself; W Comae) and flat-spectrum radio quasars with even lower X-ray peaks could be discovered, e.g., 3C 279 (Albert et al., 2008a), and PKS 1222+22 (Aleksić et al., 2011).

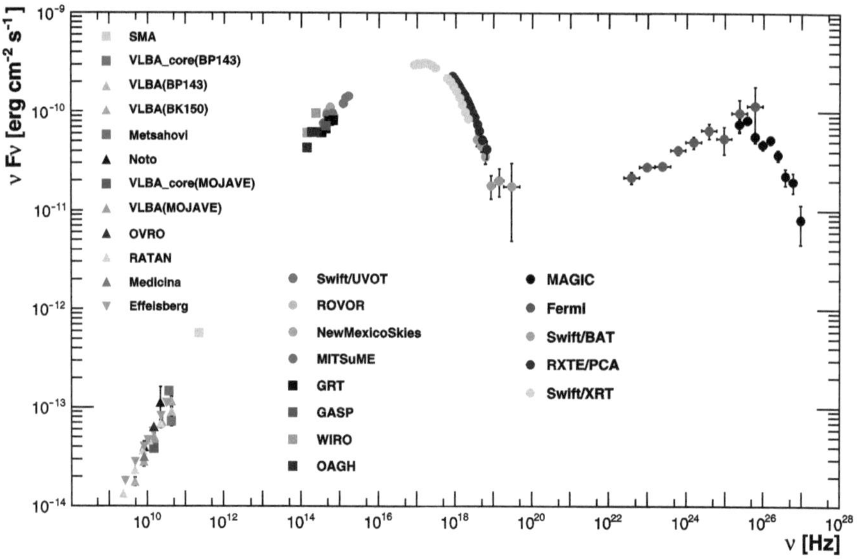


**Fig. 6.22** A strictly simultaneously measured spectral energy distribution of the blazar Mkn 421 (Abdo et al., 2011). The low-energy peak is believed to represent synchrotron radiation off a population of relativistic electrons, while the origin of the second, high-energy peak is debated. It may be due to inverse-Compton radiation of the same electron and photon population ("self-synchrotron Compton" emission), external photons scattering off the electrons ("external Compton" emission), or it may be of hadronic origin. High-energy gamma-ray observations play a crucial role in discriminating possible scenarios due to their sensitivity to time variations and the spectral shape of the SED at around GeV/TeV energies

Recently, also close-by radio galaxies like M 87 and Centaurus A have also been identified as gamma-ray emitters. Those objects are close-by and have jets misaligned to the line of sight. This allows spatial studies of the jets and the regions within them responsible for the particle acceleration, particularly by combining high-resolution radio observations and TeV light curves (Acciari et al., 2009a; Harris et al., 2011; Abramowski et al., 2012).

### 6.7.7 Starburst Galaxies

Galaxies with entirely no activity in their central engine, like M 82 and NGC 253, could be identified as TeV emitters (Acero et al., 2009; Acciari et al., 2009b). In those objects, strong stellar winds created by high supernova activities are responsible for particle acceleration that leads to gamma-ray emission up to TeV energies.

### 6.7.8 Galaxy Clusters

In addition clusters of galaxies, which in some sense represent small ecosystems of the Universe itself, have been observed by ground-based Cherenkov instruments, and recently the central galaxy of the Perseus Cluster was detected in TeV gamma rays. The emission seen so far, however, is compatible with what is expected from the galaxy itself; no extended, inter-cluster emission could be claimed (Aleksić et al., 2012a).

### 6.7.9 Gamma-Ray Bursts

Gamma-ray bursts (GRB) are transient extragalactic sources of high-energy gamma-ray emission that occur randomly and unpredictably. While gamma rays of energies of as high as 30 GeV could be detected by space-borne instruments, the specific difficulty in detecting them from ground is their presumably short time of activity. Fireball models describe the emission as being produced by relativistic shocks and predict "prompt" and "delayed" emission up to TeV energies both by leptonic and hadronic processes. Upon a trigger from space-borne, all-sky monitoring instruments, a ground-based detector has to slew very fast to the GRB direction. All ground-based instruments have GRB programs and while no GRB gamma rays have yet been detected, upper limits were reported, with observations starting up to 40 seconds after the GRB onset only. In one occasion, a (rather soft) GRB occurred while a ground-based instrument was accidentally pointing in its direction (Aharonian et al., 2009a).

### 6.7.10 Astroparticle Physics and Fundamental Physics

Besides studying of individual astrophysical objects, observations of very-high energy gamma rays are also used in the indirect search for dark matter. Some dark-matter candidate particles, namely the lightest supersymmetric particles, the neutralinos, may decay into photons (which, however, is a disfavored channel) or into quark-antiquark pairs, which would undergo further reactions producing gamma rays. A recently found channel for the production of gamma rays in DM decay processes is the "internal bremsstrahlung" process, with a gamma-ray enhancement near the kinematical limit. For some regions of the dark-matter parameter space, those gamma rays will be in the GeV to TeV energy range and may be detectable from the directions of astrophysical objects in which the dark-matter density is high. Such locations comprise the center of our Galaxy, intermediate-mass black holes, but also any object with a high mass-to-luminosity ratio, e.g. dwarf galaxies. Dark-matter signatures have not been found yet, neither from observations of the center of our Galaxy (Abramowski et al., 2011c) nor from other candidate objects; the best exclusion limit so far comes from observations of the dwarf galaxy Segue 1 (Aliu et al., 2012).

Another domain of fundamental physics in which ground-based gamma-ray astronomy can help is the search for violation of Lorentz invariance, which is predicted, most notably, by theories of quantum gravity. Qualitatively speaking, the

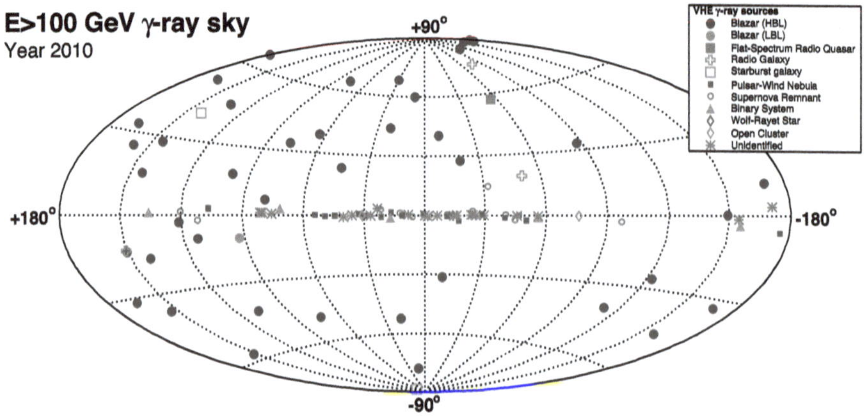


**Fig. 6.23** The VHE ($E > 100$ GeV) sky map in the year 2010

vacuum is considered to be interacting with traversing particles depending on their energy. Thus, observing delays of photon arrival times from strong AGN flares or pulsar emission constitute time of flight measurements. Any revealed relative delays of gamma rays with different energies, however, may also originate in the gamma-ray production mechanism. Thus, to demonstrate that Lorentz invariance is at work, an universal signature in many observations of objects at different distances needs to be found. Up to now, from strong flares in the AGN Mkn 501 and PKS 2155-304, only upper limits have been derived (Abramowski et al., 2011b; Albert et al., 2008b) that reach few percent of the Planck energy scale, which is the natural scale expected at which quantum-gravity effects are expected to become apparent.

## 6.8 The VHE Sky Map at the 98th Year of Cosmic-Ray Studies

The first decade of the new millennium saw a large expansion of discoveries after the large Cherenkov observatories became fully operational. Nearly every month a new source was discovered. Figure 6.23 shows the $E > 100$ GeV sky map in the year 2010 with over 110 sources. About 60 % of all sources are located in the galactic plane, while about 40 % of the sources are of extragalactic origin. The central part of the galactic plane is well visible from the H.E.S.S. observatory site while only the outer wings of the galactic plane are visible to the two northern Cherenkov observatories, MAGIC and VERITAS. Also, some sources are detected by the "tail catcher" detector Milagro.

Currently, the productivity of the three large telescope installations is high. In the 100th year of CR physics the number of 150 discovered sources is being approached. The two Northern installations have a sensitivity of about 0.8–1 % of the Crab nebula flux for a 5-$\sigma$ signal within 50 h observation time, while H.E.S.S. has a sensitivity close to 0.7 % of the Crab nebula flux. Still, most of the extragalactic area has not been scanned.

## References

Abdo, A.A., et al. (Milagro collaboration): TeV gamma-ray sources from a survey of the galactic plane with Milagro. Astrophys. J. Lett. **664**, L91–L94 (2007)

Abdo , A.A., et al.: Fermi-LAT observations of Markarian 421: the missing piece of its spectral energy distribution. Astrophys. J. **736**, 131–152 (2011)

Abramowski, A., et al. (H.E.S.S. collaboration): VHE gamma-ray emission of PKS 2155-304: spectral and temporal variability. Astron. Astrophys. **520**, A83 (2010)

Abramowski, A., et al. (H.E.S.S. collaboration): Revisiting the Westerlund 2 field with the H.E.S.S. telescope array. Astron. Astrophys. **525**, A46 (2011a)

Abramowski, A., et al. (H.E.S.S. collaboration): Search for Lorentz invariance breaking with a likelihood fit of the PKS 2155-304 Flare data taken on MJD 53944. Astropart. Phys. **34**, 738–747 (2011b)

Abramowski, A., et al. (H.E.S.S. collaboration): Search for a dark matter annihilation signal from the galactic center halo with H.E.S.S. Phys. Rev. Lett. **106**, 161301 (2011c)

Abramowski, A., et al. (H.E.S.S. collaboration, MAGIC collaboration, VERITAS collaboration): The 2010 very high energy gamma-ray flare & 10 years of multi-wavelength observations of M87. Astrophys. J. **746**, 151–169 (2012)

Acciari, V.A., et al. (VERITAS collaboration, MAGIC collaboration, H.E.S.S. collaboration): Radio imaging of the very-high-energy gamma-ray emission region in the central engine of a radio galaxy. Science **325**, 444–448 (2009a)

Acciari, V.A., et al. (VERITAS collaboration): A connection between star formation activity and cosmic rays in the starburst galaxy M82. Nature **462**, 770–772 (2009b)

Acciari, V.A., et al. (VERITAS collaboration): Discovery of TeV gamma ray emission from Tycho's Supernova Remnant. Astrophys. J. Lett. **730**, L20 (2011)

Acero, F., et al. (H.E.S.S. collaboration): Detection of gamma rays from a starburst galaxy. Science **326**, 1080–1082 (2009)

Acero, F., et al. (H.E.S.S. collaboration): First detection of VHE $\gamma$-rays from SN 1006 by HESS. Astron. Astrophys. **516**, A62 (2010)

Acero, F., et al. (H.E.S.S. collaboration): A multiwavelength view of the flaring state of PKS 2155-304 in 2006. Astron. Astrophys. **539**, A149 (2012)

Actis, M., et al. (CTA consortium): Design concepts for the Cherenkov telescope array CTA: an advanced facility for ground-based high-energy gamma-ray astronomy. Exp. Astron. **32**, 193–316 (2011)

Aharonian, F., et al.: Cherenkov imaging TeV gamma-ray telescope. In: Very High Energy Gamma Ray Astronomy: Proceedings of the International Workshop, Crimea, USSR, April 17–21, 1989, p. 36 (1989)

Aharonian, F.A., Akhperjanian, A.G., Kankanian, A.S., Mirzoyan, R.G., Stepanian, A.A., Müller, N., Samorski, M., Stamm, W., Bott-Bodenhausen, M., Lorenz, E., Sawallisch, P.: A system of air Cherenkov telescopes in the HEGRA array. In: Proceedings of the 22nd International Cosmic Ray Conference, vol. 2, Dublin, Ireland, pp. 615–618 (1991)

Aharonian, F., et al. (HEGRA collaboration): The temporal characteristics of the TeV gamma-radiation from MKN 501 in 1997. I. Data from the stereoscopic imaging atmospheric Cherenkov telescope system of HEGRA. Astron. Astrophys. **342**, 69–86 (1999)

Aharonian, F., et al. (H.E.S.S. collaboration): Very high energy gamma rays from the direction of Sagittarius A*. Astron. Astrophys. **425**, L13–L17 (2004)

Aharonian, F., et al. (H.E.S.S. collaboration): Discovery of the binary pulsar PSR B1259-63 in very-high-energy gamma rays around periastron with HESS. Astron. Astrophys. **442**, 1–10 (2005a)

Aharonian, F., et al. (H.E.S.S. collaboration): Discovery of very high energy gamma rays associated with an X-ray binary. Science **309**, 746–749 (2005b)

Aharonian, F., et al. (H.E.S.S. collaboration): The H.E.S.S. survey of the inner galaxy in very high-energy gamma-rays. Astrophys. J. **636**, 777–797 (2006a)

Aharonian, F., et al. (H.E.S.S. collaboration): 3.9 day orbital modulation in the TeV gamma-ray flux and spectrum from the X-ray binary LS 5039. Astron. Astrophys. **460**, 743–749 (2006b)

Aharonian, F., et al. (H.E.S.S. collaboration): Detection of extended very-high-energy gamma-ray emission towards the young stellar cluster Westerlund 2. Astron. Astrophys. **467**, 1075–1080 (2007)

Aharonian, F., et al. (H.E.S.S. collaboration): HESS very-high-energy gamma-ray sources without identified counterparts. Astron. Astrophys. **477**, 353–363 (2008)

Aharonian, F., et al. (H.E.S.S collaboration): H.E.S.S. observations of the prompt and afterglow phases of GRB 060602B. Astrophys. J. **690**, 1068–1073 (2009a)

Aharonian, F., et al. (H.E.S.S. collaboration): Discovery of gamma-ray emission from the shell-type supernova remnant RCW 86 with HESS. Astrophys. J. **692**, 1500–1505 (2009b)

Aharonian, F., et al. (H.E.S.S. collaboration): Simultaneous observations of PKS 2155-304 with H.E.S.S., Fermi, RXTE and ATOM: spectral energy distributions and variability in a low state. Astrophys. J. **696**, L150 (2009c)

Aharonian, F., et al. (H.E.S.S. collaboration): Very high energy gamma-ray observations of the binary PSR B1259-63/SS2883 around the 2007 Periastron H.E.S.S. collaboration. Astron. Astrophys. **507**, 389–396 (2009d)

Aiso, S., et al.: The detection of TeV gamma rays from Crab using the telescope array prototype. In: Proceedings of the 25th International Cosmic Ray Conference, vol. 3, Durban, South Africa, pp. 177–180 (1997)

Akerlof, C.W., et al.: Granite, a new very high energy gamma-ray telescope. Nucl. Phys. B, Proc. Suppl. **14**, 237–243 (1991)

Albert, J., et al. (MAGIC collaboration): Variable very-high-energy gamma-ray emission from the microquasar LS I+61 303. Science **23**, 1771–1773 (2006)

Albert, J., et al. (MAGIC collaboration): Discovery of VHE gamma radiation from IC443 with the MAGIC telescope. Astrophys. J. Lett. **664**, L87–L90 (2007)

Albert, J., Jordí, A., et al. (MAGIC collaboration): Very high energy gamma rays from a distant quasar: how transparent is the Universe? Science **320**, 1752–1754 (2008a)

Albert, J., et al. (MAGIC collaboration): Probing quantum gravity using photons from a flare of the active galactic nucleus Markarian 501 observed by the MAGIC telescope. Phys. Lett. B **668**, 253–257 (2008b)

Aleksić, J., et al. (MAGIC collaboration): MAGIC discovery of VHE emission from the FSRQ PKS 1222+21. Astrophys. J. **730**, L8 (2011)

Aleksić, J., et al. (MAGIC collaboration): Detection of very-high energy $\gamma$-ray emission from NGC 1275 by the MAGIC telescopes. Astron. Astrophys. **539**, L2 (2012a)

Aleksić, J., et al. (MAGIC collaboration): Phase-resolved energy spectra of the Crab pulsar in the range of 50–400 GeV measured with the MAGIC telescopes. Astron. Astrophys. **540**, A69 (2012b)

Aliu, E., et al. (MAGIC collaboration): Observation of pulsed $\gamma$-rays above 25 GeV from the crab pulsar with MAGIC. Science **322**, 1221–1224 (2008)

Aliu, E., et al. (VERITAS collaboration): Detection of pulsed gamma rays above 100 GeV from the Crab pulsar. Science **334**, 69–72 (2011)

Aliu, E., et al. (VERITAS collaboration): VERITAS deep observations of the dwarf spheroidal galaxy Segue 1. Phys. Rev. D **85**, 062001 (2012)

Auger, P., Ehrenfest, P., Maze, R., Daudin, J., Fréon, R.A.: Extensive cosmic-ray showers. Rev. Mod. Phys. **11**, 288–291 (1939)

Augustin, J.-E., et al.: Discovery of a narrow resonance in $e^+e^-$ annihilation. Phys. Rev. Lett. **33**, 1406–1408 (1974)

Bagge, E.R., Samorski, M., Stamm, W.: A new air shower experiment at Kiel. In: Proceedings of the 15th International Cosmic Ray Conference, vol. 12, Plovdiv, Bulgaria, pp. 24–29 (1977)

Baixeras, C., et al. (MAGIC collaboration): The MAGIC telescope. Nucl. Phys. B, Proc. Suppl. **114**, 247–252 (2003)

Barrau, A., et al. (CAT collaboration): The CAT imaging telescope for very-high-energy gamma-ray astronomy. Nucl. Instrum. Methods Phys. Res., Sect. A, Accel. Spectrom. Detect. Assoc.

Equip. **416**, 278–292 (1998)

Blackett, P.M.S.: A possible contribution to the night sky from the Cherenkov radiation emitted by cosmic rays. In: The Emission Spectra of the Night Sky and Aurorae, International Conference, London, July, 1947, p. 34. The Physical Society, London (1948)

Borione, A., et al.: High statistics search for ultrahigh energy $\gamma$-ray emission from Cygnus X-3 and Hercules X-1. Phys. Rev. D **55**, 1714–1731 (1997)

Bowden, C.C.G., et al.: The University of Durham Mark V Composite Gamma Ray Telescope. In: Proceedings of the 22nd International Cosmic Ray Conference, vol. 2, Dublin, Ireland, pp. 626–629 (1991)

Bradbury, S., et al. (HEGRA collaboration): Detection of gamma-rays above 1.5 TeV from Mkn 501. Astron. Astrophys. Lett. **320**, 5–8 (1997)

Cao, Z., et al. (LHAASO collaboration): The ARGO-YBJ experiment progresses and future extension. Int. J. Mod. Phys. D **20**, 1713–1721 (2011)

Cherenkov, P.A.: Visible emission of clean liquids by action of $\gamma$ radiation. Dokl. Akad. Nauk SSSR **2**, 451 (1934)

Chi, X., Wdowczyk, J., Wolfendale, A.W.: The clustering of the arrival directions of the highest-energy cosmic rays. J. Phys. G, Nucl. Part. Phys. **18**, 1867–1868 (1992) and further references therein

Chudakov, A.E., Dadykin, V.L., Zatsepin, V.I., Nesterova, N.M.: On the high energy photons from local sources. In: Proceedings of the 8th International Cosmic Ray Conference, vol. 4, Jaipur, India, pp. 199–203 (1963)

Cocconi, G.: An air shower telescope and the detection of $10^{12}$ eV photon sources. In: Proceedings of the 2nd International Cosmic Ray Conference, vol. 2, Moscow, Russia, pp. 309–312 (1959)

Enomoto, R., et al.: A search for sub-TeV gamma rays from the vela pulsar region with CANGAROO-III. Astrophys. J. **638**, 397–408 (2006)

Fazio, G.G., Helmken, H.F., Rieke, G.H., Weekes, T.C.: An experiment to search for discrete sources of cosmic gamma rays in the $10^{11}$ to $10^{12}$ eV region. Can. J. Phys. **46**, S451–S456 (1968)

Fomin, V.P., et al.: Comparison of the imaging gamma-ray telescopes at the Crimea and Whipple observatories. In: Proceedings of the 22nd International Cosmic Ray Conference, vol. 2, Dublin, Ireland, pp. 603–606 (1991)

Galbraith, W., Jelley, J.V.: Light pulses from the night sky associated with cosmic rays. Nature **171**, 349–350 (1953)

Goret, P., et al.: ASGAT: a fast timing VHE gamma-ray telescope. In: Proceedings of the 22nd International Cosmic Ray Conference, vol. 2, Dublin, Ireland, pp. 630–633 (1991)

Gould, R.J., Schréder, G.: Opacity of the Universe to high-energy photons. Phys. Rev. Lett. **16**, 252–254 (1966)

Grindlay, J.E., Helmken, H.F., Brown, R.H., Davis, J., Allen, L.R.: Results of a southern-hemisphere search for gamma-ray sources at energies of at least 300 GeV. Astrophys. J. **201**, 82–89 (1975)

Harris, D.E., Massaro, F., Cheung, C.C., Horns, D., Raue, M.: Stawarz, Ł., Wagner, S., Colin, P., Mazin, D., Wagner, R., Beilicke, M., LeBohec, S., Hui, M., Mukherjee, R. An experiment to locate the site of TeV flaring in M87. Astrophys. J. **743**, 177 (2011)

Helfand, D.J., Gotthelf, E.V., Halpern, J.P., Camilo, F., Semler, D.R., et al.: Discovery of the putative pulsar and wind nebula associated with the TeV gamma-ray source HESS J1813-178. Astrophys. J. **665**, 1297–1303 (2007)

Hess, V.F.: Über Beobachtungen der durchdringenden Strahlung bei sieben Freiballonfahrten. Phys. Z. **13**, 1084–1091 (1912)

Hillas, A.M.: Cherenkov light images of EAS produced by primary gamma rays and by nuclei. In: Proceedings of the 18th International Cosmic Ray Conference, vol. 3, La Jolla, USA, pp. 445–449 (1985)

Hofmann, W.: Status of the H.E.S.S. project. In: Proceedings of the 27th International Cosmic Ray Conference, vol. 7, Hamburg, Germany, pp. 2785–7288 (2001)

Holder, J., et al. (VERITAS collaboration): The first VERITAS telescope. Astropart. Phys. **25**, 391–401 (2006)

Huang, J., et al. (Tibet-AS collaboration): The complex EAS hybrid arrays in Tibet. Nucl. Phys. B, Proc. Suppl. **196**, 147–152 (2009)

Kifune, T.: The energy threshold of imaging Cherenkov technique and 3.8 m telescope of CANGAROO. In: Conference Proceedings: Towards a Major Atmospheric Cherenkov Detector for TeV Astroparticle Physics, pp. 229–237 (1992)

Kranich, D.: Temporal and spectral characteristics of the active galactic nucleus Mkn 501 during a phase of high activity in the TeV range. Ph.D. thesis, Technische Universität München (2002)

Kolhörster, W.: Messungen der durchdringenden Strahlung in Freiballons in größeren Höhen. Phys. Z. **14**, 1153–1156 (1913)

Kosack, K., et al.: TeV gamma-ray observations of the galactic center. Astrophys. J. Lett. **608**, L97–L100 (2004)

Koul, R., et al.: The Himalayan gamma ray observatory at Hanle. In: Proceedings of the 29th International Cosmic Ray Conference, vol. 5, Pune, India, pp. 243–246 (2005)

Joshi, U., et al. (TACTIC collaboration): Coordinated TeV gamma-ray and optical polarization study of BL Lac object Mkn 501. Bull. Astron. Soc. India **28**, 409–411 (2000)

Lloyd-Evans, J., Coy, R.N., Lambert, A., Lapikens, J., Patel, M., Reid, R.J.O., Watson, A.A.: Observation of gamma rays with greater than 1000 TeV energy from Cygnus X-3. Nature **305**, 784–787 (1983)

Lorenz, E.: High-energy astroparticle physics. Nucl. Instrum. Methods Phys. Res., Sect. A, Accel. Spectrom. Detect. Assoc. Equip. **567**, 1–11 (2006)

Lemoine-Gourmard, M., et al. (H.E.S.S. collaboration): HESS observations of the supernova remnant RX J0852.0-4622: shell-type morphology and spectrum of a widely extended VHE gamma-ray source. In: Proceedings of the 30th International Cosmic Ray Conference, vol. 2, Merida, Mexico, 667–670 (2007)

Maier, G., Skilton, J. (VERITAS collaboration, H.E.S.S. collaboration): VHE Observations of the binary candidate HESS J0632+057 with H.E.S.S. and VERITAS. In: Proceedings of the 32nd International Cosmic Ray Conference, Beijing, China (2011). Available at arXiv:1111.2155 [astro-ph]

Marshak, M.L., et al.: Evidence for muon production by particles from Cygnus X-3. Phys. Rev. Lett. **54**, 2079–2082 (1985)

Merck, M., et al. (HEGRA collaboration): Search for steady and sporadic emission of neutral radiation above 50 TeV with the HEGRA array. In: Proceedings of the 22nd International Cosmic Ray Conference, vol. 1, Dublin, Ireland, 261–264 (1991)

Merck, M.: Suche nach Quellen untrahochenergetischer kosmischer Strahlung mit dem HEGRA-Detektor. Ph.D. thesis, Ludwig-Maximilians-Universität München (1993)

Millikan, R.A., Cameron, G.H. New results on cosmic rays. Nat. Suppl. **121**, 19–26 (1928)

Mirabel, I.F.: Very energetic gamma-rays from microquasars and binary pulsars. Science **312**, 1759–1760 (2006)

Morrison, P.: On gamma-ray astronomy. Nuovo Cimento **7**, 858–865 (1958)

Nikolsky, S.I., Sinitsyna, V.G.: Investigation of gamma-sources by mirror telescopes. In: Very High Energy Gamma Ray Astronomy: Proceedings of the International Workshop, Crimea, USSR, April 17–21, 1989, p. 11 (1989)

Nolan, P.L., et al. (Fermi-LAT collaboration): Fermi large-area telescope second source catalog. Astron. J. Suppl. Ser. **199**, 31–77 (2012)

Parsignault, D.R., Schreier, E., Grindlay, J., Gursky, H.: On the stability of the period of Cygnus X-3. Astrophys. J. Lett. **209**, L73–L75 (1976)

Pfeffermann, E., Aschenbach, R.: ROSAT observation of a new supernova remnant in the constellation Scorpius. In: Proceedings of International Conference on X-ray Astronomy and Astrophysics: Röntgenstrahlung from the Universe, pp. 267–268 (1996)
Penzias, A., Wilson, R.W.: A measurement of excess antenna temperature at 4080 Mc/s, Astrophys. J. Lett. **142**, 419–421 (1965)
Punch, M., et al. (Whipple collaboration): Detection of TeV photons from the active galaxy Markarian 421. Nature **358**, 477–478 (1992)
Quinn, J., et al. (Whipple collaboration): Detection of Gamma Rays with $E > 300$ GeV from Markarian 501. Astrophys. J. Lett. **456**, L83–L86 (1996)
Quinn, J., et al. (Whipple collaboration): The flux variability of Markarian 501 in very high energy gamma rays. Astrophys. J. **518**, 693–698 (1999)
Remillard, R.A., Levine, M.L.: The RXTE all sky monitor: first year of performance. In: Proceedings All Sky X-Ray Observations in the Next Decade, RIKEN, Japan, p. 29 (1997)
Salazar, H.: The HAWC observatory and its synergies at Sierra Negra Volcano. In: Proceedings of the 31st International Cosmic Ray Conference, Łódź, Poland (2009)
Samorski, M.: private communication
Samorski, M., Stamm, W.: Detection of $2 \times 10^{15}$ to $2 \times 10^{16}$ eV gamma-rays from Cygnus X-3. Astrophys. J. Lett. **268**, L17–L21 (1983)
Sinnis, G.: Cosmic-ray physics with the Milagro gamma-ray observatory. J. Phys. Soc. Jpn. Suppl. A **78**, 84–87 (2009)
Stepanian, A.A., Fomin, V.P., Vladimirsky, B.M.: A method to distinguish the gamma-ray Cherenkov flashes from proton component of cosmic rays. Izv. Krimskoi Astrofiz. Obs. **66**, 234–240 (1983)
Vandenbroucke, J.: AGIS: a next-generation TeV gamma-ray observatory. Bull. Am. Astron. Soc. **41**, 909 (2010)
Vladimirsky, B.M., Zyskin, Yu.L., Neshpor, Yu.I., Stepanian, A.A., Fomin, V.P., Shitov, V.G.: Cherenkov gamma-telescope GT-48 of the Crimean astrophysical observatory of the USSR Academy of Sciences. In: Very High Energy Gamma Ray Astronomy: Proceedings of the International Workshop, Crimea, USSR, April 17–21, 1989, p. 11 (1989)
Wagner, R.: Synoptic studies of seventeen blazars detected in very high-energy gamma-rays. Mon. Not. R. Astron. Soc. **385**, 119–135 (2008)
Watson, A.A.: High-energy astrophysics: is Cygnus X-3 a source of gamma rays or of new particles? Nature **315**, 454–455 (1985)
Weekes, T.C., Turver, K.E.: Gamma-ray astronomy from 10–100 GeV: a new approach. In: ESA Recent Advances in Gamma-Ray Astronomy, pp. 279–286 (1977)
Weekes, T.C.: A fast large aperture camera for very high energy gamma-ray astronomy. In: Proceedings of the 17th International Cosmic Ray Conference, vol. 8, Paris, France, pp. 34–37 (1983)
Weekes, T.C., et al. (Whipple collaboration): Observation of TeV gamma rays from the Crab nebula using the atmospheric Cherenkov imaging technique. Astrophys. J. **342**, 379–395 (1989)
Weekes, T.C.: Very High Energy Gamma-Ray Astronomy. Institute of Physics, Bristol (2003)

# Chapter 7
# Search for the Neutrino Mass and Low Energy Neutrino Astronomy

**Kai Zuber**

## 7.1 Understanding Radioactivity – The "Invention" of the Neutrino

Historic details and most references of this section can be found in Pais (1986). The hidden entrance of neutrinos into our understanding of nature started with the discovery of radioactivity by Becquerel in 1896. He recognised that photographic plates got "fogged" while being close to uranium salts and he called this strange phenomenon *les rayons uranique* – uranium rays. Around this time Rutherford was studying ionisation of gases due to X-rays, just discovered shortly before by Roentgen. He realised that Becquerel has found a quite similar behaviour while studying his uranium rays in air. This lead him into a two-year systematic investigation of absorption features of the uranium rays and he could resolve two components of the rays. According to their penetration ability he called them $\alpha$-rays (easily absorbed) and $\beta$-rays being more penetrating (an even more penetrating radiation was discovered by Villard in 1900, nowadays called $\gamma$-rays). In another 10 years effort Rutherford and Geiger could show that the emitted $\alpha$-particle is mono-energetic and "after losing his positive charge is like a helium atom", observations which ultimately brought Rutherford to his concept of atomic nuclei.

In parallel also investigations of $\beta$-rays continued. By 1900 it was known that they are negatively charged and after Becquerel measured the e/m-ratio of these rays he obtained a similar value as the one observed 1897 in cathode rays by J.J. Thomson, strongly suggesting the electron to be the emitted particle in beta decay. It was Kaufmann in 1902 who convincingly showed that $\beta$-rays are electrons by placing a radium source into electric and magnetic fields. Only about 5 years later the question was raised by experiments whether $\beta$-rays are mono-energetic like $\alpha$-rays. Hahn and Meitner picked up this issue in 1907 while working in Berlin (actually

K. Zuber (✉)
Inst. for Nuclear and Particle Physics, Zellescher Weg 19, 01069 Dresden, Germany
e-mail: zuber@physik.tu-dresden.de

B. Falkenburg, W. Rhode (eds.), *From Ultra Rays to Astroparticles*,
DOI 10.1007/978-94-007-5422-5_7, 

they had to start their work in a carpenters workshop as women were not allowed in the Chemistry Institute; this situation changed two years later) and very first indications using photographic plates indeed supported the idea of mono-energetic electrons. However, von Baeyer in the Physics Institute was building one of the first beta magnetic spectrometers and by collaborating with him they could show in 1911 that $\beta$-rays are continuous. Nevertheless, the idea that mono-energetic electrons are emitted, which are losing energy by "secondary causes", remained. Independently, Chadwick, while working with Geiger in Berlin, picked up the investigation using a new counter system (nowadays known as Geiger counters) instead of photographic plates. He used a magnetic spectrometer with a small slit and counted the electrons arriving at his counter. By tuning the spectrometer he was able to measure the energy spectrum. In 1914 he confirmed a continuous spectrum with a few lines superimposed and could explain why there might be misleading interpretations from photographic plate data. Nowadays, the lines are known to be internal conversion lines, due to gamma rays knocking out electrons in inner atomic shells.

After recovery from World War 1 the investigation was revived in 1921, but people were looking at the problem from a different point of view. In the meantime Rutherford had convincingly proved the concept of an atomic nucleus and in 1913 Bohr invented his first quantum theory of atoms, with electrons orbiting the massive but tiny nucleus in quantised states. Using energy arguments it became obvious that $\beta$-rays are not resulting from the atomic shells, but have their origin in the nucleus. Furthermore, the mass of the nucleus was typically estimated a factor two too small if just counting all the necessary protons to compensate the charges of the electrons in the shells to make a neutral atom. Thus, the preferred model of the nucleus was one with twice the number of protons, half of them electrically neutralised by a corresponding number of electrons inside the nucleus, the remaining by the atomic shell electrons. In 1925 Wooster and Ellis went out to solve the issue of mono-energetic electrons in beta decay calorimetrically. If electrons are really mono-energetic, a calorimetric measurement should always measure the full transition energy between the two involved nuclei, independent of any "secondary causes". However, if electrons are really emitted with a continuous energy spectrum, the average energy measured in the calorimeter should be much smaller than the maximal value allowed. By using $^{210}$Bi (at that time called radium-E) they measured in 1928 an average energy of 0.35 MeV, much smaller than the maximal energy of 1.16 MeV and thus clearly proved a continuous spectrum. Meitner and Orthmann repeated the measurement and confirmed the result in 1929.

Meanwhile a second independent problem had arisen in the context of beta decay. In 1925 Uhlenbeck and Goudsmit discovered that the electron has a spin of $1/2$ and Dennison was proposing the same spin for the proton. However, it was impossible to explain a measurement performed in 1929 by Rasetti, namely that the spin of $^{14}$N is one. According to the accepted nuclear models this nucleus would contain 14 protons and 7 electrons, thus 21 spin $1/2$ objects, which results in a non-integer total spin. Also the decay of $^{14}$C into $^{14}$N cannot be explained by the emission of a single spin $1/2$ electron, as this nuclear transition is characterised as $0^+ \rightarrow 1^+$.

Given these severe puzzles rather desperate solutions were discussed, in 1930 Bohr even considered that the conservation of energy in beta decay might only be guaranteed on a statistical basis. The rescuing suggestion came in form of a famous open letter, dated 4. Dec. 1930, to the "group of radioactives" at the Gauvereins-Tagung in Tübingen, which translates as the following (after Pauli, 1961):

> Dear Radioactive Ladies and Gentlemen,
> As the bearer of these lines, to whom I graciously ask you to listen, will explain to you in more detail, how because of the "wrong" statistics of the N and Li6 nuclei and the continuous beta spectrum, I have hit upon a desperate remedy to save the "exchange theorem" of statistics and the law of conservation of energy. Namely, the possibility that there could exist in the nuclei electrically neutral particles, that I wish to call neutrons, which have spin $1/2$ and obey the exclusion principle and which further differ from light quanta in that they do not travel with the velocity of light. The mass of the neutrons should be of the same order of magnitude as the electron mass and in any event not larger than 0.01 proton masses. The continuous beta spectrum would then become understandable by the assumption that in beta decay a neutron is emitted in addition to the electron such that the sum of the energies of the neutron and the electron is constant... I agree that my remedy could seem incredible because one should have seen those neutrons very earlier if they really exist. But only the one who dare can win and the difficult situation, due to the continuous structure of the beta spectrum, is lighted by a remark of my honoured predecessor, Mr Debye, who told me recently in Bruxelles: "Oh, It's well better not to think to this at all, like new taxes". From now on, every solution to the issue must be discussed. Thus, dear radioactive people, look and judge. Unfortunately, I cannot appear in Tübingen personally since I am indispensable here in Zurich because of a ball on the night of 6/7 December. With my best regards to you, and also to Mr Back.
> Your humble servant
> W. Pauli

He suggested a new particle called neutron (the real neutron was not known and discovered by Chadwick 1932, even though Bothe and collaborators had already seen them in 1930 as an "unusual gamma radiation"). A first public mentioning of the idea of the "neutron" by Pauli was in his presentation on *Problems of Hyperfine structure* at a 171st regular meeting of the American Physical Society at Caltech in Pasadena on June 16th, 1931 (Pauli, 1931), which made it to the New York Times the next day in an article named "Dance of electrons heard by scientists". However, by no means the issue was settled and the debate whether a new particle or violation of energy conservation is responsible for the energy spectrum in beta decay continued for another five years.

The whole situation changed dramatically in 1932 when the real neutron was discovered. Chadwick and coworkers were searching already for quite some time for a "neutron" in the nucleus. After Joliot-Curie reported the observation of protons ejected from paraffin while shooting alpha-particles on beryllium, within three weeks Chadwick was able to repeat the experiment and claimed the observation of a neutron via the reaction $\alpha +^9 \mathrm{Be} \rightarrow^{12} \mathrm{C} + \mathrm{n}$. Joliot-Curie assumed high energy gammas to be responsible for the proton emission. However, in her further studies she discovered positron decay as well as delayed radioactive decays. The neutron itself was still considered to be a bound ep-state, hence the electron was still part of the nucleus but the discovery of the neutron completely changed the view of the atomic nucleus as became apparent in the famous Solvay Conference in 1932. Taking Pauli's idea for granted by using radium source it was revealed in 1933 that the

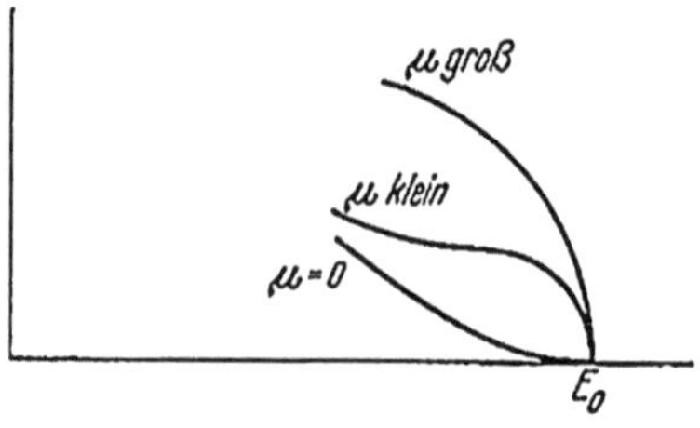


**Fig. 7.1** Sketch from Fermi's original paper on weak interaction in 1934 of the effect of a neutrino mass $\mu$ being massless, klein (small) and groß (large) (from Fermi, 1934a,b)

mean free path of "the neutron" is at least 150 km in nitrogen at 75 atmospheric pressure, by far the most penetrating object known at that time and also today. Shortly after, a radical new view and one of the most insightful papers in modern physics came up with a first sophisticated theory to explain all these new nuclear effects. It was Enrico Fermi who in 1933/1934 developed a first theory in particle physics using quantised spin 1/2 particles, published in Italian and German (Fermi, 1934a,b). His description of reactions assuming a point interaction of four particles is still valid today in the MeV range. In his seminal paper he also mentioned the neutrino appearing in beta decay and he renamed Pauli's neutron into neutrino (Italian for "small neutron"). Furthermore, he discussed the impact of a non-vanishing neutrino mass on the endpoint of a beta spectrum (Fig. 7.1) and from existing beta decay data he concluded that the best estimate is zero. Inspired by Fermi's work Bethe and Peierls calculated the cross section for neutrinos interacting with a nucleus producing an electron or positron (Bethe and Peierls, 1934). They estimated a mean free path of about $10^{16}$ km$^3$ in solid matter for 2–3 MeV antineutrinos and concluded:

> It is therefore absolutely impossible to observe processes of that kind with the neutrinos created in nuclear transformations.

Thus, for a long time nuclear recoil measurements seemed to be the only option to prove its existence. In 1935 M. Goeppert-Mayer estimated the life time of double beta decay (Goeppert-Mayer, 1935), the simultaneous decay of two neutrons in a single nucleus, resulting in values of around $10^{20}$ year, making this process extremely rare and hard to detect. Another theoretical masterpiece was published in 1937 as Majorana proposed a two component theory of neutrinos (Majorana, 1937), resulting in a neutrino being its own antiparticle (see Sect. 7.3). Immediately Racah (1937) and two years later Furry (1939) realised that this would allow for the process of neutrino-less double beta decay, just the emission of two electrons and provide an alternative neutrino mass measurement.

## 7.2 Weak Interactions – Collecting Information About the Neutrino

After the Second World war scientists started thinking about the detection of the neutrino. Especially Fred Reines at Los Alamos Laboratory, being a member of the Manhattan project and continuing with atomic bomb tests in the Pacific, came up

**Fig. 7.2** The Project Poltergeist group, on the left Clyde Cowan and on the right Fred Reines. The detector called "Herr Auge" (German for "Mr. Eye") in form of a liquid scintillator barrel can be seen in the middle (with kind permission of Los Alamos Science)

with the idea of using such an explosion for neutrino detection. The detection reaction is inverse beta decay $\bar{\nu}_e + p \rightarrow e^+ + n$. They estimated that observation of a few events via positron annihilation into two 511 keV photons would require a liquid scintillation detector of several tons if the bomb is about 50 m away. As the newly developed technology of liquid scintillators built so far had volumes of only one liter, the new detector was called "El Monstro". The detector was considered to be in free fall in a shaft, triggered by the explosion, ending up in a bath of feathers and foam rubber. The project was granted approval by Los Alamos and Reines; Cowan and others began designing and building the detector in 1951. However, in Fall 1952, once urged again to consider the option of using a nuclear power plant instead, they realised that detecting the neutron as well, by using a capture reaction producing high energy gamma rays and forming a short time coincidence of less than 100 ms between the positron and neutron, would dramatically reduce backgrounds and guarantee a much better experimental environment. The activity was called Project Poltergeist and a 300 liter detector "Herr Auge" was built (Fig. 7.2). First measurements were performed at the reactor in Hanford, Washington in 1953, which showed some indication for neutrinos but were not significant enough (Reines and Cowan, 1953). Hence a new detector was designed consisting of three scintillator tanks with 1400 liters each, interleaved by two water tanks containing dissolved cadmium chloride (the "target tanks"). The neutrino reaction was supposed to happen on the protons in the water and the thermalised neutron was captured on $^{113}$Cd, which has a huge cross section for that. The scintillators had to record all the gammas. The Los Alamos group went out to the new Savannah River Plant in South Carolina and proved the existence of the neutrino in 1956 (Cowan et al., 1956), see also Los Alamos (1997) for more details.

In this period of time a lot of further fundamental experiments were performed, some are shortly listed here. First of all, Madame Wu and collaborators discovered parity violation in weak interactions (Wu et al., 1957). Subsequent investigations actually showed that parity is maximally violated. This basically defined the structure of the weak interaction to be of vector minus axial-vector type (V-A interaction) allowing only left-handed neutrinos and right-handed antineutrinos to participate. To prove that the helicity of the neutrino is always left-handed Goldhaber performed his deeply insightful experiment measuring its helicity (Goldhaber et al., 1958). A next milestone was the proof that neutrinos emitted in pion decay are not identical to the ones from beta decay, thus at least two different neutrinos exist (Danby et al., 1962). Finally, experiments at the LEP accelerator at CERN in the 1990s showed pretty clearly that there are only three light neutrinos with mass below 45 GeV by studying the decay of the Z-boson. The first direct detection of the tau neutrino via the produced tau-lepton occurred relatively recently in 2000 by the DONUT experiment at Fermilab (Kodama et al., 2001).

## 7.3 Direct Neutrino Mass Searches

Single and neutrino-less double beta decay are two options to measure neutrino masses, however, the latter requires the neutrino to be its own antiparticle. After the Second World War people were addressing both options but not all experiments performed can be mentioned here. In beta decay the effect of a non-vanishing neutrino mass would appear as a distortion of the electron energy spectrum close to the endpoint, determined by the Q-value of the transition. Furthermore, the endpoint will be reduced by $m_\nu c^2$.

For various reasons, one is the low Q-value of 18.6 keV, tritium is the preferred beta decay candidate for neutrino mass searches. In a set of measurements Curran et al. could show in 1949 that any neutrino mass should be between 0–1 keV given the uncertainties of the Q-value at that time (Curran et al., 1949). This was improved by Langer and Moffat, showing in 1952 that the mass is smaller than 250 eV or less than 0.05 % of the electron mass (Langer and Moffat, 1952). This was one of the main reasons why the neutrino was implemented in the Standard Model of Particle Physics as a massless particle. A major step forward in improving beta decay sensitivities was the usage of new electrostatic-magnetic spectrometers. With this kind of apparatus Bergkvist in 1972 was able to set a new upper limit on the neutrino mass, of 55–60 eV (Bergkvist, 1972). As quite a surprise came the announcement of a non-zero neutrino mass by the ITEP Moscow group in 1980, claiming a value between 14 and 46 eV with 99 % confidence level (Lubimov et al., 1980). This triggered a lot of beta decay activities but no confirmation was found. Nevertheless, kinks in the energy spectrum of the electrons would indicate possible heavy neutrino admixture and these were searched for. Indeed for some time a kink linked

**Fig. 7.3** Evolution of the neutrino mass (more accurately $m_\nu^2$ is used as a fit parameter to the spectrum) as deduced from various beta decay experiments within the last two decays. The observed negative values could be explained as unknown systematic effects, which only disappeared in the very last experiments performed (from Otten and Weinheimer, 2008)

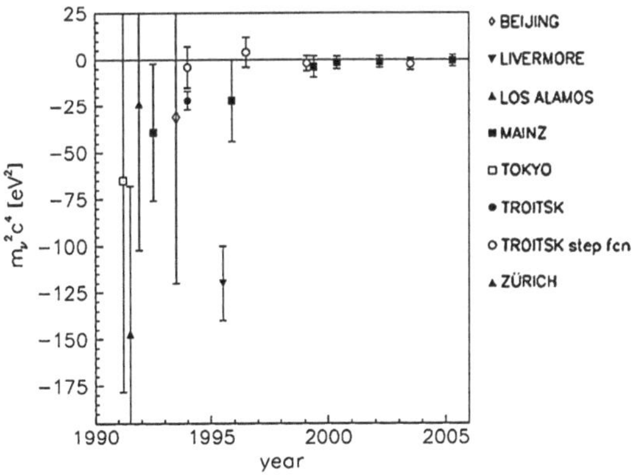


to a mysterious 17 keV neutrino appeared. This kink was observed by a fraction of experiments but not in all. Later the effect could be traced back to be due to some unrecognised experimental effects. In the latest beta decay experiments performed at Troitzk and Mainz using even more sophisticated spectrometers a further improving of sensitivity could be done and the groups were able to reduce the upper neutrino mass limit down to about 2 eV (Lobashev, 2003; Kraus et al., 2005). A next generation experiment called KATRIN is about to start soon and will have an enhanced sensitivity down to 0.2 eV. A time evolution of the improvement of the neutrino mass limits from beta decay endpoint searches of the last two decades can be seen in Fig. 7.3. In neutrino-less double beta decay the mass measurement is linked to the half-life. The longer the half-life of the decay the smaller is the neutrino mass. In 1949 a first measurement of double beta decay became available. Two tin samples were studied, one enriched in the double beta emitter $^{124}$Sn and one depleted in $^{124}$Sn and a coincidence search with four counters was performed. Apparently two coincidences per hour more were observed in the enriched sample and interpreting them as double beta events a half-life between 4–9 $\times$ $10^{15}$ years was claimed (Fireman, 1949). However, due to a lack of knowledge about nuclear structure the expectation was completely different from what we know nowadays. For this paper the neutrino-less mode was calculated with an expectation of $10^{16}$ years and $10^{24}$ years for the neutrino accompanied mode, simply based on phase space arguments, which is orders of magnitude larger if you emit only two instead of four particles. But neither the Q-value of the decay was known, which has a huge impact on the phase space nor nuclear transition matrix elements have been considered which completely changes the picture. Nowadays the two neutrino mode is expected and observed around $10^{19}$–$10^{21}$ years and the neutrino-less one is depending on the neutrino mass. Thus much longer half-lives must be considered. The result was questioned already a year later using geochemical and radiochemical experiments. The geochemical approach relies on isotopic anomalies in billion years old ores, searches are mostly done for Te (isotopes $^{128,130}$Te) and Se (isotope $^{82}$Se) double beta decay, because the daughter product is a noble gas. Hence the selective production of isotopes $^{128,130}$Xe or $^{82}$Kr will lead to an enhancement of both isotopes with respect to the natural abundance measurable by very sensitive noble gas mass spectrometry. The radiochemical approach was to search for double beta decay of $^{238}$U into $^{238}$Pu which decays with the emission of a characteristic $\alpha$-particle. Both searches came up with much longer half-lives limits then the claimed observation (Inghram and Reynolds, 1949; Levine et al., 1950). It is worthwhile to mention that in 1951 a geochemical $^{130}$Te half-life was claimed of $1.4 \times 10^{21}$ years, not too far away from the current laboratory value. A repetition of a tin experiment in 1952 with improved equipment could not confirm the original evidence (Fireman and Schwartzer, 1952). More sophisticated experiments based on the geochemical approach were performed in the 1960s and showed first serious evidence for double beta decay of $^{130}$Te (Kirsten et al., 1967, 1968). It should be mentioned that this kind of approach cannot discriminate between the decay modes as only the daughter isotope is detected, hence the signal is expected to be dominated by the neutrino accompanied decay.

The first laboratory observation of this decay mode occurred in 1987 by using selenium-foils in a TPC within a magnetic field. In this way the group at UC Irvine could announce a positive signal based on 36 events (Elliott et al., 1987). By now this decay mode has been observed for about a dozen isotopes but the important neutrino-less mode is still awaiting its detection. However, in 2001 there was a claim of observation (Klapdor-Kleingrothaus et al., 2001) in the isotope $^{76}$Ge, which is still awaiting confirmation. The next generation of large-scale detectors using huge amounts of isotopically enriched detectors is about to start, already in 2011 three projects in form of GERDA ($^{76}$Ge) and EXO, KamLAND-Zen ($^{136}$Xe) have started.

It should be mentioned that in the last decade also bounds on the sum of all neutrino masses from cosmology became more and more stringent due to much better data, leading to bounds in the region of an eV and below as well.

## 7.4 History of Stellar Energy Generation

The ideas about energy production inside the Sun and stars has a long and changing history. An extensive discussion can be found in Longair (2006). In the 19th century thermodynamic arguments were used to explain the heat production by accretion. A bombardment by meteorites was considered but the rate needed would be very large and in conflict with observed meteoritic impacts on Earth. The Sun itself was considered to be a liquid sphere gradually contracting and cooling on time scales of $10^7$ years, already causing conflicts with geological age determinations of the Earth. In 1869/1870 Lane was exploring whether the Sun could be gaseous and he was the first who found the correct hydrodynamic equations and the principle of mass conservation. He was also the first to show that if a star loses energy by radiation it contracts and that in this process the temperature actually increases and not decreases as expected from the virial theorem. This model was supported and refined later by more sophisticated work of Ritter and Emden. Independently, Lockyer in the 1880s attempted to link the spectral classes of stars with an evolutionary sequence based on the mentioned meteoritic hypothesis. In his opinion a cloud of meteorites evaporates by common collisions producing a gaseous nebula. The gas will contract to form a hot star. Its further development was an expected cooling down to become a compact red star.

During the following decades, observations and investigations of colours and luminosities of stars and their proper motion improved significantly which ultimately culminated in the Hertzsprung–Russell diagram. In 1911 Hertzsprung could reveal the main sequence based on photometry of the Plejades and Hyades clusters, because all stars in the cluster have the same age and distance. He found a wide variation in stellar luminosities but a relatively small one in mass. Independently Russell focused on precise parallax measurements. In 1914 Pickering provided him with magnitudes and spectra and he was able to obtain a similar diagram as Hertzsprung by plotting luminosity versus colour. His vision was that red giants are the earliest phase of stellar evolution, which during contraction will heat up and join the upper end of the main sequence. Somehow this follows Lockyers idea that a star cools down and ends up as a compact red star at the lower end of the main sequence. Thus, the main sequence was considered to be a cooling sequence not an evolutionary sequence as we see it today. A relic of this view of Russell is the fact that stars on the upper part of the main sequence are still called early-type and those at the lower end late-type stars.

The key player in establishing equations of stellar evolution and introducing nuclear transformation into the discussion was Sir Arthur Eddington. After Eddington proposed the basic assumptions for stellar structure and applied it successfully to red giants he did the same for the Sun. As elemental abundances were not really known at that time he assumed that the mass of the Sun is composed of atoms with an average atomic mass of 54. Jeans proved him wrong very quickly as part of a year long debate between both. Eddington changed to an average mass of 2 and found that matter would be fully ionised, the perfect gas law is applicable and any reactions can happen at much higher temperatures and densities than on Earth. In 1917 he announced the annihilation of matter as an inexhaustible source of energy, being aware of all the developments in atomic and nuclear physics at that time. In 1920 he concluded that even without any idea how this could actually happen, but just from energetic arguments, it is very attractive with an eye on Einstein's energy–mass relation (Eddington, 1920):

> Certain physical investigations in the past year... make it probable to my mind that some portion of this sub-atomic energy is actually being set free in stars... Aston has further shown conclusively that the mass of the helium atom is less than the sum of the masses of the 4 hydrogen atoms... Now mass cannot be annihilated and the deficit can only represent the mass of the electrical energy set free in the transmutation... If 5 per cent of the star's mass consists initially of hydrogen atoms, which are gradually being combined to form more complex elements, the total heat liberated will more than suffice for our demands, and we need look no further for the source of a star's energy.

A next milestone was Gamow's discovery of quantum mechanical tunnelling and its application to $\alpha$-decay in 1928. Already a year later Atkinson and Houtermans studied proton penetration through a potential barrier assuming a Maxwell–Boltzmann (MB) distribution for the protons. They concluded that this process

**Fig. 7.4** Bruno Pontecorvo, sitting in his office at the Joint Institute for Nuclear Research (JINR) in Dubna in 1983, on the day of his 70th birthday. He was the first to come up with the idea of using the chlorine–argon method for neutrino detection. In further essential work on neutrino physics, he was the first one to discuss neutrino oscillations (with kind permission from S. Bilenky)

would work most effectively for nuclei with small charge as the Coulomb barrier is smaller and that preferentially particles in the high energy tail of the MB-distribution will succeed. Two immediate consequences from this are that nuclear reactions might occur at lower temperature than thought and that the stellar luminosity should strongly rise as a function of temperature, due to the exponential dependence of the tunnelling probability on energy. Furthermore, in 1931 it became more or less clear that hydrogen is the most abundant of all elements in the Universe. Atkinson used this information and proposed that heavier elements could be created by successively adding protons on nuclei until they become too massive for nuclear stability and emit an $\alpha$-particle. This is a kind of precursor idea of the CNO-cycle proposed by Bethe and Weizsäcker in 1938 as a source of energy generation (Weizsäcker, 1937, 1938; Bethe, 1939). With the discovery of the neutron and Fermi's theory about weak interactions (see Sect. 7.1) it finally became possible to calculate reaction rates for $^3$He and $^4$He which ultimately lead to the proposal of the pp-chain by Bethe and Critchfield (Bethe and Critchfield, 1938; Bethe, 1939). They also found that the energy of the pp-chain is sufficient to explain the solar luminosity and the scaling laws for the rate of energy production $\varepsilon$ as a function of temperature ($\varepsilon \propto T^4$ for the pp-chain and $\varepsilon \propto T^{17}$ for the CNO cycle). In his papers Bethe safely ignored neutrinos as the initial reaction for fusion: he wrote $\mathrm{p} + \mathrm{p} \rightarrow \mathrm{d} + \mathrm{e}^+$. Thus, the major energy source for stars on the main sequence performing hydrogen burning into helium in hydrostatic equilibrium was finally found.

## 7.5 Solar Neutrinos

After the world recovered from the Second World War Bruno Pontecorvo (Fig. 7.4) in 1946 studied radiochemical methods to detect solar neutrinos if the mentioned nuclear processes are really occurring inside the Sun (Pontecorvo, 1946). One of his suggestions was the usage of the chlorine-argon method, based on the reaction $\nu_e + {}^{37}Cl \to {}^{37}Ar + e$, requiring neutrinos of at least 814 keV energy. However, notice that at that time the subscript e was not existing and the neutrino was not even observed. Thus only one neutrino was assumed. Luis Alvarez picked up the idea in 1948 and performed a rather detailed discussion of all the experimental issues involved but did not try to convert the idea into an experiment. Fortunately there was Raymond Davis Jr., a radiochemist hired by the newly opened Brookhaven National Laboratory. Inspired by a review article of Crane (1948) on neutrinos he started to convert the chlorine idea into reality. First he built a 200 l detector but was unable to detect anything at the Brookhaven High Flux reactor, considered to be a neutrino source. Later on, he was upgrading his tank to 3 800 l buried 5.8 m underground. Based on the data obtained he was able to give an upper limit on solar neutrinos of 40 000 SNU (see below), which is about 15 000 times higher than he measured later which forced a response from the referee:

> Any experiment such as this, which does not have the requisite sensitivity, really has no bearing on the question of the existence of neutrinos. To illustrate my point, one would not write a scientific paper describing an experiment in which an experimenter stood on a mountain and reached for the moon, and concluded that the moon was more than eight feet from the top of the mountain (from Davis, 2003).

After recognising the too low flux of the reactor, the experiment was moved to Savannah River and measured simultaneously to the Project Poltergeist (see Sect. 7.3). While Reines and Cowan discovered the antineutrino, again Davis did not see any signal. He concluded that the cross section for the neutrino capture was a factor 5 smaller than predicted by theory and he refined it within a few years to about a factor of 20 smaller. This was the prove that neutrinos and antineutrinos are different (reactors emit only antineutrinos and the chlorine–argon method is insensitive to them), but this non-observation did not get much attention. In 1958 all of a sudden a solar neutrino measurement seemed to be feasible. Until this time it was considered that the Sun is basically using the pp-I chain resulting in solar neutrinos too low in energy to be detected with ${}^{37}Cl$. In this year Holmgren and Johnston measured the cross section for the reaction ${}^{3}He + {}^{4}He \to {}^{7}Be + \gamma$ and it turned out to be 1 000 times greater than thought (Holmgren and Johnston, 1959). This opened the road to higher energetic neutrinos resulting from the electron capture of ${}^{7}Be$ and also from ${}^{8}B$ decay (Fig. 7.5). Immediately Fowler and Cameron contacted Davis to convince him to consider an experiment and he put his 3 800 l prototype in a limestone mile in Ohio. Unfortunately he was not able to detect a signal. In 1960 Kavanagh showed that the branching into ${}^{8}B$ (the pp-III chain) is very small, hence a potential neutrino flux would be small. On the positive side Bahcall in 1963 recalculated the capture cross section of solar neutrinos on ${}^{37}Cl$ including the isobaric analogue state in ${}^{37}Ar$ and found it 20 times higher than before (Bahcall, 1964). This finally triggered a larger experiment and the estimate was an event rate of 4–11 neutrino captures per day in 378 000 liters of $C_2Cl_4$ (Davis, 1964). This kind of enormous ratio (finding a few atoms in about $10^{30}$ other ones) was the beginning of low-level (often called low-background) physics. $C_2Cl_4$ was chosen as it is a bleaching agent and was available in huge amounts at that time. Based on the experience from the

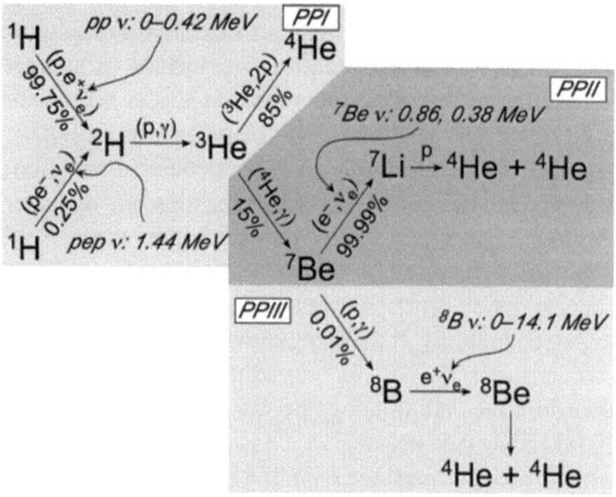


**Fig. 7.5** The pp-cycle of nuclear hydrogen burning in stars. The *upper left part*, called pp-I, was the only one considered seriously when R. Davis Jr. started working on the chlorine experiment. It was the work of Holmgren and Johnston that removed the bottleneck and opened the road to higher energetic neutrinos in the pp-II and pp-III chains (Courtesy of Brookhaven National Laboratory)

Ohio measurements in terms of $^{37}$Ar production via other processes, Davis was able to calculate the depth he needed to actually observe solar neutrinos and found out that at least 4 000 feet were necessary. At the end the decision was made to build the device at the Homestake Gold Mine in Lead, South Dakota. The expected solar neutrino spectrum based on the reaction chain and solar models is shown in Fig. 7.6.

The principle of the experiment was the counting of $^{37}$Ar which could only be produced by solar neutrino capture on $^{37}$Cl. For that the tank was left alone for 2–3 months to accumulate a few $^{37}$Ar atoms. This is feasible because the half-live of $^{37}$Ar is 35.04 days. After that via an eductor system helium was bubbled into the perchloroethylene to remove the volatile Ar (Fig. 7.7). Performed in a closed gas circuit the Ar was trapped on the surfaces in a cooled charcoal trap. After this procedure was finished a second, independent gas circuit was opened, the charcoal trap was heated and the released Ar atoms were guided into a miniaturised proportional counter. There the electron capture decay of $^{37}$Ar was measured using the released Auger electrons. This is possible, as argon–methane mixtures are commonly used counter gases. The experiment was installed in 1965/1966 and data taking started in 1967 (Fig. 7.8). Already in the first run it became apparent that less neutrinos than expected were observed, which was published in 1968 (Davis et al., 1968). This is the origin of the problem of missing solar neutrinos. Further improvements could be made in 1971 by introducing pulse shape information (a clever new way of reducing background by looking at the charge collection as a function of time). In this way single, localised charge creation like electrons can be distinguished from various other forms of charge productions. Moving the counting station for $^{37}$Ar decays from Brookhaven to the underground location in Homestake reduced the background even further (Fig. 7.9). The experiment was running for about 30 years and the initially observed deficit remained persistent over the whole period. In the final publication in 1998 an average measured rate of $2.56 \pm 0.32$ SNU is quoted while the expected rate from Standard Solar Models was $8.5 \pm 1.8$ SNU (Cleveland et al., 1998). Thus, a deficit of about a factor three was seen. The SNU is an acronym for

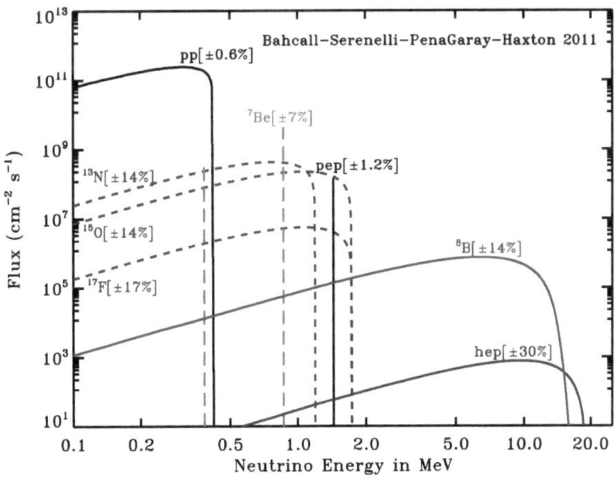


**Fig. 7.6** A modern day solar neutrino spectrum, showing the individual contributions discussed in the text (with kind permission from A. Serenelli)

Solar Neutrino Unit, used in radiochemical experiments and corresponding to $10^{-36}$ captures per target atom per second. It was first mentioned in Bahcall (1969). A lot of details can also be found in Bahcall (1989); Bahcall et al. (1994); Davis (2003); Lande (2009).

Despite giving the pioneering Homestake experiment and Ray Davis Jr. the major credit, it is apparent that $^{37}$Cl is not the only isotope which can be used in a radiochemical approach. During the 1960s also the $^7$Li–$^7$Be system was considered as a good detector. Due to the well understood nuclear structure the theoretical predictions are rather solid. Reines and Woods were planning for a 570 kg metallic lithium detector and the apparatus was ready for assembly when the first Homestake results were announced. They stopped and the detector was never used for solar neutrino searches. A third isotope became very popular when Kuzmin proposed in 1966 to use $^{71}$Ga to even measure the fundamental pp-neutrinos (Kuzmin, 1966) via the reaction $^{71}\mathrm{Ga} + \nu_e \rightarrow {}^{71}\mathrm{Ge} + e^-$. Rate estimates showed that about an order of magnitude more gallium than the world production rate at this time would be needed for solar neutrino detection. In this decade also the Russians started major initiatives in cosmic ray research and neutrino astrophysics. Without any deep mine available, they started to build a laboratory in Baksan Valley in the North Caucasus mountains by drilling a 4 km long horizontal tunnel into Mt. Andyrchi. The original idea was to install three radiochemical experiments based on $^{37}$Cl, $^7$Li and $^{71}$Ga to disentangle the individual pp-, $^7$Be and $^8$B fluxes. At the end only $^{71}$Ga was realised, nevertheless groups in Russia are still exploring a metallic $^7$Li experiment. In the early 1970s increased Al manufacturing produced a large amount of Ga as a byproduct. Hence, the Brookhaven group including R. Davis borrowed some 50 kg of gallium to explore extraction schemes of $^{71}$Ge, the expected isotope due to neutrino capture, and its detection in proportional counters like the ones for the chlorine experiment. For that two methods were explored, from a $GaCl_3$ solution and from metallic Ga. Both method seemed to work and were finally used by the two major experiments GALLEX and SAGE.

**Fig. 7.7** Ray Davis near a part of the argon gas extraction system in 1978 (Courtesy of Brookhaven National Laboratory)

The Gallium Experiment (GALLEX) was a dominantly European collaboration using 30.3 tons of Ga in form of 110 tons of $GaCl_3$ in the Italian Underground Laboratory Gran Sasso (Fig. 7.10). It was running from 1991 until 1996 and after some maintenance work continued as the Gallium Neutrino Observatory (GNO) until 2002 (Fig. 7.11). The Soviet–American Gallium Experiment (SAGE) started around 1990 with initially 30 tons and later upgraded to 60 tons (Fig. 7.12). The experiment is still running today. It was quite a surprise when SAGE released their first results being in agreement with zero or less than 79 SNU with 90 % Confidence Level, while the expectation was 132 SNU (Abazov et al., 1991). Hence, the first real positive detection of pp-neutrinos belongs to GALLEX, who published a number of $83 \pm 27$ SNU a year later (Anselmann et al., 1992). It turned out to be a good choice to perform two experiments with different extractions for this delicate and important measurement. In the meantime the SNU numbers measured by both experiments are in good agreement and about 67 SNU, definitely much less than the solar model prediction (Hampel et al., 1999; Altmann et al., 2005; Abdurashitov et al., 2009). Recently, a reanalysis of the GALLEX data has been performed (Kaether et al., 2010).

As beautiful as these results are, there is one major disadvantage of the radiochemical method. Neither information on the time of the neutrino capture nor the incoming neutrino energy (besides the fact that it has to be above threshold) can be deduced. Hence, a detection principle with real-time and energy information is desirable. Fortunately this was already realised before the gallium experiments in large water Cherenkov detectors. The idea of this type of detector is to use neutrino–electron scattering $\nu + e^- \rightarrow \nu + e^-$ where the recoiling electron is measured. If the speed of the electron is larger than the speed of light in the medium (in the case of relevance here the medium is water), it will start to emit coherent Cherenkov light in analogy to Mach cones in acoustics due to supersonic speed. Notice that the radio-

**Fig. 7.8** As it is pretty warm underground, R. Davis Jr. took a swim in the water shielding of the chlorine experiment. This shielding was built to reduce external backgrounds and moderate high energy neutrons (Courtesy of Brookhaven National Laboratory)

chemical experiments are only sensitive to $\nu_e$ and the water Cherenkov detectors are dominated by $\nu_e$ interactions as well, because the cross section of electron scattering for the remaining two flavours is about a factor 6 smaller. However, due to the energy of the electron required to emit significant Cherenkov light, this method can only be used for neutrinos of several MeV, i.e. only the $^8$B and hep neutrinos (Fig. 7.6). At energies involved in solar neutrino searches, the recoiling electron more or less follows the direction of the incoming neutrino. The water tank is equipped with photomultipliers; the energy can be reconstructed from the number of struck phototubes and the measured amount of light. Nowadays this technique is also used for neutrino telescopes. So far so good, but nobody would have built a detector for solar neutrinos alone, fortunately another fundamental physics topic was on the bench, which triggered the effect of building large-scale water detectors.

For decades people are desperate to find new physics which is not covered by our current understanding of Particle Physics. In the environment of Grand Unified Theories attempts are made to merge the four fundamental interactions of nature. The most promising way discussed around 1980 had as a major prediction the fundamental instability of matter due to a possible decay of the proton. The favourite decay mode was into a positron and a neutral pion. As water is a cheap source of a large number of protons and the potential decay products all end up in light, the idea of water Cherenkov detectors came up. At the end no proton decay was observed, but solar and supernova neutrinos were. In a kind of competition in finding proton decay two experiments based on this technology were built, one in the

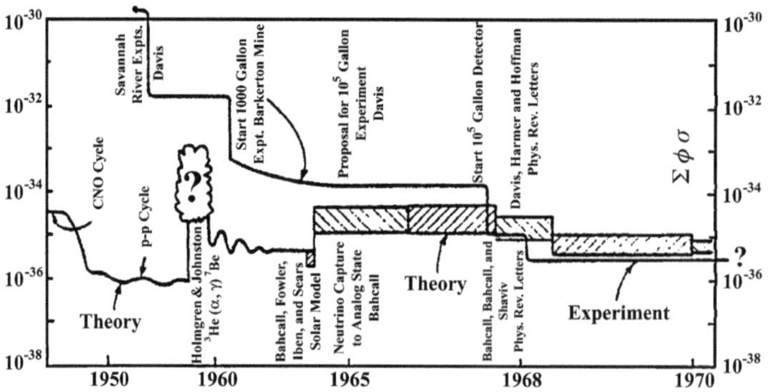


**Fig. 7.9** Slide of a talk from R. Davis Jr. shown in 1971 sketching the evolutionary timelines of experiments and theory of solar neutrinos (Courtesy of Brookhaven National Laboratory)

US called the Irvine-Michigan-Brookhaven detector (IMB) and one in Japan called the Kamioka Nucleon Decay Experiment (KamiokaNDE). As only the latter measured solar neutrinos the following description will be of this experiment. The detector was installed in the Kamioka mine 1 000 m underground and consisted out of 3 000 t of water surrounded by specially developed large photomultipliers covering 20 % of the detector surface. The experiment started in 1983 and besides the fact of non-observation of proton decay it became apparent that energies as low as about 10 MeV could be achieved.

Thus, at the International Conf. on BAryon Non-conservation in 1984 (ICOBAN'84) in Park City, Utah (USA) an upgrade to Kamiokande-II was announced to tackle the region of solar neutrinos as well as a new future project called Super-Kamiokande using 50 000 tons of water. After major upgrade work and attracting new collaborators the experiment started again in January 1986. The energy threshold of the detector was initially 7.6 MeV and could be reduced down to 6.1 MeV during the running period until April 1990. The first observation of solar neutrinos released again showed a deficit with respect to solar model expectation by about a factor of two (Hirata et al., 1990). Hence, in all experimental approaches performed until 1995, a significant deficit in all solar neutrino experiments with respect to expectation has been observed.

Two solutions are evident: either the Sun is producing less neutrinos than predicted or neutrinos have new unknown properties. While almost all solutions regarding solar properties require a change in the core temperature of the Sun by several percent, all neutrino solutions require a non-vanishing rest mass. The latter is not foreseen in the Standard Model of Particle Physics where neutrinos are considered to be massless. Explanations based on the second option included among others properties like neutrino decay, a magnetic moment (motivated by a claimed anticorrelation of solar neutrinos flux with sunspots in the Homestake experiment) or neutrino oscillations. Neutrino oscillations rely on the fact that the flavour eigenstates $\nu_e, \nu_\mu$ and $\nu_\tau$ participating in the weak interaction are not identical to the mass eigenstates used to describe the propagation $\nu_1, \nu_2$ and $\nu_3$. This leads to a

**Fig. 7.10** Ray Davis Jr. (*left*) and Till Kirsten (*right*) on a Sunday afternoon in the Gardens of Schwetzingen Castle (near Heidelberg, Germany) during the Second International GALLEX planning meeting end of Sep. 1979. T. Kirsten became Principal Investigator and spokesperson of GALLEX (with kind permission of D.D. Clayton)

mixing effect, i.e. even if the Sun emits only $\nu_e$ in a certain distance you have a mixture of all three neutrino flavours. The probability for a new flavour to show up in a distance L from the source depends, in the simple case that only two states contribute, on a mixing angle $\theta$ (determining the amplitude of the oscillation), the energy E of the neutrino, the distance L from the source to the detector and the quantity $\Delta m_{ij}^2 = m_j^2 - m_i^2$ with $i, j = 1, 2, 3$, the difference of two involved mass eigenvalues squared. Thus, at least one of the mass eigenstates has to be non-zero, otherwise this phenomenon cannot occur. The case described would imply that neutrinos oscillate on the way from the Sun to the Earth, hence it is called vacuum oscillation. L. Wolfenstein, A. Smirnov and S. Mikheyev realised around 1980 that a conversion can occur within the Sun already, now called the matter effect or MSW-effect (Wolfenstein, 1978; Mikheyev and Smirnov, 1986). The physics behind this phenomenon relies on the fact that all neutrinos can interact with the electrons in the solar interior via the exchange of Z-bosons (so-called weak neutral currents), while only electrons can additionally interact via W-boson exchange (called weak charged currents). This singles out electron neutrinos and provides them with an "additional effective mass" which is proportional to the electron density. Thus in the solar core, where the density is highest, the $\nu_e$ is heavier than the other neutrinos which get inverted as neutrinos leave the Sun. Hence, somewhere on the way out, the involved states are about equal in mass and there is a good chance for the conversion of one

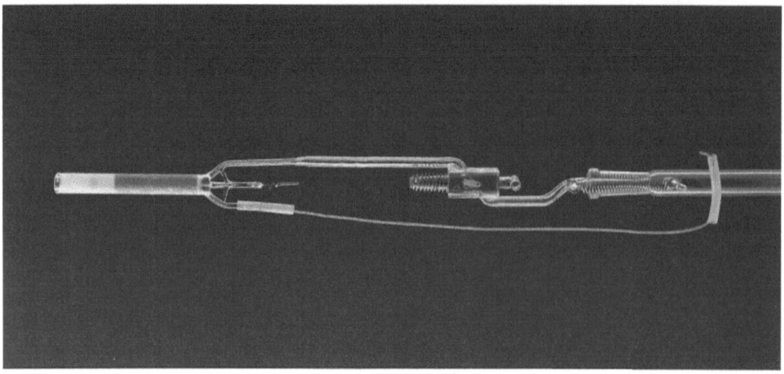

**Fig. 7.11** Proportional counter from the GALLEX experiment to detect the $^{71}$Ge decay. It is a modified and improved version of the counter used by the Homestake experiment. A similar counter is also used in SAGE (Courtesy of Max-Planck Institute for Nuclear Physics, Heidelberg)

flavour into the other. In this way it is also possible to describe the three observations and it allows for different energy dependent oscillation probabilities. Interestingly, for certain parameter values of $\theta$ and $\Delta m^2$ there could be some regeneration effect for $\nu_e$ if neutrinos have to travel through the Earth due to the same effect. Hence, the Sun in neutrinos could be brighter during the night than at daytime. The effect is called day–night asymmetry. However, no day–night effect has been seen yet.

The next player in the game is Super-KamiokaNDE starting data taking in 1996 and still operational. They were able to reduce the threshold below 5 MeV and accumulated a high statistics with far more than 20 000 events of $^8$B neutrinos. This detector was the first one claiming the observation of neutrino oscillation based on a deficit of upward going muons due to atmospheric neutrino interactions.

Already in 1984 an approach was suggested by H. Chen at a Solar Neutrino Conference in Homestake to solve the problem of missing neutrinos independently of any solar models by using heavy water $D_2O$. On deuterium two reactions are possible, the flavour sensitive reaction $\nu_e + D \rightarrow p + p + e$ (charged current) and the flavour blind reaction $\nu_x + D \rightarrow \nu_x + p + n$ (neutral current). The latter is possible for all neutrinos with an energy above 2.2 MeV and additionally measuring the total solar neutrino flux (Chen, 1985). Of course, also neutrino–electron scattering is possible but plays a minor role. Hence by measuring the number of electrons and neutrons a discrimination of the two solutions can be found. If both rates are down by a factor 2–3 with respect to expectation, the Sun indeed would be producing less neutrinos, however, if only the electron events are reduced but the neutron reaction corresponds to expectation then neutrinos oscillate. Already in the 1960s at times when the chlorine experiment was considered, groups at Case Western Reserve University, including F. Reines, were exploring various kinds of solar neutrino detector. As one of the investigated approaches T.L. Jenkins and his student F. Dix built a 2 000 liter heavy water Cherenkov detector which suffered from a high background rate but could place a limit on the $^8$B flux which is more than two orders of magnitude higher than expected from the models. G. Ewan and H. Chen started a

**Fig. 7.12** Vladimir Gavrin (*left*), spokesperson of the SAGE experiment, with John N. Bahcall (*right*), who was starting and refining Standard Solar Models and the resulting neutrino fluxes over decades, during a meeting in Princeton in summer 1988 (with kind permission of V. Gavrin)

project which became later the Sudbury Neutrino Observatory (SNO). After a feasibility study published in 1985 a full proposal of a 1 000 ton heavy water Cherenkov detector was released in 1987 (Fig. 7.13). The heavy water was borrowed from the Canadian Atomic Energy Commission and installed in a nickel mine close to Sudbury (Ontario, Canada) 2 km underground. The experiment started data taking in 1998 and finished end of 2006. Three different phases were performed: A first running period with pure $D_2O$ which allowed a good measurement of the charged current reaction producing electrons but also had some sensitivity to neutral current reactions. To enhance the sensitivity for the latter, 2 tons of salt were added to the detector, because $^{35}$Cl has a much higher neutron capture cross section. In a third phase $^{3}$He filled proportional counters were added which allowed the detection of neutrons on an event-by-event basis. As the first neutral current data were released in 2001 it became obvious that the total neutrino flux is in agreement with solar model expectations, thus neutrinos must be responsible (Ahmad et al., 2001). Furthermore, the charged current reaction was down by about a factor 3, confirming the decades long claim of the Homestake experiment. Finally one of the longest standing problems in particle astrophysics, the one of missing solar neutrinos, was solved (Fig. 7.14). The solution is that there are no missing solar neutrinos but 60–70 % are passing the earth in a wrong flavour, i.e. not as $\nu_e$. One year after the release of the first SNO results the Nobel prize was awarded to R. Davis Jr. and M. Koshiba for the detection of cosmic neutrinos. With all the data obtained it became very likely that indeed matter effects are the solution.

However, this is not the end of the story. Now being equipped with real-time and spectroscopic measurements of $^{8}$B neutrinos above 4 MeV and integral capture rates starting from 814 keV (from Homestake) and 233 keV (from GALLEX/GNO/SAGE) there is still information to be deduced about the solar interior and the neutrino spectrum and it was always a desire to measure neutrinos in real-time below 1 MeV. Obviously the maximal information is available if the full solar neutrino spectrum can be measured in real time. As Cherenkov detectors do not allow for this, scintillation detectors are considered like the one F. Reines used but bigger. This was the aim of the BOREXINO experiment installed at the Gran

**Fig. 7.13** Picture from one of the first SNO collaboration meetings. Among others there is Herb Chen (*fifth from right*), George Ewan (*third from right*) and Art McDonald (*right*) as the long term spokesperson of SNO (with kind permission of the SNO collaboration)

Sasso Laboratory in Italy (the name is a relic from the original idea using a boron loaded scintillator with much larger target mass proposed as BOREX. Neutral current excitations and charged current reactions on $^{11}$B were considered. However, at the end only a scintillator without any boron was used and the detector became a "small Borex", i.e. Borexino). It contains 300 tons of a liquid scintillator surrounded by photomultipliers. Already proposed in 1991 a lot of effort had to be spent on purifying the scintillator and the used materials from any kinds of radioactive contaminant to an incredibly low level. However, the effort paid off, because already shortly after the start of the experiment in 2007 a first real-time detection of mono-energetic $^7$Be neutrinos at 0.862 MeV could be announced (Arpesella et al., 2008). In the meantime also the $^8$B spectrum with a threshold of 2.8 MeV has been measured and a first evidence for the mono-energetic pep neutrinos at 1.44 MeV has been claimed. The experiment is still taking data. Another large-scale scintillator experiment called KamLAND using 1 000 tons was installed in the Kamioka mine to use nuclear power plants for an independent prove that matter effects are the solution of the solar neutrino deficit. If matter effects occur in the Sun, the required $\Delta m^2$ would lead to oscillation effects in vacuum (consider the earth atmosphere as vacuum) on a baseline of roughly 100–200 km. Surrounded by a large amount of reactors in a reasonable distance from KamLAND, the experiment was able to show that the neutrino energy spectrum from reactors is distorted as one would expect from oscillation results (Eguchi et al., 2003). This finally pinned down the oscillation solution and determined the $\Delta m^2$ involved precisely. In 2011 KamLAND also released their first $^8$B solar neutrino measurement.

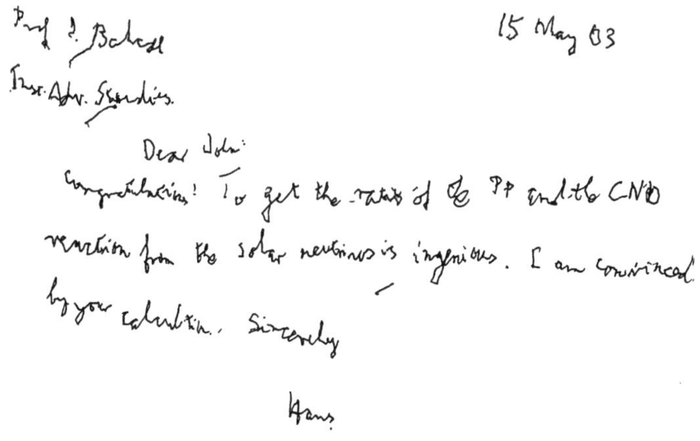

Prof J. Bahcall
Inst. Adv. Studies.

15 May 03

Dear John:
Congratulations! To get the ratio of the pp and the CNO reaction from the solar neutrinos is ingenious. I am convinced by your calculation. Sincerely

Hans

**Fig. 7.14** A letter of the late Hans Bethe to John Bahcall on solar modelling originating from 2003 (with kind permission of C. Pena-Garay)

## 7.6 Neutrinos from Stellar Collapse

Further sources of low energy astrophysical neutrinos besides the Sun are supernova explosions. Unfortunately the world-wide data set relies on only one event from 1987 which is discussed now in some detail. Stars with initial masses larger than about eight solar masses are considered to continue fusion processes all the way up to iron group elements. Beyond that no further energy gain is possible by fusion. The star has developed on onion-like structure with the inner iron core being made out of iron. The following describes a very simplified version of a very complex phenomenon which is still under intense research. The pressure of the degenerate electron gas in the core balances gravitation. The usual "trick" of a star of contraction $\rightarrow$ pressure increase $\rightarrow$ temperature increase $\rightarrow$ ignition of new fusion fails in the last two steps. First of all the pressure of a degenerate electron gas is no function of temperature, hence pressure increase does not lead to temperature rise. Secondly iron group elements cannot do fusion anymore, because they have the highest binding energy per nucleon. Even worse, energy and electrons are taken away by electron capture processes on protons and heavy nuclei, and photodisintegration of iron group elements occurs, which is strongly endothermic. As a consequence the iron core becomes unstable and collapses quickly. At densities beyond $10^{12}\ \mathrm{g\,cm^{-3}}$ matter becomes opaque even for neutrinos and they start to diffuse, probably the only known scenario in the Universe where this can happen. As the iron core contracts further it will be compressed to densities beyond nuclear density (about $10^{14}\ \mathrm{g\,cm^{-3}}$). This will convert the collapse into an explosion as the density of the atomic nucleus is highly incompressible and thus bounces back. The resulting shock wave is travelling outwards and dissociates the still infalling iron nuclei. If it successfully leaves the iron core then the outer shells are not really a hurdle and they will be blown away in a giant explosion, which is called supernova.

**Fig. 7.15** Masatoshi Koshiba, spokesperson of the Kamiokande-experiment and proposer of Super-Kamiokande. He won the Nobel prize in 2002 for the discovery of cosmic neutrinos together with Ray Davis Jr. (with kind permission of M. Koshiba)

So what about neutrinos? Well, even being a spectacular optical event, the energy released is minor compared to neutrinos. About 99 % of the released binding energy is actually emitted in neutrinos! The expected neutrino spectrum, details are still a hot topic of research, basically consists of two parts: First the so-called deleptonisation burst consisting of the trapped $\nu_e$ emitted in the electron capture process which are released within milliseconds as the outgoing shock wave reaches sufficiently low densities which allows them to escape immediately. They have piled up behind the shock wave as in the dissociated material the mean free path for neutrinos is much larger. The second contribution is from the Kelvin–Helmholtz cooling phase of the protoneutron star. This will emit neutrinos of all flavours and last for about 10 s or so. Average neutrino energies are in the region of 10–30 MeV, thus enables all solar neutrino detectors to detect them. So what happened in 1987? A lively account can be found in Koshiba (1992, 2003). On February 23rd, 1987, the brightest supernova since the Kepler supernova in 1604 was discovered in the Large Magellanic Cloud, about 150 000 light years away. It is named Supernova 1987A and was first announced in the IAU Circular 4 316 on February 24th. As is clear from the previous section, several proton decay detectors were online but it took them a while to loop through the data to find the supernova. Nevertheless on February 28th in IAUC 4323 the Mont Blanc Neutrino Observatory (Large Scintillator Detector, LSD) announced the observation of five events above 7 MeV within 7 seconds, an unusual high rate. The detection time was corrected to be 5 minutes earlier in IAUC 4332 on March 6th. The detector was a 60 ton Liquid Scintillation detector distributed in 72 counter units. In circular IAUC 4329, dated March 3rd, J. Bahcall, A. Dar and T. Piran made a first attempt to estimate the number of events in the various detectors based on supernova models by Wilson and collaborators. Homestake was considered to see about one Ar atom while the water Cherenkov detectors should see about a dozen. The chlorine experiment did an immediate extraction but could not find any $^{37}$Ar as announced in IAUC 4339 on March 10. The full announcement from the Homestake experiment, never released in this form (but see Davis 1994), reads

> Immediately after Supernova 1987A was announced, a special run was made with the Homestake detector to look for electron neutrinos. The preceding solar neutrino extraction, run 92, was fortunately completed on February 13th, just 9.1 days before the neutrinos from SN1987A arrived at the earth. The extraction for the supernova-produced 37Ar began on 1 March and was finished on 3 March. The run was counted

for 134 days (120 live-days) and only two 37Ar-like events were detected. These events occurred 35 and 119 days after the time of extraction. These times of occurrence are best fit by attributing both of these events to counter background, and yield an upper limit (one sigma) for the number of detected 37Ar atoms of less than two.
If we assume that this limiting number of atoms was produced by the supernova, then, taking into account the 92 % extraction efficiency, the 40 % counting efficiency, the 8.6 day delay between the time of neutrino arrival and the time of extraction, and the dead time during counting, the limit of less than two detected atoms implies that less than 7.3 atoms of 37Ar were produced in the C2Cl4 absorber by the neutrinos from the supernova. This limit is consistent with the number expected to be produced – approximately one. A supernova at 150 kiloparsecs distance is simply out of range for the Homestake detector.
Alternatively, if we assume the 37Ar production rate was constant during this run, the upper limit of less than two detected atoms implies a production rate of less than 0.5 37Ar atoms per day, not inconsistent with what is expected to be produced by the average solar neutrino flux during the 17.7 day exposure period.

However, in the circular before (IAUC 4338) Kamiokande-II reported the detection of 11 events within 13 s, but this event was about 4.7 hours later than the reported Mont Blanc detection. Evidently this caused a lot of discussion.... At the time of the Mont Blanc detection Kamiokande-II did not observe anything unusual. Kamiokande was lucky to observe it for two reasons: First of all two minutes before the supernova signal there was a 105 second dead time in data taking due to gain check monitoring clearly visible in Fig. 7.15. Secondly the mine was on a substitute holiday and hence the experiment in holiday trigger scheme. Under normal conditions 5 mins before the supernova event the shifters would have started to change the data tapes with a fair chance that the experiment would have missed the event. One day later in IAUC 4340 the IMB-3 collaboration announced their discovery of SN1987A in the form of eight events in 6 seconds at the same time as Kam-II. Unfortunately, due to a failure of one of the four high voltage power supplies 25 % of the photomultipliers were not operational. IMB-3 was an 8 000 ton rectangular water Cherenkov detector using 2 048 photomultipliers located in the Morton Thiokol salt mine near Cleveland, Ohio. Last but not least, the Baksan Scintillator Telescope experiment with a total mass of 330 tons investigated their data and could not find anything around the time of LSD but five events in 9 seconds around $\pm 1$ min of the Kam-II observation. To summarise, three experiments observed 25 events in coincidence while one detector has seen five events about 5 hours before (Fig. 7.16). The number of publications based on these events is at least an order of magnitude larger. For example from the observed time spread of the events a neutrino mass limit could be set which was as good as the ones from beta decay at that time. Astonishingly the relative time between the events is quite accurately known but the absolute time is not. The events observed in around 7:35 UT are shown in Fig. 7.17. This event marks the birth of neutrino astrophysics, the first observation from astrophysical neutrinos besides the Sun.

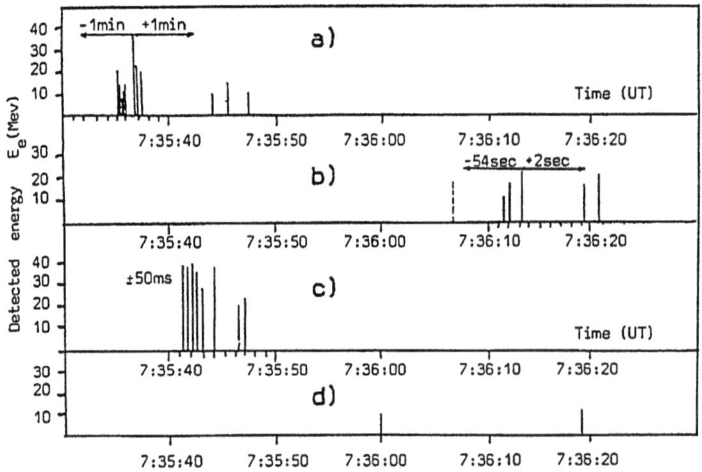


**Fig. 7.16** Compilation of event of all four experiments around 7:35 UT (from Alekseev et al., 1988, with kind permission of Elsevier)

## 7.7 Outlook

Neutrino physics made major progress over the last two decades, especially establishing a non-vanishing neutrino mass via the phenomenon of neutrino oscillations. The so-called flavour eigenstates participating in weak interactions are not identical to the mass eigenstates describing the propagation. Hence different flavours can occur even if the source is producing only one kind of neutrino, and more precise measurements are ongoing or planned. The long standing problem of missing solar neutrinos was finally solved to be due to these effects of converting neutrinos within the sun, called matter effects. However, still a lot can be explored. The improvements in helioseismological observations and new abundance determinations from the photosphere would make a measurement of CNO neutrinos desirable as this will put one of the fundamental assumptions of stellar evolution, the homogeneous distribution of elements, on the bench. After a first glimpse on the pep neutrinos a precise measurement would be desirable to reduce the error on the involved parameter $\theta$ from neutrino oscillations. Both issues might be addressed by SNO+, a new liquid scintillator detector using the SNO infrastructure. Last but not least, the ultimate goal would and should be a real-time measurement of solar pp-neutrinos, which is by far the dominant flux. This would allow to study the dynamics of energy production within the Sun, and some detector concepts exist to perform such a measurement.

Also, scientists are better prepared now for the next supernova explosion. There are larger (Super-Kamiokande) and more detectors available (ICECUBE, Borexino, KamLAND, LVD and in the near future HALO and SNO+) to detect supernova neutrinos. In the era of GPS also the absolute synchronisation of time will not be an issue anymore. Furthermore, the detectors are connected within a Supernova Early Warning System (SNEWS) to warn astronomers about a potential supernova. This is

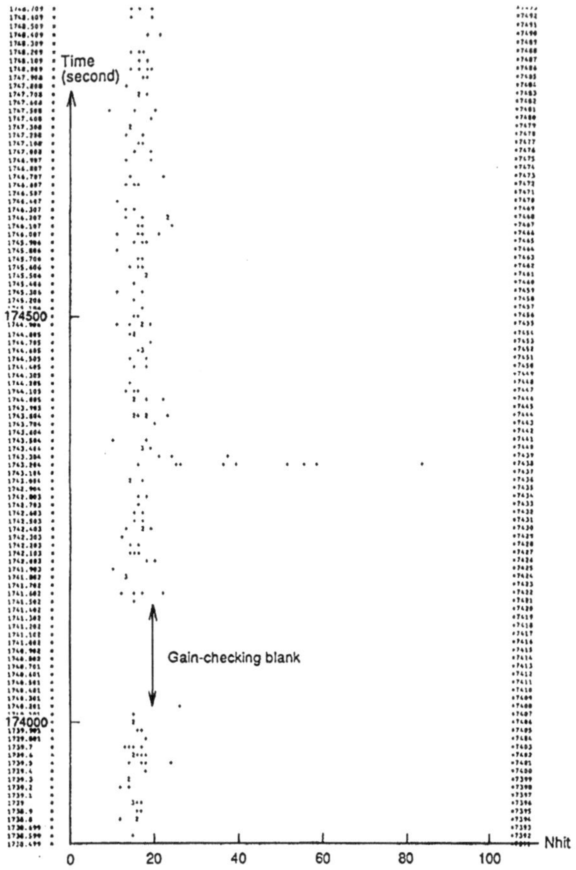


**Fig. 7.17** Printout of the SN87A observation from Kamiokande-II. On the $y$-axis the measuring time in seconds since the start of the run is shown and on the $x$-axis the number of hit photomultipliers, Nhit, a measure of energy. Clearly seen is the spike in the middle of the plot caused by SN 1987A (from Koshiba 1992, with kind permission of Elsevier)

possible as neutrinos leave the star significantly before the actual explosion. Additionally some of the mentioned detectors have already placed significant bounds on a diffuse supernova neutrino background (DSNB) resulting from a summation of all supernova explosions over the history of the Universe. A detection still remains to be done. The same is true for the Cosmic neutrino background originating from the Big Bang. Like the cosmic microwave background (CMB) there should be a 1.95 K relic neutrino background, whose detection is extremely difficult.

Absolute mass measurements are facing interesting times as well. The KATRIN experiment will perform a new tritium endpoint measurement down to 0.2 eV, another order of magnitude with respect to existing limits and supposed to start data taking soon. Four new large-scale double beta decay experiments (GERDA, EXO, Kamland-Zen and Candles) started data taking in 2011, several others are in the preparation stage. Also these experiments are probing absolute neutrino masses in the sub-eV range and are complementary to the tritium measurements. Last but not least, the progress in cosmology over the last decade was extremely rapid, so that based on large-scale structure and CMB data, limits on the mass density $\Omega_\nu$ of neutrinos in the universe and hence on the sum of the total neutrino masses could be strongly improved. With some cosmological model dependence and correlation among all the involved parameters again values in the sub-eV range are preferred. Hence we are looking into a very bright near term future, especially because neutrino oscillations guarantee that neutrinos have a non-vanishing rest mass. But the ultimate fascination from studying neutrino properties stems from the fact that one has always to expect the unexpected, as in history neutrinos rarely behaved in experiments as was thought before.

## 7.8 Nobel Prizes

The field of neutrino physics already attracted various Nobel prizes (see also Appendix B). The ones of most relevance for this article are those to Raymond Davis Jr. and Masatoshi Koshiba in 2002 for the detection of astrophysical neutrinos. In addition, Fred Reines got the 1995 Nobel prize for the discovery of the neutrino, Jack Steinberger, Melvin Schwartz and Leon Lederman got the 1988 Nobel Prizes for showing that there are at least two different neutrinos. Hans Bethe got the 1967 Nobel prize for his work on energy production in stars.

**Acknowledgements** I would like to thank B. Cleveland for valuable information and for providing the SN1987A note of the Homestake experiment not published in this form before.

## References

Abazov, A.I., et al.: Phys. Rev. Lett. **67**, 3332–3335 (1991)
Abdurashitov, J.N., et al.: Phys. Rev. C **80**, 015807 (2009)
Ahmad, Q.R., et al.: Phys. Rev. Lett. **87**, 071301 (2001)
Alekseev, E.N., et al.: Phys. Lett. B **205**, 209 (1988)
Altmann, M., et al.: Phys. Lett. B **616**, 174 (2005)
Anselmann, P., et al.: Phys. Lett. B **285**, 376–389 (1992)
Arpesella, C., et al.: Phys. Lett. B **658**, 101 (2008)
Bahcall, J.N.: Phys. Rev. Lett. **12**, 300–302 (1964)
Bahcall, J.N.: Phys. Rev. Lett. **23**, 251 (1969)
Bahcall, J.N.: Neutrino Astrophysics. Cambridge University Press, Cambridge (1989)
Bahcall, J.N., Davis, R. Jr., Parker, P., Smirnov, A., Ulrich, R. (eds.) Solar Neutrinos – The First Thirty Years. Addison Wesley, Reading (1994)
Bergkvist, K.E.: Nucl. Phys. B **39**, 317 (1972)
Bethe, H., Peierls, R.: Nature **133**, 532 (1934)
Bethe, H.A., Critchfield, C.L.: Phys. Rev. **54**, 248 (1938)
Bethe, H.A.: Phys. Rev. **55**, 434 (1939)
Chen, H.H.: Phys. Rev. Lett. **55**, 1534–1536 (1985)
Cleveland, B.T., et al.: Astrophys. J. **496**, 505 (1998)
Cowan, C., et al.: Science **124**, 103 (1956)
Crane, H.R.: Rev. Mod. Phys. **20**, 278 (1948)
Curran, S.C., Angus, J., Cockcroft, J.L.: Phys. Rev. **76**, 853 (1949)
Danby, G., et al.: Phys. Rev. Lett. **9**, 36 (1962)
Davis, R. Jr.: Phys. Rev. Lett. **12**, 303–305 (1964)

Davis, R. Jr., Harmer, D.S., Hoffman, K.C.: Phys. Rev. Lett. **20**, 1205–1209 (1968)
Davis, R.: Prog. Part. Nucl. Phys. **32**, 13 (1994)
Davis, R.: Rev. Mod. Phys. **75**, 985–994 (2003)
Eddington, A.: Observatory **43**, 341–358 (1920)
Eguchi, K., et al.: Phys. Rev. Lett. **90**, 021802 (2003)
Elliott, S.R., Hahn, A.A., Moe, M.: Phys. Rev. Lett. **59**, 2020 (1987)
Fermi, E.: Z. Phys. **88**, 161 (1934a)
Fermi, E.: Nuovo Cimento **11**, 1 (1934b)
Fireman, E.L.: Phys. Rev. **75**, 323 (1949)
Fireman, E.L., Schwartzer, D.: Phys. Rev. **86**, 451 (1952)
Furry, W.H.: Phys. Rev. **56**, 1184 (1939)
Goeppert-Mayer, M.: Phys. Rev. **48**, 512 (1935)
Goldhaber, M., Grodzin, L., Sunyar, A.W.: Phys. Rev. **109**, 1015 (1958)
Hampel, W., et al.: Phys. Lett. B **447**, 127 (1999)
Hirata, K.S., et al.: Phys. Rev. Lett. **65**, 1297–1300 (1990)
Holmgren, H.D., Johnston, R.L.: Phys. Rev. **113**, 1556–1559 (1959)
Inghram, M.G., Reynolds, J.H.: Phys. Rev. **76**, 1265 (1949)
Kaether, F., et al.: Phys. Lett. B **685**, 47 (2010)
Kirsten, T., Gentner, W.: Schaeffer, O.A.: Z. Phys. **202**, 273 (1967)
Kirsten, T., Gentner, W.: Schaeffer, O.A.: Phys. Rev. Lett. **20**, 1300 (1968)
Klapdor-Kleingrothaus, H.V., et al.: Mod. Phys. J. A **12**, 147 (2001)
Kodama, K., et al.: Phys. Lett. B **504**, 218 (2001)
Koshiba, M.: Phys. Rep. **220**, 229–381 (1992)
Koshiba, M.: Rev. Mod. Phys. **75**, 1011–1020 (2003)
Kraus, C., et al.: Eur. Phys. J. C **40**, 447 (2005)
Kuzmin, V.A.: Sov. Phys. JETP **22**, 1051 (1966)
Lande, K.: Annu. Rev. Nucl. Part. Sci. **59**, 21–39 (2009)
Langer, L.M., Moffat, R.J.D.: Phys. Rev. **88**, 689 (1952)
Levine, C.A., Ghiorso, A., Seaborg, G.T.: Phys. Rev. **77**, 296 (1950)
Lobashev, V.M.: Nucl. Phys. A **719**, 153 (2003)
Los Alamos: Celebrating the neutrino, Los Alamos Science, vol. 25 (1997)
Longair, M.: The Cosmic Century. Cambridge University Press, Cambridge (2006)
Lubimov, V.A., et al.: Phys. Lett. B **94**, 266 (1980)
Majorana, E.: Nuovo Cimento **14**, 171 (1937)
Mikheyev, S.P., Smirnov, A.Y.: Nuovo Cimento **9**, 17 (1986)
Otten, E.W., Weinheimer, C.: Rep. Prog. Phys. **81**, 086201 (2008)
Pais, A.: Inward Bound. Oxford University Press, Oxford (1986)
Pauli, W.: Phys. Rev. **38**, 579 (1931)
Pauli, W.: In: Aufsätze und Vorträge über Physik und Erkenntnistheorie, p. 156. Vieweg, Braunschweig (1961)
Pontecorvo, B.: Chalk River Laboratory Report PD-205 (1946)
Racah, G.: Nuovo Cimento **14**, 322 (1937)
Reines, F., Cowan, C.: Phys. Rev. **92**, 830 (1953)
v. Weizsäcker, C.F.: Phys. Z. **38**, 176 (1937)
v. Weizsäcker, C.F.: Phys. Z. **39**, 633 (1938)
Wolfenstein, L.: Phys. Rev. D **17**, 2369 (1978)
Wu, C.S., et al.: Phys. Rev. **105**, 1413 (1957)

# Chapter 8
# From Particle Physics to Astroparticle Physics: Proton Decay and the Rise of Non-accelerator Physics

**Hinrich Meyer**

## 8.1 Introduction

In this chapter, we report on the way leading from accelerator laboratories to underground physics, which paradoxically enough turned out to studying cosmic rays. The standard model of particle physics established in the early 1970s (see Riordan, 1987) had an unexpected consequence for astroparticle physics. Its symmetry would have required that matter and antimatter annihilated in the early universe, so that no world made up of 'matter' could have formed. In 1968, Andrei Sacharov showed that the matter–antimatter asymmetry might have formed in a state of the universe far from thermal equilibrium (such as obviously given in big bang cosmology), together with the P- and CP-violations (which today are well-confirmed and further investigated e.g. in the LHC experiment LHCb), and proton decay. The latter phenomenon, however, could be only investigated in large none-accelerator experiments. The size of the first generation of such experiments depended on the idea of unifying the fundamental forces beyond the standard model. In the middle of the 1980s, the most simple extension of the standard model, the SU(5) theory, implied a proton lifetime of about $10^{29}$ years. With detectors consisting of 1 000 tons of matter and hidden from the cosmic radiation as deep as possible under the Earth surface, such as the water Cherenkov detectors Kamiokande and IBM or the French–German Fréjus iron calorimeter, one expected to detect several proton decays per year.

The passage from accelerator physics to cosmic ray studies by means of underground detectors began in the 1980s (for the following see Meyer, 1990). It was based on detailed knowledge about the formation of the primordial light nuclei and baryon number violation and had its first striking successes in the measurements of the solar neutrino flux and the neutrino signals emitted by supernovae (see Chap. 7).

H. Meyer (✉)
DESY, Hamburg, Germany
e-mail: meyerw@mail.desy.de

B. Falkenburg, W. Rhode (eds.), *From Ultra Rays to Astroparticles*,
DOI 10.1007/978-94-007-5422-5_8, 

Non-accelerator particle physics is a field of high energy physics that exploits the sources of energy and related particle beams as provided by nature. It works on grand scales of space, time and energy by considering e.g. the matter content of the whole universe, single events of huge energy release like the big bang, supernova explosions and the process of energy production in stars as well as acceleration of protons and electrons up to energies that are still far beyond the capabilities of man-made accelerators. It is the photons and neutrinos that carry very important information on cosmic events to us observers on earth and may even give new insights on the basic properties of those particles. The atmospheric neutrino and muon flux turned out to be particularly relevant.

## 8.2 Big Bang Nucleosynthesis

Let us review the stage of knowledge around 1990. It was well known then that one of the most successful fields of non-accelerator particle physics concerns the formation of the primordial light nuclei, D, $^3$He, $^4$He and $^7$Li in the environment of an expanding gas of protons, neutrons, electrons, photons and neutrinos (Boesgaard and Steigman, 1985; Yang et al., 1984). In addition, other (very) weakly interacting stable or unstable particles may have been present in large numbers and it was suspected that they may consist of the ubiquitous dark matter present on all scales in the universe (Trimble, 1987). Experimental information relevant for this problem has become available in the late 1980s, namely a precise measurement of the neutron lifetime, a safe limit on the number of different light neutrinos from studies of the $Z^0$ particle in $e^+e^-$-annihilation and finally a better estimate of the amount of primordial $^4$He from the $^4$He abundance as seen today. It had been possible to use these data to reassess the ratio of baryons to photons, $\eta = n_{\mathrm{b}}/n_{\mathrm{y}}$, which is estimated to be in the range of $10^{-9}$–$10^{-10}$. Measurements of the neutron lifetime using very different techniques have been performed in the late 1980s at the reactor in Grenoble. The results for the neutron lifetime $\tau_{\mathrm{n}}$ are $877 \pm 10$ sec using a magnetic neutron storage ring (Paul et al., 1989) and $887.6 \pm 3$ sec from storing ultra cold neutrons in a glass box coated with reflecting oil (Mampe et al., 1989). Since the amount of $^4$He produced in the early universe, $Y_{\mathrm{p}}$, depends on the uncertainty $\Delta t$ of the neutron lifetime like $Y_{\mathrm{p}} = 0.24 \pm (2 \times 10^{-4} \times \Delta t\ (\mathrm{sec}))$, the small error in the neutron lifetime is no longer of great concern and its influence on the $Y_{\mathrm{p}}$ is about a factor of ten smaller than a change in the number of different neutrino flavors by 1.

The measurements of the width of the $Z^0$ essentially rule out four neutrino flavors and are in very good agreement with three different neutrino species, the well known electron-, muon- and tau-neutrinos. From the measurements of $^7$Li abundance in old stars and a careful reanalysis of the amount of primordial $^4$He[6] one obtains as nominal values for $\eta_{10}$ ($\eta$ measured in units of $10^{-10}$) a value of 2.2 from $^7$Li (choosing the lower of two possible values) and 1.6 from $^4$He. This would seem to be in mild conflict with the lower limit on $\eta_{10} = 2.6$ from the upper limit on the D and $^3$He abundance levels. Putting more weight on the measurements of $^4$He and $^7$Li led to an estimate for $\eta_{10}$ of $\sim 2$. This can be converted into a value for the baryonic density normalized to the critical density of $\Omega_b = \rho/\rho_{crit} \times 1/H^2 = 0.0077/H^2$ using 2.75 K for the temperature of the primordial photon background. This is not too far from the density of the visible matter $\Omega_{vis} = 0.005$ using $H = 1$, where $H$ is the Hubble constant in units of 100 km/sec/Mpc. Around 1990, the exact value of $H$ was still a matter of debate with a tendency to lower values, $H < 0.65$. The uncertainty in $H$ is indeed one of the main problems for a reliable estimate of the possible amount of invisible baryonic matter. Based on the arguments given above it was suspected that in fact no dark baryonic matter may be required at all if only the lower limit on $\eta_{10}$ from the upper limit on primordial deuterium was relaxed by about a factor of 2.

## 8.3 The Search for Proton Decay

In 1967, A.D. Sacharov suggested a very general reason to expect the instability of nucleons. Based on the existence of P- and CP-violations together with the absence of significant amounts of antimatter in a uniformly expanding universe, he argued that nucleons should decay. In the 1970s, much more specific and detailed arguments came up in the context of attempts to achieve grand unification of the fundamental interactions (Pati and Salam, 1973; Georgi and Glashow, 1974). The most specific prediction used SU(5) as the unification group, with the result for the proton lifetime

$$\tau_p = 10^{28\pm1}\left(M_X/2 \times 10^4\right)^4 \text{ years}$$

In 1987, the evolution of the gauge couplings with energy as the basis of this prediction was known with impressive precision (Amaldi et al., 1987), however, a common unification mass was apparently missing, and furthermore the lower limit on proton decay reached values incompatible with the SU(5) prediction for the dominant decay mode $p \to e^+\pi^0$. (See Fig. 8.1.)

### 8.3.1 *Early Considerations and First Results*

After World War II physicists started to consider the extraordinary stability of protons a problem that required an explanation. Many nuclei are known to be unstable to alpha emission, beta decay and even fission. But protons seem to stay there forever and seem to require an extra law of conservation of protons. It then seemed of interest to demonstrate to what extent protons are stable. If indeed protons decayed, then so much energy would be released that any nucleus would be destroyed. This was not observed and a first estimate of a lower limit of proton lifetime was estimated of about $10^{20}$ years (Wick et al., 1949; Wigner, 1952; Segre, 1952; Seaborg

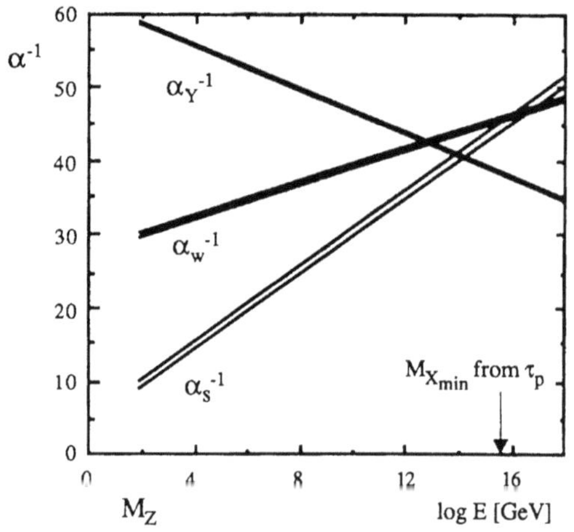


**Fig. 8.1** The energy dependence of coupling constants from the $Z^0$ mass into the GUT region. The value for $M_{X_{\rm min}}$ is determined from the lower limit on $p \rightarrow e^+\pi^0$ (Meyer, 1990, p. 138)

et al., 1953; Feinberg and Goldhaber, 1959). Reines et al. (1954) then started with an experiment where a large piece of scintillator was placed under about 100 feet of rock to reduce the counting rate from penetrating cosmic rays that exceed an energy deposition of about 15 MeV. It provided an improved lower limit of proton lifetime of $10^{22}$ years from the non-observation of destroyed nuclei.

Cosmic ray interactions were considered the only source of the observed counts entirely due to muons when going deep enough. Therefore deeper and deeper sites were chosen in e.g. mines and tunnels.

Backenstoss et al. (1960) went into the *Loetschberg* tunnel getting more than 800 meters of rock on top to dramatically reduce counts due to cosmic rays to reach a limit of $10^{26}$ years, depending somewhat on the assumed decay mode of the proton.

Reines and collaborators (Gurr et al., 1967) then decided to go even deeper into a South African gold mine at a depth of 3 200 meters. Here another cosmic ray issue came into consideration, finding neutrino events, first of all those ones created in air showers in the earth atmosphere. Even the detection of neutrinos from extraterrestrial sources was discussed. A similar experiment at about the same depth was placed in a gold mine in Southern India in the Kolar Gold Field (Achar et al. 1965a, 1965b). Both experiments succeeded with neutrino detection, but for proton decay only limits could be established at more than $10^{28}$ and later up to $10^{30}$ years (Reines and Crouch, 1974).

### 8.3.2 More Definite Motivations

We think that all we observe around us was created in the Big Bang Event long time ago when everything was very hot and dense and uniform. While it expands structure can develop, very rich structure indeed. Everything consists of matter (protons) and very likely not of antimatter although all basic interactions seem to be symmetrical between matter and antimatter. Why is it that only matter has survived? What (if at all) would matter decay to? Photons and neutrinos in the end, of course, and indeed as a result of the hot early universe, neutrinos and photons fill and dominate (by number) the universe today.

A. Sacharov in the 1960s (Sacharov, 1967) started to wonder about this situation that apparently only protons were surviving from the big bang environment. Could it be that the just discovered CP-violation was *involved* so that antiprotons decayed in the very early big bang event and only some protons survived?

This idea could not gain ground. At the same the discovery of the whole zoo of strongly interacting particles and of a rather strange theory, quantum chromodynamics (QCD), ruling them, allowed to propose a new hypothesis, called Grand Unification (GUT) (Georgi and Glashow, 1974; see also Pati and Salam, 1973) which had the surprising consequence of the existence of proton decay e.g. to photons and neutrinos and with lifetime predictions just an order of magnitude more than the best experimental limit. Also simple decay modes were predicted, like $e^+\pi^0$. This was very challenging for experimenters and in particular for those working in cosmic ray physics since one had to be deep underground, with very large size and with good resolution. However, the existing experimental installations were not sufficient, both in size and resolution to get convincing evidence for proton decay.

### *8.3.3 Experimental Considerations*

Right from the outset it is clear that a proton lifetime is very long indeed: the early results tell, somehow beyond $10^{30}$ years otherwise it would not have escaped attention. Then cosmic rays interactions start to be a strong, in fact the dominating background to a search for a possible proton decay. Now cosmic rays on the surface of earth dominantly consist of muons observed to be rather penetrating through any material. Going into deep mines, however, offers sufficient shielding. The gold mines of the Kolar Gold Field in India and in Witwatersrand in South Africa, with (by modern standards) rather simple setups consisting of scintillator detectors and drift tubes could only be used to set limits on proton decay a little more than $10^{30}$ years with simple considerations on possible decay modes.

With definite predictions for lifetimes and decay modes, in particular from the ideas of Grand Unifications (GUT) it became clear that specific designs needed to be realized that had significant efficiencies for the observation of proton decay candidates and that were placed deep enough underground to provide sufficient shielding against cosmic ray events. Part of cosmic rays are neutrinos, which cannot be shielded and their sure observation deep underground was indeed one of the main considerations to go deep enough. What material should be used? It is considered that using any chemical element is safe for proton decay searches given the large energy release as compared to binding energies of nucleons in heavier nuclei. The scintillator is generically CH($n$) and is to be preferred because of the free proton content, and so is water ($H_2O$). But also iron or argon are considered well suitable.

In addition a size of more than a few hundred tons was needed because of simple rate considerations.

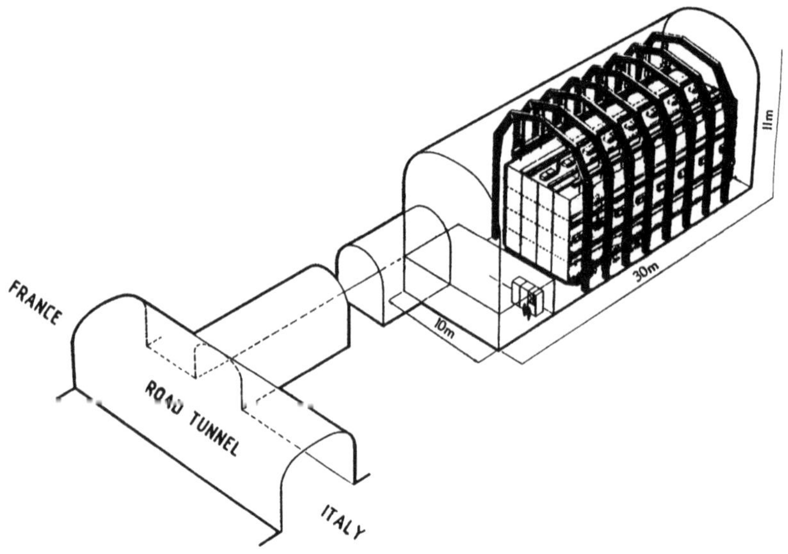


**Fig. 8.2** The Fréjus detector (FREJUS EXP 87-01)

## *8.3.4 Dedicated Experiments*

In the mid-1970s new great efforts started and in the first half of the 1990s new experiments came into operation using water, scintillator and iron as the detector material, with photomultipliers to detect Cherenkov light in water, scintillator light and in iron detectors with tracking elements like flash tubes and drift tubes. The detectors based on water were the largest, with IMB (3300 tons) (Seidel et al., 1988), Kamiokande (800 tons) (Hirata et al., 1989).

The experiments based on iron are somewhat less massive between 120 tons (Kolar, Krishnaswamy et al., 1986 and NUSEX, Battistoni et al., 1983, and 900 tons Fréjus, Berger et al., 1991). All experiments observed atmospheric neutrinos. The masses of the detectors were large enough so that all particles from p-decay as well as from neutrinos would be contained and total energies could be well established. Also the separation into muon neutrino and electron neutrino events could be well achieved. Although several candidates for proton decay were discussed at times, however, with increasing exposures no candidate was confirmed and only limits on proton decay lifetimes could be established. With values above $10^{33}$ years, the predictions of Grand Unification were exceeded by about a factor of 100!

However, in other possible decay modes of nucleons the experimental search for evidence continued. This has been done notably by the three large experiments IMB, Kamiokande II and Fréjus (see Fig. 8.2) (Seidel et al., 1988; Hirata et al., 1989; Berger et al., 1989a; Phillips et al., 1989). The experiments came close to the unavoidable background level due to the interactions of atmospheric neutrinos with the nuclei of the detector material. The limits obtained then mainly depend on detection efficiencies for a given channel and ranged from few times $10^{31}$ years for modes involving neutrinos in the final state up to few times $10^{32}$ years for modes with only electrons and photons to be detected. Although IMB and Kamiokande II continued taking data, they achieved only small improvements. A big step forward could only be achieved by construction of a really huge detector, even bigger than Superkamiokande (Kajita et al., 1989). In order to be able to suppress the neutrino background it had to have an energy resolution and spatial resolution much better than the detectors running then.

Baryon number violations were also suspected to manifest through the $\Delta B = 2$ process of $nn$ oscillations (Mohapatra and Marshak, 1980). Two very different experimental approaches were used to search for $nn$ transitions. In the first method a very cold neutron beam generated in a reactor, carefully shielded against the earth magnetic field passes through a target. Antineutrons created along the flight path would annihilate with neutrons and protons in the target to produce a multipion final state that has to be separated against the cosmic ray interactions in the target (Bressi et al., 1989; Baldo-Ceolin et al., 1990). Secondly the underground experiments to search for nucleon decay also yield very interesting limits on $nn$ oscillations (Jones et al., 1984; Takita et al., 1986). The transition of a neutron to an antineutron would occur inside a nucleus ($^{16}$O or $^{56}$Fe) with subsequent $nN$ annihilation. The only background is from atmospheric neutrino interactions in the underground detectors. No evidence for annihilation events has been seen in any of the experiments. The best limits obtained were $\tau_{\text{n}} > 1\text{–}2 \times 10^8$ sec from the nucleon decay experiments (Jones et al., 1984; Takita et al., 1986) and $\tau_{\text{n}} > 10^7$ sec from the experiment at the reactor in Grenoble (Bressi et al., 1989; Baldo-Ceolin et al., 1990).

The relevance of these experimental limits for the underlying baryon number violating mechanism is not easily assessed, since the theoretical frame work still has many unknowns. However, in the long run the water detectors were very successful as far as the observation of neutrinos is concerned (see Chap. 7).

In February 1987 the two big water detectors Kamiokande and IMB were online just at the right moment to receive signals from a flash of low energy neutrinos originating from a supernova of type II that occurred in the large Magellanic Cloud, a galaxy next to our own (SN 1987A), see e.g. Koshiba (1988). Furthermore, only a year later Kamiokande also saw the neutrinos from the sun, but at a rate lower than predicted by solar models in agreement with other experiments indicating the possibility of neutrino oscillation. Further evidence for neutrino oscillations was obtained by increasing the statistic concerning the ratio of electron neutrinos to muon neutrinos coming from the atmosphere, which also deviated from expectation. What was believed to be background for the search for p-decay proved to be the source of great discoveries. Thus the original motivation to go deep underground into mines and tunnels, namely a quantitative determination of fluxes of neutrinos, was well met. One of the key problems of cosmic ray physics was solved.

Since the water detector in Kamioka, Japan, was earlier increased in mass by another order of magnitude, through the construction of Superkamiokande while keeping the resolution, one could continue the hunt for the elusive proton decay. But again only limits could be obtained, reaching a new lower limit about $10^{34}$ years for the favored decay modes and a few factors less for other decay modes (Nishino et al., 2012; Regis et al., 2012). Now, 40 years after the GUT prediction, limits are a factor of 100 larger than predicted have been achieved. Should one really dare to go even further, say, by another order of magnitude? Probably neutrino events set a natural limit.

## 8.4 Atmospheric Neutrinos

An expected source of neutrinos is cosmic ray interactions in earth atmosphere and in fact the muon neutrino energy spectrum could be firmly determined rather soon while for electron neutrinos it took considerable longer because of lower flux. The source of atmospheric neutrinos is pion, kaon and muon decay in the upper atmosphere. The pions are produced in interactions of primary nucleons and nuclei with atmospheric nuclei, the muons are from pion decay. The neutrino energy spectra are very steep, $dN/dE \sim E^{-a}$, with $a \sim 3.7$, and the yield of neutrinos to antineutrinos is about 1.3 and the ratio of $\nu_\mu/\nu_e \sim 2$. For the lower energies, of the order of 1 GeV, that we want to consider here, many calculations of the atmospheric neutrino flux have been performed, but in all cases with some simplifying assumptions (Volkova, 1980; Gaisser et al., 1983, 1988; Bugaev and Naumov, 1987; Lee and Bludman, 1988). A complete calculation is very involved, it has to take account a rather detailed pion and kaon production model for nucleon–nucleus (mostly $p + {}^{14}N$) collisions, a transport equation of the pions through the atmosphere with pion and muon decay including muon polarization and the energy loss of muons in the atmosphere, and finally geomagnetic effects (including the influence of solar spot number cycle) on the flux of primary cosmic rays. The absolute yield of atmospheric neutrinos is at this stage assumed to be systematically uncertain by about $\pm$ 20 %, however, in all computations it is found that the ratio of the electron (anti)neutrino flux to muon (anti)neutrino flux is rather precisely know (to better than 5 %), only the influence of muon polarization needs to taken care of, as was done in the late 1980s (Barr et al., 1988, 1989; Honda et al., 1990).

Atmospheric neutrinos can be observed in well shielded detectors in the deep underground, however, the rate is very low, even in very big detectors since the rate is only about one event/day/3 000 tons. For electron neutrino detection, volume detectors are required, to fully contain an event it takes a volume of $\sim 3\ m^3$. On the other hand charged current muon neutrino interactions can also be detected as upward going muon events even if the vertex is far outside the detector, at a rate of about one event/day for 500 $m^2$ detector area (Gaisser and Stanev, 1984; Gaisser and Grillo, 1987).

It was realized long ago that due the long flight path through the earth of $\sim 10^4$ km atmospheric neutrino observations allow searches for neutrino oscillations in the region $\Delta m^2 > 10^{-4}\ eV^2$ but only at rather large mixing angles due to the low rate (Ayres et al., 1984; Carlson, 1986; Auriemma et al., 1988). The region of smaller mixing angles, $\sin^2 2\theta > 10^{-3}$, becomes accessible only with very large detectors ($\sim 10^5$ tons) before the limit due to systematic uncertainties of the flux calculations is reached.

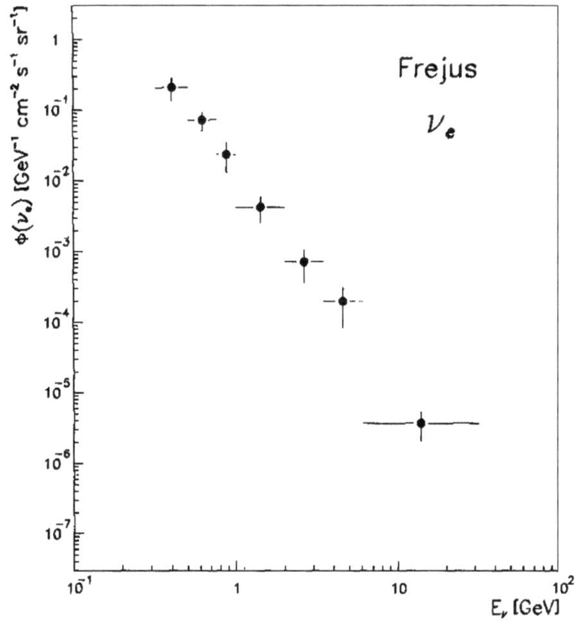


**Fig. 8.3** The absolute $\nu_e$-spectrum obtained from the contained $\nu_e$-interactions in the Fréjus detector averaged over the zenith angle (Daum et al., 1995, p. 422)

The absolute rates of neutrino events as observed by a number of underground experiments were in very good agreement with the available calculations (Berger et al., 1989b; LoSecco et al., 1987a; Aglietta et al., 1989), but due to the large uncertainty of the absolute flux the oscillation of atmospheric neutrinos could only be doubtless detected in the late 1990s by Superkamiokande.

Two of the large nucleon decay detectors, Kamiokande II and Fréjus (and with of smaller size the NUSEX detector) had sufficient quality to distinguish muon neutrino from electron neutrino induced events (see Fig. 8.3) to perform a much more sensitive search for $\nu_e$–$\nu_\mu$ and $\nu_\mu$–$\nu_\tau$ oscillations. No result on $\nu_e$–$\nu_\tau$ oscillations could be obtained in the 1980s, due to the small number of events available for analysis. Kamiokande II had reported a low energy ($< 1$ GeV) muon neutrino event deficit (Hirata et al., 1988), which was interpreted as evidence for neutrino oscillations (Learned et al., 1988; Barger and Wishnant, 1988; Hirata et al., 1988; Bugaev and Naumov, 1989; Honda et al., 1990). The Fréjus experiment did not find such an effect and the data were used to define exclusion regions in $\Delta m^2$ vs $\sin^2\theta$ that include the central values allowed by the Kamiokande II data.

The new limits for $\Delta m^2$ were better by more than a factor 10 compared to previous accelerator results, with only about 200 events available for analysis. This showed the great potential of underground experiments for neutrino oscillation searches, given sufficient size and quality. Crucial to achieve such results was the ability to distinguish electrons and muons with high reliability and that can be done very safely in the Fréjus detector since it was calibrated with electron and muons and pions at accelerators (Berger et al., 1987). Kamiokande on the other hand had to rely on Monte Carlo simulation techniques. The Fréjus results (see Fig. 8.4) were

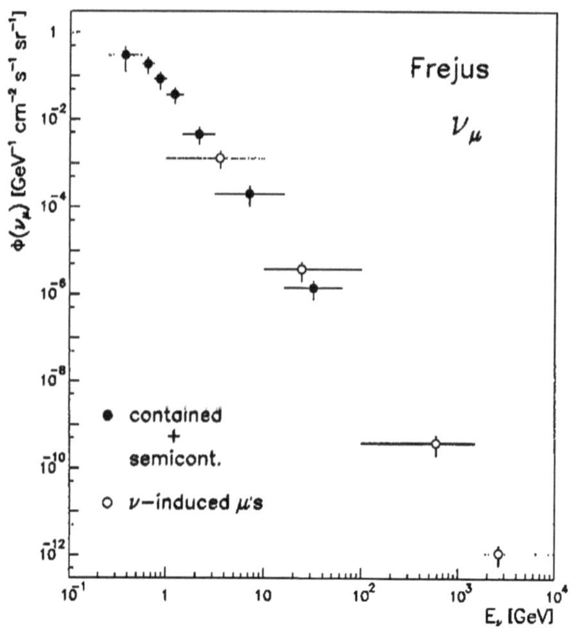


**Fig. 8.4** The absolute $\nu_\mu$-spectrum obtained from the total neutrino induced event sample averaged over the zenith angle (Daum et al., 1995, p. 424).

supported, although with less statistics, by NUSEX (Aglietta et al., 1989), and finally, by Superkamiokande. The latter experiment, however, detected the neutrino oscillations in a parameter space inaccessible to the former generation of experiments.

## 8.5 Signals from Dark Matter Annihilation

It seems rather certain that our universe is filled with matter not accounted for by detectable electromagnetic radiation. This matter of fact was recognized through gravitational effects only, very clearly in flat rotation curves of galaxies throughout the whole range of Hubble classifications and secondly by the application of the virial theorem to clusters of galaxies (Trimble, 1987). About 10–100 times more matter is seen than accounted for by the emission of light. This is the famous dark matter problem, which is most likely due to non-baryonic dark matter (Hegyi and Olive, 1989).

From the particle physics point of view, the lightest supersymmetric particle (LSP) has been proposed as a dark matter candidate (Pagels and Primack, 1982). They can be produced in the big bang without spoiling the successful predictions of nucleosynthesis. They would constitute the dark matter also in our galaxy and both the sun and the earth could capture them into Keplerian orbits well inside their main bodies (Gilliland et al., 1986; Gould, 1987). Above well defined particle masses of $\sim 3$ GeV for the sun and $P \sim 10$ GeV for the earth evaporation and capture become balanced and the particles build up an equilibrium abundance (Silk et al., 1985).

Particles and antiparticles could annihilate into ordinary quarks and leptons. The quarks hadronize to produce high energy neutrinos from their subsequent decay. Since inside the sun and the earth one has a beam dump situation a neutrino signal would mainly come from charm and bottom particle decay (Ritz and Seckel, 1988; Gaisser et al., 1986; Ng et al., 1987; Ellis et al., 1987; Hagelin et al., 1987). The neutrinos can be observed in detectors deep underground while waiting for nucleon decay to occur. The expected neutrino rates are very low but so is the background from atmospheric neutrinos within the angular acceptance from the direction of the sun (earth). None of the big underground detectors reports an excess in the neutrino flux from the sun (LoSecco et al., 1987b; Totsuka, 1989; Daum, 1989), and for the Fréjus detector also limits on the neutrino flux from the direction of the center of the earth are available (Kuznik, 1989).

It may be convenient to convert these limits into limits on the abundance of galactic dark matter for several species of dark matter particles. This can be done on the basis of quantitative calculations of capture and annihilation rates for the heavy neutrinos from supersymmetry of Majorana $\nu_M$, Dirac-$\nu_D$ and s-neutrino $\nu_e$, $\nu_\mu$ type. For Dirac type neutrinos the limit at lower particle masses ($< 10$ GeV) covers the interesting region not excluded by the direct scattering experiments using Ge detectors. The pure (unmixed) photino has been proposed as a viable candidate for the LSP but no useful abundance limit could be obtained (Goldberg, 1983; Kraus, 1983). It seems more appropriate, however, to consider the more general case of a mixed particle the neutralino (Barbieri et al., 1989) at the expense of more parameters that are unknown. Limiting regions in this parameter space on the basis of several sources of experimental limits have been discussed (Olive and Srednicki, 1989). The dark matter particles could as well annihilate in the galactic halo, with photons, antiprotons and positrons as annihilation products to be detected (Ellis et al., 1988; Turner and Wilczek, 1989; Tylka, 1989; Bergstrom, 1989).

### 8.5.1 *High Energy Photons from Galactic Sources*

The primary energy spectrum of cosmic rays as observed near earth extends up to about $10^{20}$ eV $= 10^{11}$ GeV, following roughly a power law with spectral index ($-2.75$) at energies $\leq 3 \times 10^{15}$ eV and ($-3.1$) to the upper end of the spectrum. The particle composition of the cosmic ray flux had been determined with balloon and satellite experiments up to about 1000 GeV/nucleon and protons seem to dominate (Jones et al., 1987; Grunsfeld et al., 1988). The basic acceleration mechanisms and their site inside or even outside the Galaxy are still not known, both large scale as well as point like sources have been proposed (Berezinsky and Prilutsky, 1978; Gaisser and Stanev, 1987). Observations of primary photons (and neutrinos) are crucial to help solving this outstanding problem (Berezinsky and Ptuskim, 1989). The most likely source of neutrinos will be $\pi^+$ (and $\mu^\pm$) decay that are produced in collisions of protons with ambient matter (Berezinsky et al., 1985; Reno and Quigg, 1988). The neutrinos are difficult to detect on earth because typical detection efficiencies are very low (between $10^{-12}$ at 1 GeV and $10^{-5}$ at 10 TeV) and also due to the high atmospheric background. Photons, however, are much easier to detect at or near earth than neutrinos.

High energy photons will in general be produced through synchrotron radiation of high energy electrons, by backward Compton scattering of low energy photons off high energy electrons and finally by high energy $\pi^0$ decay. Below about 100 GeV detectors have to be located above the atmosphere on satellites. At larger energies $> 1\,000$ GeV primary photons generate large air showers detectable in counter arrays at mountain altitudes and alternatively through air Cherenkov light detection in clear moonless nights using open photomultipliers (Weekes, 1988).

In the low energy range from 50 MeV to 5 GeV two satellite experiments, SASII (Fichtel et al., 1975) and COSB (Bignami et al., 1975), had provided a fascinating first look on the high energy gamma ray sky. Mostly galactic $\gamma$-ray sources had been detected (Mayer-Hasselwander and Simpson, 1988), with the possible exception of the quasar 3C273. The strongest point sources are pulsars, Vela, Crab and PSR 1820-11, identified by the characteristic time structure of photon emission. The Crab nebula has also been detected at $> 700$ GeV using the air Cherenkov technique with very high significance (Weekes et al., 1989). Most of the observed photon sources, however, found no definite identification so far (Mayer-Hasselwander and Simpson, 1988), mainly because the large positional error boxes contained too many astronomical objects as possible candidates.

The photon flux from the galaxy was believed to be generally dominated by so-called diffuse emission (Bloemen, 1989). It indeed originates from $\pi^0$ production in pp collisions. The interstellar hydrogen (both in atomic and molecular form) constitutes the target protons, the proton beam is provided by cosmic rays assumed to have the same intensity and spectral shape as we detect it near earth. The column density of the proton gas along a given line of sight through our galaxy is rather well known on the basis of 21 cm line observations (atomic hydrogen) and also of $H_2$ using CO (carbon monoxide) as a tracer. This simple model provides a quantitative description of the observed diffuse photon flux (Mayer-Hasselwander and Simpson, 1988; Bloemen, 1989). A comparison of the energy dependence of the photon flux from the inner and the outer galaxy reveals a flatter energy spectrum, (by about 0.4–0.5 in the spectral index) for photons from the outer part of the galaxy (Bloemen, 1987) and specifically at moderate galactic latitudes (Bloemen et al., 1988). The reason was considered to be a flatter energy spectrum of the primary protons in the outer part of the Galaxy.

Driven by these motivations, it was decided to extend these measurements to much higher energies, to approach the energy where the spectral index of the primary cosmic ray energy spectrum changes, $E = 3 \times 10^{15}$ eV, and where significant leakage out of the galaxy is expected. This was attempted in a number of air shower array experiments. In the 1990s, an increasing number of particle physicists took the decision to join one of the astroparticle telescopes.

### 8.5.2 Further Remarks

After about 50 years, and on the basis of the great success of the big bang hypothesis, and given the breathtaking developments in particle physics it is time to remember that gravitation still presents us (since already 80 years) with the enigma that there seems to be much more matter in the cosmos whose nature is completely open. But it would outnumber protons by a factor of about 10, provided it exists at all. I expect that when a solution can be found to include gravitation into a combined description with Grand Unification, this way an understanding of the deviations from Newtonian gravitation can be found. This would most likely involve a new view onto the fundamental question of the stability of protons.

## References

Achar, C.V., et al.: Phys. Lett. **18**, 196 (1965a)
Achar, C.V., et al.: Phys. Lett. **19**, 78 (1965b)
Aglietta, M., et al.: Europhys. Lett. **8**, 611 (1989)
Amaldi, U., et al.: Phys. Rev. D **36**, 2191 (1987)
Auriemma, G., et al.: Phys. Rev. D **37**, 665 (1988)
Ayres, D.S., et al.: Phys. Rev. D **29**, 902 (1984)
Backenstoss, G.K., et al.: Nuovo Cimento **16**, 749 (1960)
Baldo-Ceolin, M., et al.: Phys. Lett. B **236**, 95 (1990)
Barbieri, R., et al.: Nucl. Phys. B **313**, 725 (1989)
Barger, V., Wishnant, K.: Phys. Lett. B **220**, 308 (1988)
Barr, S., et al.: Phys. Lett. B **214**, 147 (1988)
Barr, G., et al.: Phys. Rev. D **39**, 3523 (1989)
Battistoni, G., et al.: Phys. Lett. B **133**, 454 (1983)
Berezinsky, V.S., Prilutsky, O.F.: Astron. Astrophys. **66**, 325 (1978)
Berezinsky, V.S., Ptuskim, V.S.: Astrophys. J. **340**, 351 (1989)
Berezinsky, V.S., et al.: Nuovo Cimento C **8**, 185 (1985)
Berger, Ch., et al..: Nucl. Instrum. Methods **A262**, 463 (1987)
Berger, C., et al.: Nucl. Phys. B **313**, 509 (1989a)
Berger, C., et al.: Phys. Lett. B **227**, 489 (1989b)
Berger, C., et al.: Z. Phys. C **50**, 385 (1991)
Bergstrom, L.: Phys. Lett. B **225**, 372 (1989)
Bignami, G.F., et al.: Space Sci. Instrum. **1**, 245 (1975)
Bloemen, J.B.G.M.: Astrophys. J. **317**, L15 (1987)
Bloemen, H.: Annu. Rev. Astron. Astrophys. **27**, 469 (1989). Full Text via CrossRef
Bloemen, J.B.G.M., et al.: Astron. Astrophys. **204**, 88 (1988)
Boesgaard, A.M., Steigman, G.: Annu. Rev. Astron. Astrophys. **23**, 319 (1985)
Bressi, G., et al.: Z. Phys. C **43**, 175 (1989)
Bugaev, E.V., Naumov, V.A.: Phys. Lett. B **232**, 391 (1989)
Bugaev, E.V., Naumov, V.A.: Sov. J. Nucl. Phys. **45**, 857 (1987)
Carlson, D.: Phys. Rev. D **34**, 1454 (1986)
Daum, H.J.: Preprint WUB89-22 Univ. of Wuppertal (1989)
Daum, K., et al.: Z. Phys. C **66**, 417–428 (1995)
Ellis, J., et al.: Phys. Lett. B **198**, 393 (1987)
Ellis, J., et al.: CERN-TH 5062/88 (1988)
Feinberg, G., Goldhaber, M.: Proc. Natl. Acad. Sci. **45**, 1301 (1959)

Fichtel, C.E., et al.: Astrophys. J. **198**, 163 (1975)
Gaisser, T.K., Grillo, A.F.: Phys. Rev. D **36**, 2752 (1987)
Gaisser, T.K., Stanev, T.: Phys. Rev. D **30**, 985 (1984)
Gaisser, T.K., Stanev, T.: Phys. Rev. Lett. **58**, 1695 (1987)
Gaisser, T.K., et al.: Phys. Rev. Lett. **51**, 223 (1983)
Gaisser, T.K., et al.: Phys. Rev. D **34**, 2206 (1986)
Gaisser, T.K., et al.: Phys. Rev. D **38**, 85 (1988)
Georgi, H., Glashow, S.L.: Phys. Rev. Lett. **32**, 438 (1974)
Gilliland, R.L., et al.: Astrophys. J. **306**, 703 (1986)
Goldberg, H.: Phys. Rev. Lett. **50**, 1419 (1983)
Gould, A.: Astrophys. J. **321**, 571 (1987)
Grunsfeld, J.M., et al.: Astrophys. J. **327**, L31 (1988)
Gurr, H.S., et al.: Phys. Rev. **158**, 1321 (1967)
Hagelin, J.S., et al.: Phys. Lett. B **180**, 375 (1987)
Hegyi, D.J., Olive, K.A.: Astrophys. J. **346**, 648 (1989)
Hirata, K.S., et al.: Phys. Lett. B **205**, 416 (1988)
Hirata, K.S., et al.: Phys. Lett. B **220**, 308 (1989)
Honda, M., et al.: ICRR-Report-208-90-1 (1990)
Jones, T.W., et al.: Phys. Rev. Lett. **52**, 720 (1984)
Jones, W.V., et al. (JACEE Coll.): Annu. Rev. Nucl. Part. Sci. **37**, 71 (1987)
Kajita, K., et al.: Solid State Commun. **70**, 1189 (1989)
Koshiba, M.: Nucl. Phys. B **3**, 441 (1988)
Kraus, L.M.: Nucl. Phys. B **227**, 556 (1983)
Krishnaswamy, M.R., et al.: Nuovo Cimento C **9**, 167 (1986)
Kuznik, B.: Thesis, University of Wuppertal (1989)
Learned, J.G., et al.: Phys. Lett. B **207**, 79 (1988)
Lee, H., Bludman, S.A.: Phys. Rev. D **37**, 122 (1988)
LoSecco, J.M., et al.: Phys. Lett. B **184**, 305 (1987a)
LoSecco, J.M., et al.: Phys. Lett. B **188**, 388 (1987b)
Mampe, W., et al.: Phys. Rev. Lett. **63**, 593 (1989)
Mayer-Hasselwander, H.A., Simpson, G.: MPE preprint 136 (December 1988)
Meyer, H.: Nucl. Phys. B **16**, 136–149 (1990)
Mohapatra, R.N., Marshak, R.E.: Phys. Lett. B **91**, 222 (1980)
Ng, K., et al.: Phys. Lett. B **188**, 138 (1987)
Nishino, H., et al.: arXiv:1203.4030v1 (19 Mar 2012)
Olive, K.A., Srednicki, M.: Preprint UMN-TH-801/89 (July 1989)
Pagels, H., Primack, J.: Phys. Rev. Lett. **48**, 223 (1982)
Pati, J.C., Salam, A.: Phys. Rev. Lett. **31**, 661 (1973)
Paul, W., et al.: Z. Phys. C **45**, 25 (1989)
Phillips, T.J., et al.: Nucl. Phys. B **224**, 348 (1989)
Regis, C., et al.: arXiv:1205.6538v1 (30 May 2012)
Reines, F., Crouch, M.F.: Phys. Rev. Lett. **32**, 493 (1974)
Reines, F., et al.: Phys. Rev. **96**, 1157 (1954)
Reno, M.H., Quigg, C.: Phys. Rev. D **37**, 657 (1988)
Ritz, S., Seckel, D.: Nucl. Phys. B **304**, 877 (1988)
Riordan, M.: The Hunting of the Quark. Simon & Schuster, New York (1987)
Sacharov, A.D.: JETP Lett. **5**(1), 32–35 (1967)
Seaborg, G., et al.: Rev. Mod. Phys. **25**, 469 (1953)
Segre, E.: Phys. Rev. **86**, 211 (1952)
Seidel, S., et al.: Phys. Rev. Lett. **61**, 2522 (1988)
Silk, J., et al.: Phys. Rev. Lett. **55**, 257 (1985)
Takita, M., et al.: Phys. Rev. D **34**, 902 (1986)
Totsuka, Y.: ICR-Report-192-9 (May 1989)
Trimble, V.: Annu. Rev. Astron. Astrophys. **25**, 425 (1987)

Turner, M.S., Wilczek, F.: FERMILAB-Pub-89/44-A (1989)
Tylka, A.J.: Phys. Rev. Lett. **63**, 840 (1989)
Volkova, L.V.: Sov. J. Nucl. Phys. **31**, 784 (1980)
Weekes, T.C.: Phys. Rep. **160**, 1 (1988). Article – PDF (10 552 K)
Weekes, T.C., et al.: Astrophys. J. **342**, 379 (1989). Full Text via CrossRef
Wick, G.C., et al.: Annu. Rev. Nucl. Part. Sci. **8**, 1 (1949)
Wigner, E.P.: Proc. Natl. Acad. Sci. **38**, 449 (1952)
Yang, J., et al.: Astrophys. J. **281**, 493 (1984)

# Chapter 9
# Towards High-Energy Neutrino Astronomy

**Christian Spiering**

## 9.1 Introduction

The search for the sources of cosmic rays is a three-fold assault, employing charged cosmic rays, gamma rays and neutrinos. The first conceptual ideas how to detect high energy neutrinos date back to the late 1950s. The long evolution towards detectors with a realistic discovery potential started in the 1970s and 1980s, by the pioneering works in the Pacific Ocean close to Hawaii (DUMAND) and in the Siberian Lake Baikal (NT200). But only now, half a century after the first concepts, such a detector is in operation: IceCube at the South Pole. We do not yet know for sure whether with IceCube we will indeed detect extraterrestrial high-energy neutrinos or whether this will remain the privilege of next generation telescopes. But whatever the answer will be: already the path to the present detectors was a remarkable journey. This chapter sketches its milestones. It focuses to the first four decades and keeps the developments of the last decade comparatively short. I refer the reader to the 2011 review of the field (Katz and Spiering, 2012) for more detailed information on actual results and plans for future detectors.

## 9.2 From First Concepts to the Detection of Atmospheric Neutrinos

The initial idea of neutrino astronomy beyond the solar system rested on two arguments: The first was the expectation that a supernova stellar collapse in our galaxy would be accompanied by an enormous burst of neutrinos in the 5–10 MeV range. The second was the expectation that fast rotating pulsars must accelerate charged particles in their Tera-Gauss magnetic fields. Either in the source or on their way to

C. Spiering (✉)
DESY, Platanenallee 6, 15738 Zeuthen, Germany
e-mail: christian.spiering@desy.de

B. Falkenburg, W. Rhode (eds.), *From Ultra Rays to Astroparticles*,
DOI 10.1007/978-94-007-5422-5_9, © Springer Science+Business Media Dordrecht 2012

Earth they must hit matter, generate pions and neutrinos as decay products of the pions.

The first ideas to detect cosmic high-energy neutrinos underground or underwater date back to the late 1950s. In the 1960 Annual Review of Nuclear Science, K. Greisen and F. Reines discuss the motivations and prospects for such detectors. In his paper entitled *Cosmic Ray Showers* (Greisen, 1960), Greisen writes:

> As a detector, we propose a large Cherenkov counter, about 15 m in diameter, located in a mine far underground. The counter should be surrounded with photomultipliers to detect the events, and enclosed in a shell of scintillating material to distinguish neutrino events from those caused by $\mu$ mesons. Such a detector would be rather expensive, but not as much as modern accelerators and large radio telescopes. The mass of the sensitive detector could be about 3000 tons of inexpensive liquid.

Later he estimates the rate of neutrino events from the Crab Nebula as one count per three years and optimistically concludes:

> Fanciful though this proposal seems, we suspect that within the next decade cosmic ray neutrino detection will become one of the tools of both physics and astronomy.

F. Reines in his article *Neutrino Interactions* (Reines, 1960) is more conservative with respect to extraterrestrial neutrinos:

> At present no acceptable theory of the origin and extraterrestrial diffusion exists so that the cosmic neutrino flux cannot be usefully predicted.

At this time, he could not be aware of the physics potential of atmospheric neutrinos and continues:

> The situation is somewhat simpler in the case of cosmic-ray neutrinos ("atmospheric neutrinos" in present language. C.S.) – they are both more predictable and of less intrinsic interest.

In the same year, on the 1960 Rochester Conference, M. Markov published his ground-breaking idea (Markov, 1960)

> ...to install detectors deep in a lake or a sea and to determine the direction of charged particles with the help of Cherenkov radiation.

This appeared to be the only way to reach detector volumes beyond the scale of $10^4$ tons.

During the 1960s, no predictions or serious estimates for neutrino fluxes from cosmic accelerators were published. Actually, many of the objects nowadays considered as top candidates for neutrinos emission were discovered only in the 1960s and 1970s (the first quasar 1963, pulsars 1967, X-ray binaries with a black hole 1972, gamma-ray bursts 1973). The situation changed dramatically in the 1970s, when these objects were identified as possible neutrino emitters, triggering an enormous amount of theoretical activity.

Different to extraterrestrial neutrino fluxes, the calculation of the flux of atmospheric neutrinos became more reliable. First serious estimates were published in the early 1960s. These pioneering attempts are described in detail in the recollections of Igor Zheleznykh (2006) who was a student of Markov. In his diploma work from 1958 he performed early estimates for the flux of atmospheric neutrinos and for the

**Fig. 9.1** Kenneth Greisen (1918–2007), Frederick Reines (1918–1998) and Moisej Markov (1908–1994)

flux of neutrinos from the Crab nebula. The real explosion of papers on atmospheric neutrinos, however, happened between 1980 and 1990 when the large underground detectors became operational and the field turned into a precision science (see the contribution of Kai Zuber).

Still, from the perspective of the 1960s and early 1970s, the study of atmospheric neutrinos appeared equally interesting as the search for extraterrestrial neutrinos (Markov and Zheleznykh, 1961). Neutrino oscillations did not yet play a role in the discussions of the late 1960s and appeared only in the 1970s on the shopping list. However, atmospheric neutrinos offered a possibility to study neutrino cross sections in an energy region, which was not accessible to accelerator experiments at that time. Using the language of the 1970s, these studies would have given information on the mass of the intermediate W-boson. Without proper guidance on the W-mass, these effects were expected to be visible already in the few-GeV range, and actually this was one of the main motivations to build the first underground neutrino detectors (Zheleznykh, 2006). More generally, the availability of neutrinos with energies beyond what could realistically be expected from accelerator beams was recognized as a tempting method to search for phenomena beyond the standard model; however, not by everybody! F. Reines notes in his summary of the Neutrino-81 conference (Reines, 1981):

> Estimates of the atmospheric flux suggest that interactions of this source of $\geq 1$ TeV neutrinos might be usefully observed, although our accelerator-based colleagues are not keen on this as a source of new information.

Actually the attitude of the broader particle physics community w.r.t. the physics potential of atmospheric neutrinos changed only together with the detection of neutrino oscillations in the 1990s.

Before studying atmospheric neutrinos they had to be detected. This was achieved in 1965, almost simultaneously, by two groups. The one was led by F. Reines (Case-Witwatersrand group, later Case-Witwatersrand-Irvine, CWI). The group operated two walls of segmented liquid scintillator in a South African Gold mine, at a depth of 8800 m water equivalent. The configuration was chosen to identify horizontal muon tracks which, at this depth, could be due only to neutrino

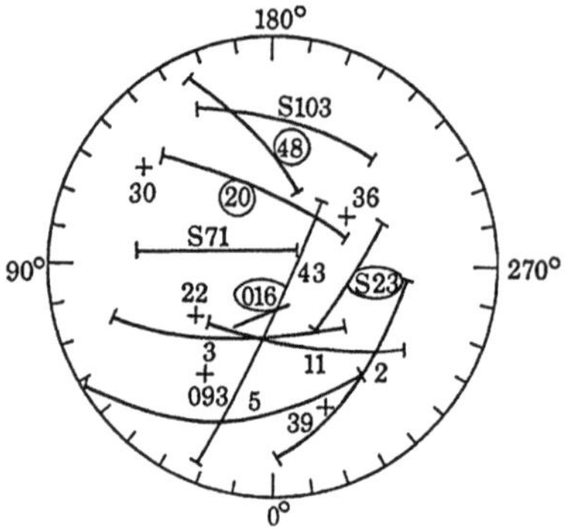


**Fig. 9.2** The first neutrino sky map with the celestial coordinates of 18 KGF neutrino events (Krishnaswamy et al., 1971). Due to uncertainties in the azimuth, the coordinates for some events are arcs rather than points. The labels reflect the numbers and registration mode of the events (e.g. "*S*" for spectrograph). Only for the ringed events the sense of the direction of the registered muon is known

interactions (the detector could measure the direction but not the sense of the direction, so that an upward moving muons could not be distinguished from downward moving muons). Between February and July 1965, seven such tracks were recorded, with a background of less than one event from muons not induced by neutrinos. It is interesting to note that the first of these tracks was recorded at February 23, 1965, exactly 22 years before the neutrinos from supernova SN1987A reached the Earth (23/2/1987). The detector of the other group (a Bombay–Osaka–Durham collaboration) was operated in the Indian Kolar Gold Field (KGF) mine, at a depth of 7 500 m water equivalent. It consisted of two walls of plastic scintillators and flash tubes. The KGF group started data taking nearly six months after the CW group, saw the first of three neutrino candidates two months later than Reines (20/4/1965), but published two weeks earlier than the CW group: KGF at August 15, 1965 (submitted 12/7/1965 Achar et al., 1965), CW at August 30, 1965 (submitted 26/7/1965 Reines et al., 1965). So, indeed Reines recorded the first cosmic neutrino ever, but the formal priority is with the KGF group. A historic race which had no losers but two winners.

With improved detectors, the two groups continued measurements for many years collecting a total sample of nearly 150 neutrino events. The KGF group was the first to release a sky map (see Fig. 9.2).

In 1978, the Baksan Neutrino Telescope (BNT) in the Caucasus started (partial) operation. It was followed by a phalanx of new detectors in the 1980s, which mostly were motivated by the search for proton decay. The largest of them was the IMB detector which produced the first neutrino sky map of reasonable quality, with 187 events from 396 live-days (Svoboda et al., 1987).

The study of atmospheric neutrinos and of MeV-neutrinos from a galactic supernova seemed to be feasible with detectors of a few hundred or thousand tons, with the main unknown being the rate at which those supernovae occur. Predictions ranged from a several per millennium up to a few per century. Therefore a

**Fig. 9.3** John Learned (born 1940), leader of the DUMAND project and a driving force of neutrino astronomy

supernova neutrino detector had better to be a multipurpose device with alternative goals, for instance atmospheric neutrinos or cosmic ray studies as reliable aims, combined with the high-risk aim to search for proton decay. Actually, the two water Cherenkov detectors which detected neutrinos from the supernova 1987A, IMB (USA) and Kamiokande (Japan), had both been funded for different primary purposes, most notably proton decay.

In contrast to investigating atmospheric and supernova neutrinos, the study of high-energy extraterrestrial neutrinos had the inherent risk that no reliable predictions for the expected fluxes could be made. Under these circumstances it appeared logical to tackle this problem with the largest devices conceivable, with underwater detectors of the kind which M. Markov had proposed in 1960. The first step from conceptual ideas to large-scale experimental efforts was done by the DUMAND project which will be described in the next section.

## 9.3 DUMAND: From the First Workshop to the DUMAND-II Proposal

The history of underwater neutrino telescopes starts with a project which eventually was cut off but left an incredibly rich legacy of ideas and technical principles: The DUMAND project. DUMAND stands for Deep Underwater Muon and Neutrino Detector. Its early history is excellently covered in a *Personal history of the DUMAND project* by A. Roberts (1992).

At the 1973 International Cosmic Ray Conference (ICRC), a few physicists including F. Reines and J. Learned (USA), G. Zatsepin (USSR) and S. Miyake (Japan) discussed a deep-water detector to clarify puzzles in muon depth-intensity curves. The puzzles faded away, but it was obvious that such a detector could also work for neutrinos.

The year 1975 saw the first of a – meanwhile legendary – series of DUMAND Workshops, this one at Washington State University (Kotzer, 1976). A survey of possible sites converged on the Pacific Ocean close to Hawaii, since it offered deep locations close to shore. A year later, a two-week workshop took place in Honolulu (Roberts, 1977). At that time, three options for a deep-sea array were discussed:

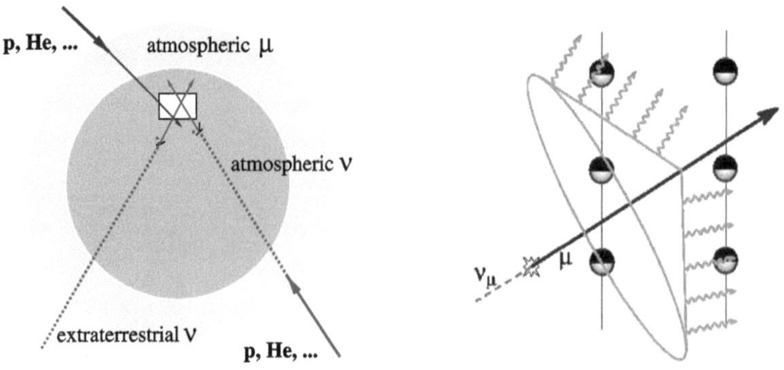


**Fig. 9.4** *Left*: Sources of muons in deep underwater/ice detectors. Cosmic nuclei – protons (p), $\alpha$ particles (He), etc. – interact in the Earth atmosphere (*light-colored*). Sufficiently energetic muons produced in these interactions ("atmospheric muons") can reach the detector (*white box*) from above. Upward-going muons must have been produced in neutrino interactions. *Right*: Detection principle for muon tracks

- UNDINE (for "UNderwater Detection of Interstellar Neutrino Emission") was intended to detect neutrinos from supernova collapses from far beyond our own Galaxy (leaving the Galactic supernovae to underground detectors). Based on overoptimistic assumptions of the neutrino energy spectrum, it was soon discarded.
- ATHENE (for "ATmospheric High-Energy Neutrino Experiment") was tailored to high-energy particle physics with atmospheric neutrinos.
- UNICORN (for "UNderwater Interstellar COsmic-Ray Neutrinos") had the primary goal to search for high-energy extraterrestrial neutrinos.

At the 1976 workshop and, finally, at the 1978 DUMAND workshop at the Scripps Institute in La Jolla (Roberts and Wilkins, 1978) the issue was settled in favor of an array which combined the last two options, ATHENE and UNICORN.

The principle of the detector was to record upward-traveling muons generated in charged current muon neutrino interactions. Neutral current interactions which produce no muons had been only discovered in 1973. They result in final-state charged particles of rather low energy and did not play a role for the design studies. The upward signature guarantees the neutrino origin of the muon since no other particle can traverse the Earth. Since the 1960s, a large depth was recognized as necessary in order to suppress downward-moving muons which may be mis-reconstructed as upward-moving ones (Fig. 9.4, left). Apart from these, only one irreducible background to extraterrestrial neutrinos remains: neutrinos generated by cosmic ray interactions in the Earth's atmosphere ("atmospheric neutrinos"). This background cannot be reduced by going deeper. On the other hand, it provides a standard calibration source and a reliable proof of principle.

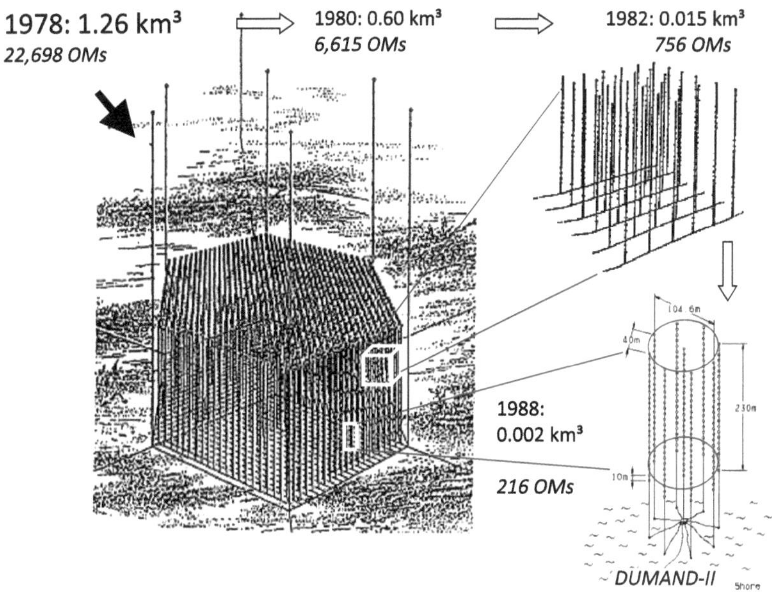


**Fig. 9.5** The originally conceived DUMAND cubic kilometer detector and the phased downgrading to the 1988 plan for a first-generation underwater neutrino telescope DUMAND-II

The DUMAND design envisaged an array of photomultiplier tubes (PMTs) housed in transparent pressure spheres spread over a cubic kilometer (see Fig. 9.5, left). The PMTs would record arrival time and amplitude of Cherenkov light emitted by muons or particle cascades. The spheres were to be attached to strings moored at the ground and held vertically by buoys. From the arrival times, the direction of the muon track can be reconstructed, and it turned out that a directional accuracy of 1 degree is achievable. This is of a similar size as the kinematic smearing between neutrino and muon direction and allows for *neutrino tracing*, i.e. for neutrino astronomy (see Fig. 9.4, right).

Naturally, the idea to construct a cubic-kilometer detector with more than 20 000 large-size photomultipliers (see Fig. 9.5) challenged technical and financial possibilities. A. Roberts remembers (Roberts, 1992):

> The 1978 DUMAND Standard Array, on closer examination, assumed more and more awesome proportions. ... 1261 sensor strings, each with 18 complex sensor modules ... to be deployed on the ocean bottom at a depth of 5 km! The oceanographers were amazed – this project was larger than any other peacetime ocean project by a factor of the order of 100. The size of the array was based on relatively scant information on the expected neutrino intensities and it was difficult to justify in detail; the general idea was that neutrino cross sections are small and high-energy neutrinos are scarce, so the detector had better be large.

Confronted with the oceanographic and financial reality, the 1.26 km$^3$ array was abandoned. A half-sized configuration (1980) met the same fate, as did a much smaller array with 756 phototubes (1982). The latter design was comparable in size to the AMANDA detector at the South Pole (see Sect. 9.7) and the ANTARES telescope in the Mediterranean Sea, close to Toulon (see Sect. 9.8). What finally emerged as a technical project was a 216-phototube version, dubbed DUMAND-II or "The Octagon' " (eight strings at the corners of an octagon and one in the center), 100 m in diameter and 230 m in height (Bosetti et al., 1988) (see Fig. 9.5). The plan was to deploy the detector 30 km off the coast of Big Island, Hawaii, at a depth of 4.8 km.

The evolution of the detector design, largely following financial and technological boundary conditions, was the one side of the story. What about the flux predictions?

At the 1978 workshop first investigations on neutron star binary systems as point sources of high-energy neutrinos were presented, specifically Cygnus X-3 (D. Eichler/D. Schramm and D. Helfand in Roberts and Wilkins, 1978). The connection to the indications for sources of TeV-$\gamma$-ray (none of them significant at that time!) was discussed by T. Weekes. At the same time, the possibilities for diffuse source detection were disfavored (R. Silberberg, M. Shapiro, F. Stecker).

The gamma-neutrino connection was discussed further by V. Berezinsky at the 1979 DUMAND Workshop in Khabarovsk and Lake Baikal (Learned, 1979). He emphasized the concept of "hidden" sources which are more effectively (or *only*) detectable by neutrinos rather than by $\gamma$ rays. Among other mechanisms, Berezinsky also investigated the production of neutrinos in the young, expanding shell of a supernova which is bombarded by protons accelerated inside the shell ("inner neutrino radiation from a SN envelope"). He concluded that a 1 000 $m^2$ detector should be sufficient to detect high-energy neutrinos from a galactic supernova over several weeks or months after the collapse. Naturally, ten years later, in 1987, much attention was given to this model in the context of SN1987. But alas! – this supernova was at about 50 kpc distance, more than five times farer than the Galactic center. Moreover all underground detectors existing in 1987 had areas much smaller than 1 000 $m^2$. Therefore the chances to see inner neutrino radiation from the envelope were rather small, and actually "only" the MeV burst neutrinos and no high-energy neutrinos have been recorded.

A large number of papers on expected neutrino fluxes was published during the 1980s. The expected neutrino fluxes were found to depend strongly

(a) on the energy spectrum of the $\gamma$-ray sources which could only be guessed since the first uncontroversial TeV-$\gamma$ observation was the Crab nebula in 1989 (Weekes et al., 1989), and
(b) on the supposed $\nu/\gamma$ ratio which depends on the unknown thickness of matter surrounding the source.

The uncertainty of expectations is reflected in the DUMAND-II proposal (Bosetti et al., 1988). Pessimistic and optimistic numbers differed by 2–3 orders of magnitude and left it open whether DUMAND-II would be able to detect neutrino sources or whether this would remain the realm of a future cubic kilometer array. Two years later, V. Berezinsky reiterated his earlier estimates that for neutrinos from a fresh neutron star a detector with an effective area of 1 000 $m^2$ (i.e. a large underground detector) would be sufficient, but that the detection of extragalactic sources would require detectors of 0.1–1.0 $km^2$ size (Berezinsky, 1990). DUMAND-II, with 25 000 $m^2$ area, fell just below these values. Again citing A. Roberts (1992):

> These calculations serve to substantiate our own gut feelings. I have myself watched the progression of steadily decreasing size ... at first with pleasure (to see it become more practical), but later with increasing pain. ... The danger is, that if DUMAND II sees no neutrino sources, the funding agencies will decide it has failed and, instead of expanding it, will kill it.

After various site surveys and technical tests, in 1983 the Department of Energy (DOE) approved the funding for DOE-supported U.S. groups to deploy the "Short Prototype String" (SPS). With additional support from NSF, ICRR in Japan and the University of Bern in Switzerland, the SPS was conceived to develop and test the basic detector techniques, to further study the environmental effects, to demonstrate that muons can be reconstructed and to measure the muon vs. depth dependence. In 1987, this 7-phototube test string was successfully deployed for some hours from a high-stability Navy vessel (Babson et al., 1990). It provided the measured muon intensity as a function of depth.

After the successful SPS test, in 1988 the DUMAND-II proposal was submitted to DOE and NSF. The collaboration groups of this proposal were: UC Irvine, CalTech, U. Hawaii, Scripps Inst. Oceanology, U. Vanderbilt, U. Wisconsin (USA), U. Kinki, ICRR Tokyo (Japan), TH Aachen, U. Kiel (Germany). DUMAND-II with its 100 m diameter and 230 m height, would have detected three down going muons per minute and about 3 500 atmospheric neutrinos per year.

A wealth of technological solutions was found within the design study for the SPS, and many remained as a legacy for other neutrino telescopes. For the first time, relevant optical properties of sea water were measured. Tests and investigations made during many cruises pushed ahead of oceanographic practice at that time. Some of the innovative solutions were only possible since basic technologies had only recently appeared on the market.

One example is the photomultiplier tube (PMT) which had to be large (to collect much light), fast (to allow for fast timing and good muon angular resolution) and to have a good amplitude resolution (allowing identification of the 1-photoelectron (PE) peak and separation from noise and possibly from the 2-PE signals). Various innovative attempts were made but eventually discarded after Hamamatsu Comp. (Japan) committed to develop a spherical 15-inch PMT R2018. This PMT fitted into a 17-inch commercial pressure sphere. The only other practicable design for a light sensor was the PHILIPS "smart" photomultiplier XP2600 (van Aller et al., 1983). I will sketch its operation principle in the context of the Baikal neutrino telescope, where a similar tube was developed.

Another example for applying brand-new techniques is the use of optical fibers for data transmission. The processed signal is sent as optical pulse through a multimode fiber cable. Fibers for use in undersea cables had been become available just in the late 1970s. This was the second fortunate event with remarkable consequences since it removed the low-data-rate barrier imposed by shore cables with copper lines of 40 km length.

**Fig. 9.6** Grigorij Domogatsky (born 1941), leader of the Baikal project, during a winter expedition to Lake Baikal

## 9.4 The Evolution of the Baikal Project

Russian participation in the DUMAND project was strong from the beginning. However, in the context of the Soviet invasion in Afghanistan, in 1980 the Reagan administration terminated the cooperation. As A. Roberts remembers (Roberts, 1992):

> The severing of the Russian link was done with elegance and taste. We were told, confidentially, that while we were perfectly free to choose our collaborators as we liked, if perchance they included Russians it would be found that no funding was available.

About the same time, however, A. Chudakov proposed to use the deep water of Lake Baikal in Siberia as the site for a "Russian DUMAND". The advantages of Lake Baikal seemed obvious: it is the deepest freshwater lake on Earth, with its largest depth at nearly 1 700 meter, it is famous for its clean and transparent water, and in late Winter it is covered by a thick ice layer which allows installing winches and other heavy technique and deploying underwater equipment without any use of ships.

In 1981, first shallow-site experiment with small PMTs started. Chair of a dedicated laboratory at the Moscow Institute of Nuclear Research, Academy of Science of USSR (INR) became G. Domogatsky, a theoretician, flanked by L. Bezrukov as leading experimentalist.

Soon a site in the Southern part of Lake Baikal was identified as suitable. It was about 30 km South-West from the outflow of Lake Baikal into the Angara river and approximately 60 km from the large city Irkutsk. A site at a distance of 3.6 km to shore and at a depth of about 1 370 m was identified as the optimal location for a detector which would be installed at a depth of about 1.0–1.1 km. Detectors could be installed in a period between late February and early April from the ice cover, and operated over the full year via a cable to shore.

In the US, these efforts were noticed but obviously not understood as a competition to DUMAND. V. Stenger, who was the leading Monte Carlo expert of the DUMAND project, repeatedly expressed his doubts that one could separate neutrinos from background in Lake Baikal: the lake was too shallow and the background of downward-going muons much too high; the necessary cuts to reject the background would inevitably also strongly diminish the signal of upward-going muons from neutrino interactions, with the exception of rather small and dense arrays.

After operation of first underwater modules with a 15-cm PMT in 1982, in the following year a small *string* was operated for several days. In 1984, a first *stationary* string was deployed (Girlanda-84) and recorded downward moving muons (Bezrukov et al., 1984). It consisted of three floors each with four PMTs in two pressure-tolerant cylinders of glass-fiber enforced epoxy. At that time, no pressure-tight glass spheres where available in the USSR. The end of the cylinders were closed by caps of plexiglass. The PMT was a Russian tube with a 15 cm flat photocathode and modest amplitude and time resolution. An electrical cable connected the string to the shore. The 1984 string was followed by another stationary string in 1986 (Girlanda-86). Data from this string were used to set stringent limits on the flux of magnetic monopoles catalyzing proton decays along their path (Domogatsky et al., 1986).

Girlanda-84 took data for a total of 50 days and then sank down due to leaking buoys which held the string in vertical position. But also the cable penetrators through the epoxy cylinders as well as the cap-to-cylinder hermetic connection tended to leak and were a notorious source of headaches. Moreover, it was clear that the PMT used was much too small and too slow for a neutrino telescope. Therefore a technology using glass spheres and a new type of photo-sensor were developed.

The Russians started with testing the “smart” XP2600 from PHILIPS (see Sect. 9.3) in Lake Baikal (Bezrukov et al., 1987). In parallel, the development of an equivalent Russian device, the QUASAR, was tackled, in cooperation with the EKRAN company in Novosibirsk. The QUASAR (Fig. 9.7) is a hybrid device similar to the PHILIPS 2600 developed for the DUMAND project. Photoelectrons from a 370 mm diameter cathode ($K_2CsSb$) are accelerated by 25 kV to a fast, high-gain scintillator placed near the center of the glass bulb. The light from the scintillator is read out by a small conventional photomultiplier (type UGON). One photoelectron from the hemispherical photocathode yields typically 20 photoelectrons in the small photomultiplier. This high multiplication factor results in an excellent 1-PE resolution, clear distinction between 1-PE and 2-PE pulses, a time jitter as small as 2 ns and negligible sensitivity to the Earth’s magnetic field (Bagduev et al., 1999).

In 1988, the Baikal experiment was approved as a long-term direction of research by the Soviet Academy of Sciences and the USSR government which included considerable funding. A full-scale detector (yet without clear definition of its size) was planned to be built in steps of intermediate detectors of growing size. In the same year 1988, our group from the East German Institute of High Energy Physics in Zeuthen (part of DESY from 1992 on) joined the Baikal experiment.

After German unification in 1990, the Zeuthen group had access to the Western market and contributed with Jena glass spheres and some underwater connectors to the strings which were deployed in 1991 to 1993 (see below). In parallel, Russian spheres were developed in collaboration with industry, as well as penetrators and

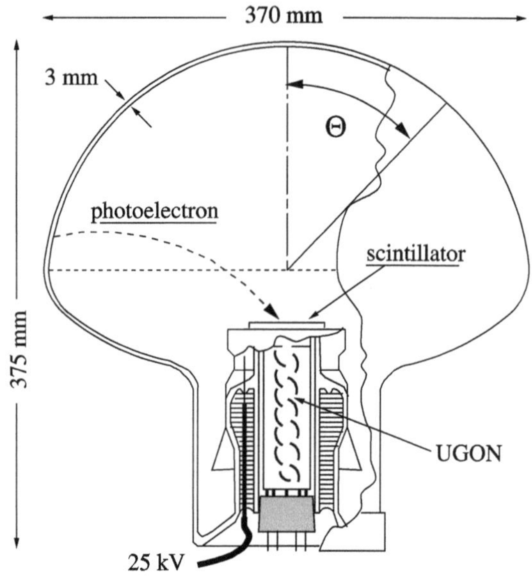


**Fig. 9.7** The QUASAR-370 phototube

connectors which tolerated water depths down to 2 km – not suitable for large-depth ocean experiments but sufficient for Lake Baikal.

In 1989, a preliminary version of what later was called the NT200 project was developed, an array comprising approximately 200 optical modules. The final version of the project description was finished in 1992 (Sokalski and Spiering, 1992). At this time, the participating groups came from INR Moscow, Univ. Irkutsk, Moscow State Univ., Marine Techn. Univ. Petersburg, Polytechnical Institutes in Niszhni Novgorod and Tomsk, JINR Dubna, Kurchatov Inst. (Moscow), Limnological Inst. Irkutsk (all Russia), DESY-Zeuthen (Germany) and KFKI Budapest (Hungary).

NT200 (Fig. 9.8, left) is an array of 192 optical modules carried by eight strings which are attached to an umbrella-like frame consisting of 7 arms, each 21.5 m in length. The strings are anchored by weights at the lake floor and held in a vertical position by buoys at various depths. The configuration spans 72 m in height and 43 m in diameter. The finely balanced mechanics of this frame, with all its buoys, anchor weights and pivoted arms is another stunning feature of the Baikal experiment. The detector is deployed (or hauled up for maintenance) within a period of about six weeks in February to April, when the lake is covered with a thick ice layer providing a stable working platform. It is connected to shore by several copper cables on the lake floor, which allows for operation over the full year.

The optical modules with the QUASAR-370 phototubes are grouped pair-wise along a string. In order to suppress accidental hits from dark noise (about 30 kHz) and bio-luminescence (typically 50 kHz but seasonally raising up to hundreds of kHz), the two photomultipliers of each pair are switched in coincidence. The time calibration is done using several nitrogen lasers in pressure-tight glass cylinders.

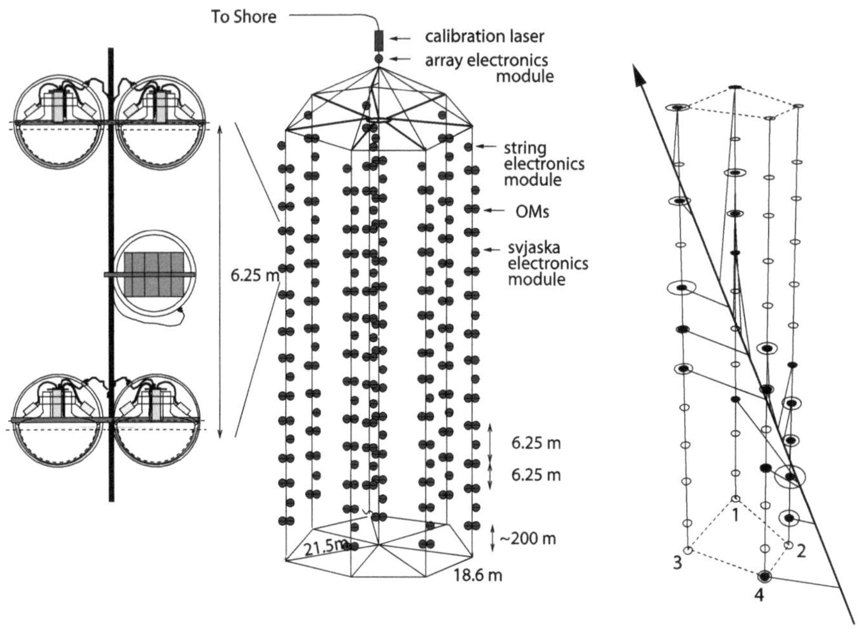


**Fig. 9.8** *Left*: The Baikal Neutrino Telescope NT200. *Right*: One of the first upward moving muons from a neutrino interaction recorded with the 4-string stage of the detector in 1996 (Balkanov et al., 1999). The Cherenkov light from the muon is recorded by 19 channels

The construction of NT200 coincided with the decay of the USSR and an economically desperate period. Members of the collaboration and even some industrial suppliers had to be supported by grants from Germany; nevertheless many highly qualified experimentalists left the collaboration and tried to survive in the private sector. Over a period of three years, a large part of the food for the winter campaigns at Lake Baikal had to be bought in Germany and transported to Siberia. Still, a nucleus of dedicated Russian physicists heroically continued to work for the project. Under these circumstances, the construction of NT200 extended over more than five years. It started with the deployment of a 3-string array (Wischnewski et al., 1993) with 36 optical modules in March/April 1993. The first two upward moving candidates were separated from the 1994 data. In 1996, a 96-OM array with four NT200 strings was operated (Balkanov et al., 1999) and provided the first textbook neutrinos like the one shown in Fig. 9.8 (right).

NT200 was completed in April 1998 and has been taking data since then. The basic components have been designed and built in Russia, notably the mechanics of the detector, the optical module, the underwater electronics and cabling. The non-Russian contributions all came from DESY: the laser time calibration system, a transputer farm in the shore station for fast data processing, an online monitoring system and a special underwater trigger tailored to register slowly moving very bright particles (as GUT monopoles), not to mention the supply of Western electronics and glass spheres.

The small spacing of modules in NT200 leads to a comparably low-energy threshold of about 15 GeV for muon detection. About 400 upward muon events were collected over 5 years. This comparatively low number reflects the notoriously large number of failures of individual channels during a year rather than what would correspond to the effective area. Still, NT200 could compete with the much larger AMANDA for a while by searching for high-energy cascades *below* NT200, surveying a volume about ten times as large as NT200 itself (Aynutdinov et al., 2006a). In order to improve pattern recognition for these studies, NT200 was fenced in 2005–2006 by three sparsely instrumented outer strings (six optical module pairs per string). This configuration is named NT200+ (Aynutdinov et al., 2006b), but suffered from several problems (from the new strings as well as from the meanwhile antiquated NT200 itself) so that no satisfying statistics and no convincing results have yet been obtained.

## 9.5 The Community Broadens

The high threshold required to get a detector working in a hostile environment such as the deep Pacific or the harsh conditions on the frozen Lake Baikal, resulted in apparently long preparatory periods of both DUMAND and Baikal. This led others to think about detectors near surface (for a review see Belotti and Laveder, 1993). The advantages seemed tempting: much easier access and a less challenging environment. Moreover, proven techniques like tracking chambers or Cherenkov techniques à la Kamiokande could be used. However, none of these projects was realized, be it by financial reasons, by the failure to convincingly demonstrate the background rejection capabilities, or since shallow lake water parameters turning out to be worse than expected.

At the same time, underground detectors moved from success to success. Remarkably, two of these successes had not been on the top priority list of the experiments: neutrino oscillations (since the trust in their existence was low in the 1980s) and neutrinos from supernova SN1987A (since Galactic or near-Galactic supernovae are rare). At the same time, neutrinos from high-energy astrophysics neutrinos did not show up. Even the data from largest detectors with about 1 000 $m^2$ area (MACRO and Super-Kamiokande) did not show any indication of an excess over atmospheric neutrinos. Seen from today, the search for sources of high-energy neutrinos with detectors of 1 000 $m^2$ or less appears to be hopeless from the beginning, with the possible exception of certain transient Galactic sources. But when these detectors were constructed, this knowledge was not common and the search for point sources appeared as a legitimate (although not priority) goal.

**Fig. 9.9** Francis Halzen (born 1944), spiritus rector of AMANDA and principal investigator of IceCube

## *9.5.1 Neutrinos in Ice?*

In this situation, a new, spectacular idea appeared on stage. In 1988, Francis Halzen from the University of Wisconsin gave a talk at the University of Kansas. At this occasion he was contacted by Ed Zeller, a Kansas glaciologist. Zeller told him about a small test array of radio antennas at the Soviet Vostok station, close to the geomagnetic South Pole. The Russians were going to test whether secondary particles generated in neutrino interactions could be detected via their radio emission. The idea that showers of charged particles would emit radio signals had been published back in 1962 by the Soviet physicist Gurgen Askaryan. Together with his colleagues Enrique Zas and Todor Stanev, Halzen realized that the threshold for this method was discouraging high (Halzen, 1995). Instead he asked himself whether the optical detection via Cherenkov light, i.e. the DUMAND principle, would be also feasible for ice. Halzen (1998) remembers:

> I suspect that others must have contemplated the same idea and given up on it. Had I not been completely ignorant about what was then known about the optical properties of ice I would probably have done the same. Instead, I sent off a flurry of E-mail messages to my friend John G. Learned, then the spokesman of DUMAND. ... Learned immediately appreciated the advantages of an Antarctic neutrino detector.

A few months later, Halzen and Learned released a paper "High energy neutrino detection in deep Polar ice" (Halzen and Learned, 1988). With respect to the light attenuation length they

> ... proceeded on the hope that a simple test will confirm the belief that is similar to the observed 25 m attenuation length for blue to mid UV light in clear water in ocean basins.

Bubble-free ice was hoped to be found at depths smaller than 1 km. Holes drilled into the ice were supposed to refreeze or, alternatively, to be filled with a non-freezing liquid.

Halzen is a theorist and Learned had its hands full with DUMAND, so both did not proceed to do an experiment. But the idea made it to Buford Price's group at University of California, Berkeley. In 1989, two young physicists of the Price group, Doug Lowder and Andrew Westphal, joined a Caltech group drilling holes in Antarctic ice and tried to measure the ice transparency using existing boreholes.

It would take, however, another year until the first successful transparency measurement of natural ice was performed – this time in Greenland. Bob Morse from the University of Wisconsin and Tim Miller (Berkeley) lowered photomultipliers into a 3 km hole drilled by glaciologists (Lowder et al., 1991).

In parallel to these first experimental steps, Buford Price, Doug Lowder and Steve Barwick (Berkeley), Bob Morse and Francis Halzen (Madison) and Alan Watson (Leeds) met at the International Cosmic Ray Conference in Adelaide and decided to propose the Antarctic Muon and Neutrino Detection Array, AMANDA.

In 1991 and 1992, the embryonic AMANDA collaboration deployed photomultipliers at various depth of the South Polar ice. Holes were drilled using a hot-water drilling technique which had been developed by glaciologists. Judging the count rate of coincidences between photomultipliers (which are due to down-going muons), the light absorption length of the ice was estimated at about 20 m and scattering effects were supposed to be negligible. It would turn out later that this was a fundamental misinterpretation of the rates. But exactly this interpretation encouraged the AMANDA physicists to go ahead with the project.

### 9.5.2 *Neutrinos in the Mediterranean Sea?*

With ongoing activities in Hawaii and at Lake Baikal and the first ideas on a telescope in polar ice, the exploration of the Mediterranean Sea as a site for an underwater neutrino telescope was natural. First site studies along a route through the Mediterranean Sea were performed in 1989 by Russian physicists who also measured the muon counting rate as a function of depth (Deneyko et al., 1991). In July 1991, a Greek/Russian collaboration led by Leonidas Resvanis from the University of Athens performed a cruise and deployed a Russian built hexagonal structure made of titanium and carrying 10 photomultipliers down to a depth of 4 100 m. The site was close to Pylos at the West coast of the Peloponnesus. They measured the vertical muon intensity and the angular distribution of down-going muons. This was the start of the NESTOR project (NESTOR: http://www.nestor.noa.gr/), which was named after the mythic king of Pylos who counseled the Greeks during the Trojan war.

The advantages of the Pylos site were obvious: The depth can be chosen down to 5 200 m, dependent on the acceptable distance to shore, deeper than at any other candidate site. This would reduce the background of mis-reconstructed downward muons faking upward muons from neutrino interactions and even allows looking above horizon. The water quality was excellent and the bio-luminescence seemed to be lower than at other Mediterranean sites.

In 1992, the collaboration included the University of Athens, the Scripps Institute for Oceanography in San Diego (USA), the Universities of Florence (Italy), Hawaii, Wisconsin (USA), Kiel (Germany) and the Institute for Nuclear Research Moscow. Naturally, more Greek institutes joined. French institutes joined and left again to pursue their own project ANTARES. More Italian institutes joined but later also decided to follow their own project, NEMO, close to Sicily. LBNL Berkeley provided

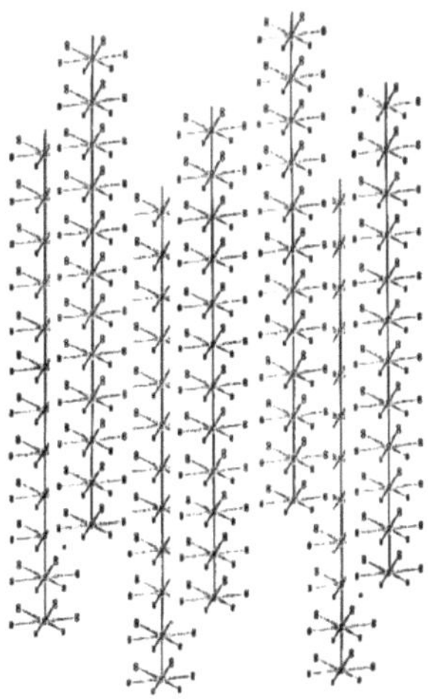

**Fig. 9.10** The original NESTOR concept of a large array of towers

essential electronics for the first stationary hexagonal floor which was deployed in 2004.

The results of the early cruises and the concept for the NESTOR detector were developed and presented during a series of Workshops in Pylos. NESTOR was conceived to consist of a hexagon of hexagonal towers covering an area of about $10^5$ m$^2$ as shown in Fig. 9.10.

A single tower should carry 168 PMTs on 12 hexagonal floors, vertically spaced by 20–30 m, each with six omni-directional modules at the end of 16 m arms and one in the center (Resvanis et al., 1994). The "omni-directional module" contained two 15-inch PMTs each in a 17-inch glass pressure sphere.

The philosophy of NESTOR was to be not only sensitive to high-energy neutrinos (therefore the large area covered by seven towers) but also to study atmospheric neutrino oscillations (hence the 5 GeV threshold inside the geometrical volume of a tower and the omni-directionality of the modules).

## 9.6 The Three-String Race and the Termination of DUMAND

In 1993 and 1994, three collaborations were going to deploy detectors with three or more strings. Three strings are the minimum to achieve full spatial reconstruction.

The DUMAND collaboration was working towards installation of the first three of the nine DUMAND-II strings. Two of these strings were to be equipped with "Japanese Optical Modules" (JOMs) containing a 15-inch PMT "R2018" from Hamamatsu, and one string with "European Optical Module" (EOMs) containing the hybrid XP2600 from Philips. This stage was christened TRIAD.

In 1992, the AMANDA collaboration was joined by the Swedish collaborators from Stockholm and Uppsala. Steve Barwick, meanwhile at University of California in Irvine, designed a four-string detector with 80 PMTs which was going to be deployed between 800 and 1 000 m depth. At Wisconsin, computer-controlled hot-water drills were developed, in close collaboration with the Polar Ice Coring Office (PICO).

However, the first three-string array (NT36) was deployed at Lake Baikal in March/April 1993. It consisted of only 18 PMT-pairs at three strings of meager 40 m length. But it served its purpose to demonstrate that 3-dimensional reconstruction of muon tracks works as expected. John Learned from DUMAND recognized the importance of the Baikal achievement and sent an E-mail to the author: *Congratulations for winning the 3-string race!*. Actually the first two neutrino candidates were isolated from the data taken with the same array in 1994.

Meanwhile, in December 1993, a first string of the TRIAD, together wit a string of environmental instruments was deployed and linked to shore via a junction box placed on the ocean bottom and a shore cable which had been laid some months earlier. However, some pressure housings developed leaks. A short circuit in the junction box (the central component for communication to shore) did not clear due to a fuse failure, and soon the communication to shore failed.

The DUMAND progress had been slow, but had shown remarkable progress compared to ocean research at that time. This impressed oceanographers but not the main funding organization, the Department of Energy (DOE), which was not used to a "try-and-try-again" mode of progress. Review committees without any ocean expertise judged the project, following criteria typical for accelerator research. Moreover, the termination of the Superconducting Super Collider (SSC) by the US congress in 1993 created a strong risk aversion in DOE. On the technical side, the reasons of the 1993 DUMAND failures had been identified and a redeployment was in preparation. But in 1995, the mentioned circumstances regrettably led to a termination of the support for DUMAND. From now on the goal to begin high-energy neutrino astronomy was carried forward at the South Pole, in the Mediterranean Sea and in Lake Baikal.

## 9.7 AMANDA

AMANDA is located some hundred meters from the Amundsen–Scott station. Holes 60 cm in diameter were drilled with pressurized hot water; strings with optical modules were deployed in the water which subsequently refreezes. Installation operations at the South Pole are performed in the Antarctic summer, November to February. For the rest of the time, two operators (of a winter-over crew of 25–40 persons in total) maintained the detector, connected to the outside world via satellite communication.

The first AMANDA array with 80 optical modules on four strings was deployed in the austral summer 1993/1994, at depths between 800 and 1 000 m (Askebjer

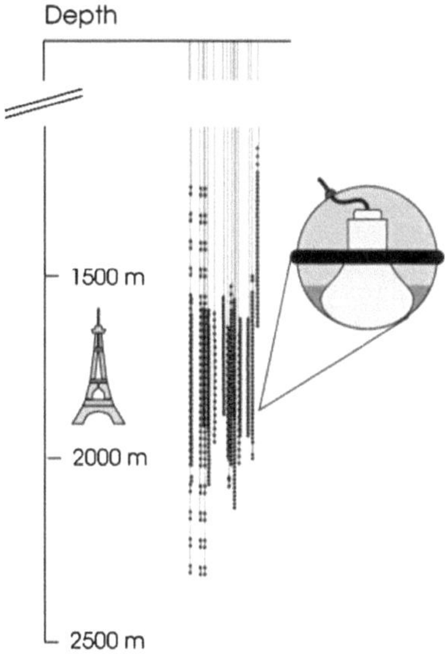


**Fig. 9.11** The AMANDA configuration. The detector consisted of 677 optical modules at 19 strings. Three of the strings have been sparsely equipped towards larger and smaller depth in order to explore ice properties, one string got stuck during deployment at too shallow depth and was not used in analyses. The Eiffel tower is shown to scale for size comparison

et al., 1995). Surprisingly, light pulses sent from on string to a neighbored string over 20 m distance, did not arrive after the expected 100 ns, but were considerably delayed. The surprise was resolved at the 1994 Venice Workshop on Neutrino Telescopes. Here, Grigorij Domogatsky informed Francis Halzen about results from an ice core extracted at the geomagnetic South Pole where the Russian Vostok station is located. The data proved that air bubbles which remain from the original firn ice at the surface did not yet disappear at 1 km depth. The delay was due to light scattering at the bubbles. Light would not travel straight but via random walk and become nearly isotropic after a few times a distance called effective scattering length. The effective scattering length was found to be between 40 cm at 830 m depth and 80 cm at 970 m. The scattering by air bubbles trapped in the ice made track reconstruction impossible.

A great story might have been over before it really got started. AMANDA seemed to be "nothing than a big calorimeter" – as Leonidas Resvanis sarcastically formulated – without any real tracking capabilities. This could have been the point to give up the project. Nevertheless, our group from DESY joined. We were encouraged by a trend seen in the AMANDA data themselves, as well as by ice core data taken at the Russian Vostok station: below 1 300 meters bubbles should disappear.

This expectation was confirmed with a second 4-string array which was deployed in 1995/1996. The remaining scattering, averaged over 1 500–2 000 m depth, corresponds to an effective scattering length of about 20 m and is assumed to be due to dust. This is still considerably worse than for water but sufficient for track reconstruction (Ackermann et al., 2006). The proof that is indeed *was* sufficient took some time, as well as the development of the suitable reconstruction methods and selection criteria (Ahrens et al., 2004a). The array was upgraded stepwise until January 2000 and eventually comprised 19 strings with a total of 677 optical modules, most

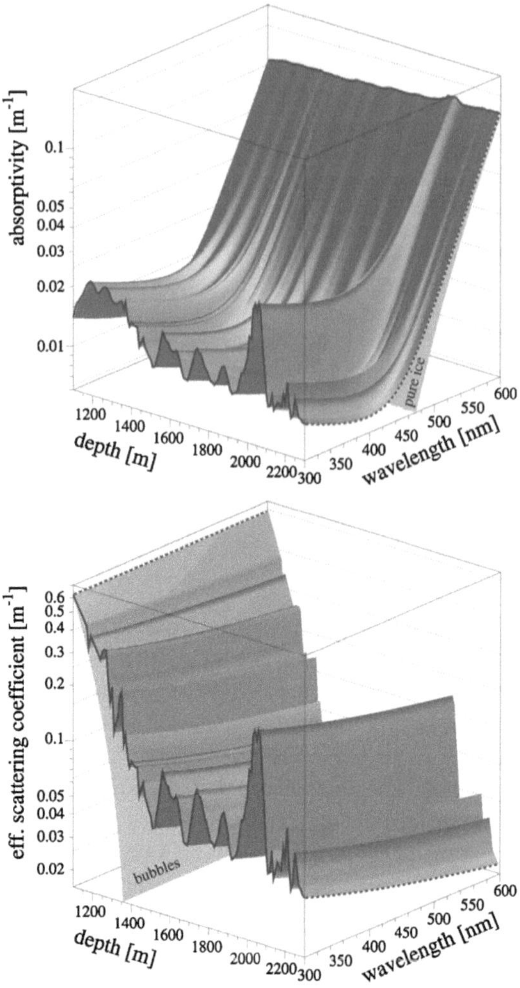


**Fig. 9.12** Absorption coefficient (*top*) and scattering coefficient (*bottom*) in the South Polar ice as functions of depth and wavelength (Ackermann et al., 2006)

of them at depths between 1 500 and 2 000 m. Figure 9.11 shows the AMANDA configuration.

In Fig. 9.12, absorption and scattering coefficients are shown as functions of depth and wavelength (Ackermann et al., 2006). The variations with depth are due to bubbles at shallow depth leading to very strong scattering and, at larger depths, to dust and other material transported to Antarctica during varying climate epochs. The quality of the ice improves substantially below a major dust layer at a depth of about 2 000–2 100 m, with a scattering length about twice as large as for the region above 2 000 m. The depth dependence of the optical properties complicates the analysis of the experimental data. Furthermore, the large delays in photon propagation due to the strong scattering cause worse angular resolution of deep-ice detectors compared to water. On the other hand, the large absorption length, with a cut-off below 300 nm instead of 350–400 nm in water, results in better photon collection.

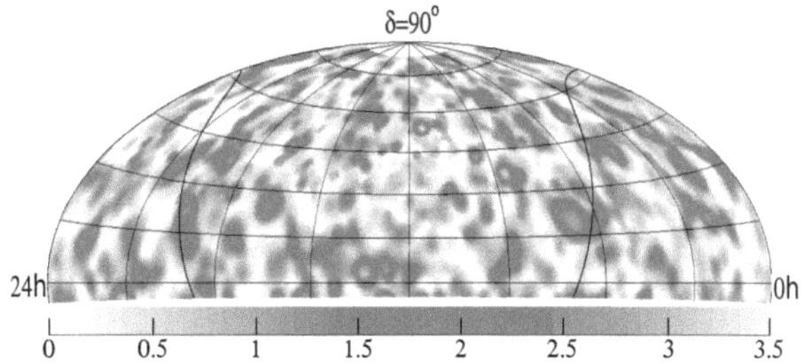


**Fig. 9.13** 7-year significance map of the Northern hemisphere derived AMANDA (Abbasi et al., 2009)

A big advantage compared to underwater detectors is the small photomultiplier noise rate, about 0.5 kHz in an 8-inch tube, compared to 20–40 kHz due to $K^{40}$ decays and bio-luminescence in lakes and oceans. The contamination of hit patterns from particle interactions with noise hits is thus small and makes hit selection much easier than in water and allows for identifying burst-like low-energy events from supernovae (see Sect. 9.9).

The angular resolution of AMANDA for muon tracks is 2°–2.5°, with a lower energy threshold around 50 GeV. Although better than for Lake Baikal (3°–4°), it is much worse than for ANTARES (<0.5°, see below). This is the result of the strong light scattering which deteriorates the original information contained in the Cherenkov cone.

AMANDA was switched off in April 2009, after more than 9 years of data taking in its full configuration. Figure 9.13 shows the skyplot derived from of 6 959 neutrinos taken in the years 2000–2007. None of the spots is statistically significant, therefore only upper limits could be derived.

AMANDA provided record limits on fluxes for cosmic neutrinos, be it for diffuse fluxes, for point sources or for transient sources like gamma-ray bursts. These limits rules out the first models on neutrino production in cosmic sources. AMANDA extended the measured spectrum of atmospheric neutrinos by nearly two orders of magnitude, from a few TeV to 200 TeV (Fig. 9.14). It also established record limits on indirect dark matter search, on the flux of magnetic monopoles, and on effects violating Lorentz invariance. It would have detected neutrinos from a supernova burst in our Galaxy (if such a burst would have appeared!), and it provided results on the spectrum and composition of cosmic rays.

What was hoped for in optimistic dreams – the discovery of an extraterrestrial source of neutrinos – did not happen. But there was one moment when the adrenaline level of some of us went up and we thought we were close to a discovery. While analyzing in 2005 the data taken from 2000–2003, five events where identified from the direction of the Active Galaxy 1ES1959+650. Interestingly, three of these came within 66 days in 2002 (Ackermann, 2006). Two of the three neutrinos were coinciding within about a day with gamma-ray flares observed by the gamma-ray telescopes HEGRA and Whipple – see Fig. 9.15. Excitingly, one of these two flares was not accompanied by an X-ray flare, a so-called "orphan flare", which one would expect for a hadron flare where the X-ray flux accompanying electron flare is absent. This result was quickly followed by two theoretical papers, one claiming that the

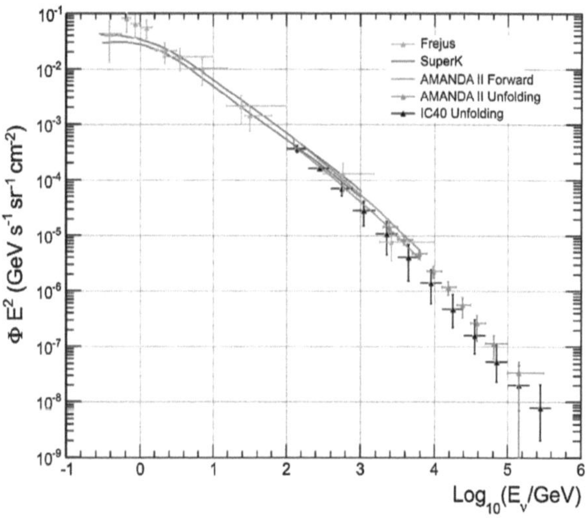


**Fig. 9.14** Energy spectrum of atmospheric neutrinos (Abbasi et al., 2011a)

corresponding neutrino flux would not fit any reasonable assumption on the energetics of the source (Reimer et al., 2005), the other claiming that scenarios yielding such fluxes were conceivable (Halzen and Hooper, 2005). Since the analysis was not a fully blind analysis, it turned out to be impossible to determine chance probabilities for this event, and actually the result was never published in a journal. However, it initiated considerations to send alerts to gamma-ray telescopes in case time-clustered events from a certain direction would appear. Such a "Target-of-Opportunity" alert is currently operating between IceCube and the gamma-ray telescopes MAGIC (La Palma) and VERITAS (Arizona).

## 9.8 Mediterranean Sea: NESTOR, ANTARES, NEMO

Whereas AMANDA and Baikal developed coherently, the Mediterranean community split in three branches which are related to three different locations and host countries.

The NESTOR collaboration presented the full concept of a "tower" in 1993 (Resvanis et al., 1994). After a long phase of tests and developments, a cable was installed to a site at 4 km depth. In 2004, a single prototype floor was deployed, connected and operated for about one month (Aggouras et al., 2005a). Then, its operation had to be terminated due to a failure of the cable to shore. However, the data taken with this prototype demonstrated the detector functionality and provided a measurement of the atmospheric muon flux (Aggouras et al., 2005b).

Comparing the original plan to deploy an array of towers within a decade (see Fig. 9.10) with the reality of a single floor operated over just a month, demonstrates the enormous challenges which these projects face. The deep-sea medium is hostile and unforgiving, the iterative approach is not what funding agencies like. A professional management and a coherent collaboration are necessary for any large

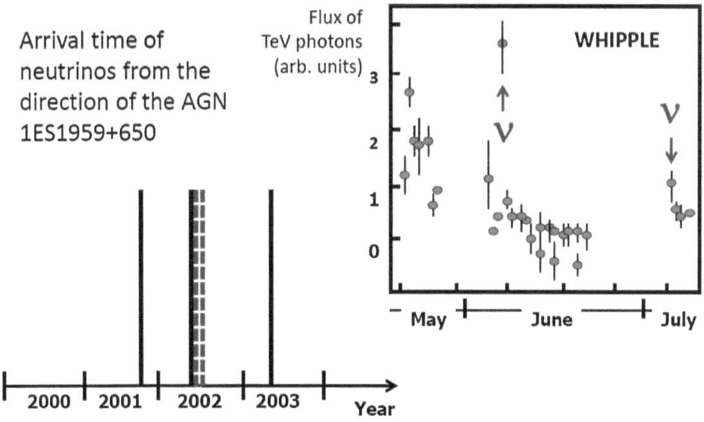


**Fig. 9.15** The "curious" coincidence of neutrino events from the direction of an AGN with gamma flares from the same source. The second and third of the three events recorded in 2002 (*dashed*) coincide within about one day with peaks seen by WHIPPLE

projects, and if problems from this corner add to the inherent problems of deep underwater projects, delays or even failure are inevitable.

Currently NESTOR is part of the KM3NeT framework which is directed towards a multi-cubic kilometer detector in the Mediterranean Sea.

French collaborators who temporarily had been members of NESTOR, pursued an independent strategy from the mid-1990s. Together with collaborators from Italy and the Netherlands they presented a full proposal for a 12-string detector in 1999 (Aslanides et al., 1999). In 2001 also a German group from Univ. Erlangen joined the experiment. ANTARES stands for Astronomy with a Neutrino Telescope and Abyss environmental RESearch (ANTARES: http://antares.in2p3.fr). This proposal was based on the operation of a demonstrator string (Blondeau, 1998; Feinstein, 1999) as well as on the results of extensive site exploration campaigns in the region off Toulon at the French Mediterranean coast, indicating that the optical background (Amram et al., 2000) as well as sedimentation and biofouling (Amram et al., 2003) are acceptable at that site. However, taken all together (depth, optical clarity, optical background and sedimentation) the site is inferior to the Greek and Italian sites.

The construction of ANTARES started in 2002 with the deployment of a shore cable and a junction box, the central element connecting the shore cable to the detector. In 2002/2003, a preproduction string was deployed and operated for a few months. Several technical problems were identified that required further studies, design modifications and the operation of a mechanical test string (Ageron et al., 2007). The detector in its final 12-string configuration was installed in 2006–2008 and has been operational since then, with a break of a few months in 2009 due to a failure of the main cable that required repair.

ANTARES consists of 12 strings, each carrying 25 "storeys" equipped with three optical modules, an electronics container and calibration devices. The optical module consists of a 17-inch glass sphere housing a hemispherical 10-inch photomul-

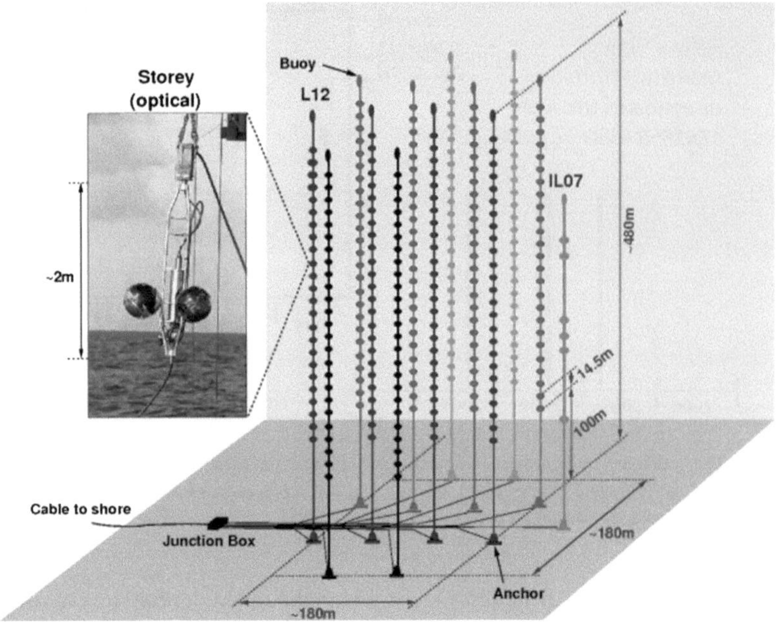


**Fig. 9.16** Schematic of the ANTARES detector. Indicated are the 12 strings and the instrumentation line in its 2007 configuration (IL07). Shown as an inset is the photograph of a story carrying three photomultipliers

tiplier (Hamamatsu R7081-20). A further string, the "instrumentation line", carries devices for environmental monitoring. The depth at the ANTARES site is 2475 m. The schematic setup is shown in Fig. 9.16, a detailed technical description can be found in (Ageron et al., 2011).

An almost background-free separation of neutrino-induced upward-going muons from the huge background of downward-going muons is the central requirement for an underwater or under-ice telescope. Baikal and AMANDA, followed by ANTARES, have mastered this challenge, even more so IceCube. Figure 9.17 (top) shows the rate of muons as a function of the zenith angle $\theta$ as measured with ANTARES. Below the horizon ($\theta < 0$) the rate is well described by the expectation for atmospheric neutrinos, above the horizon by that for atmospheric muons. Figure 9.17 (bottom) shows the 1-year skymap of neutrino candidates – as to be expected without a clear source signal.

The newest Mediterranean project is NEMO (NEutrino Mediterranean Observatory) (NEMO: http://nemoweb.lns.infn.it). It was launched in 1998 after Italian groups left NESTOR. From the beginning, their objective was to study the feasibility of a cubic kilometer detector, to develop corresponding technologies and to identify and explore a suitable site, in their case close to Sicily rather than to build a separate detector of medium size.

The basic unit of NEMO are towers composed by a sequence of floors. Different to NESTOR, floors consist of horizontal "bars", originally foreseen to be 15 m

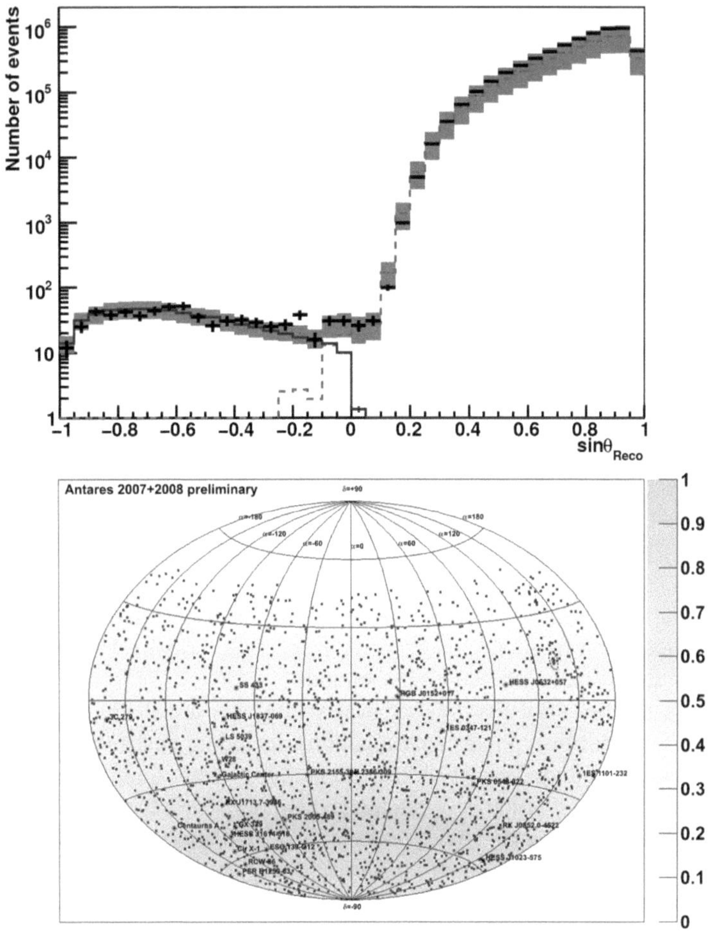


**Fig. 9.17** *Top*: Number of reconstructed muons in the 2008 ANTARES data, as a function of the reconstructed zenith angle (*horizontal bars*). Also indicated are the simulation results for atmospheric muons (*dashed*), and muons induced by atmospheric neutrinos (*full line*). The *shaded band* indicates the systematic uncertainties. Figure taken from Aguilar et al. (2011). *Bottom*: Equatorial skymap of neutrino-induced muon events from 295 days of ANTARES data in 2007/2008. The *background color scale* indicates the sky visibility in percent of the time. The most significant accumulation of events, marked with a *circle*, is fully compatible with the background expectation (Eberl, 2011)

long and each equipped with four 10-inch PMs. The floors are tilted against each other and form a three-dimensional structure (Capone et al., 2009). A tower can be folded together and deployed to the sea floor as a compact object that is subsequently unfurled. Contrary to single strings and similarly to the NESTOR concept, the 3-dimensional arrangement of photomultipliers per tower allows for local reconstruction of muon directions.

A suitable site at a depth of 3.5 km, about 100 km off Capo Passero on the South-Eastern coast of Sicily has been identified and investigated during various campaigns. During the first prototyping phase, a cable to a test site near Catania at a depth of 2 km was installed and equipped with a junction box. In 2007, a "mini-tower" with 4 bars was deployed, connected and operated for several weeks. Although the data taking period was limited to a few months due to technical problems, the mini-tower provided the proof of concept for the technologies and most of the components employed.

The setup of a second phase (Taiuti et al., 2011) includes shore infrastructure at Capo Passero and a 100 km long cable to the site at 3.5 km depth; both are currently in place. A remotely operated vehicle (ROV) is available for the deep-sea operations. A mechanical test tower of limited size was successfully deployed and unfurled in early 2010. The plans to deploy a full-size prototype tower will be pursued in the KM3NeT framework.

## 9.9 IceCube

With IceCube (IceCube: http://www.icecube.wisc.edu/), the idea of a cubic-kilometer detector was finally realized (Ahrens et al., 2004b). However, the way towards the first installation was all but smooth.

Actually, the first initiative beyond AMANDA was a concept called *DeepIce*, a proposal for multi-disciplinary investigations, including neutrino and cosmic ray astrophysics, glaciology, glacial biology, seismology and climate research. As an example, we note the relation between the layered impurities from dust and climatic effects or volcano eruptions (Ackermann et al., 2006). concluded that a neutrino detector was sold under the flag of multi-disciplinary research, (mis)using the NSF funding model for multi-disciplinary centers. The advice was to go ahead with a dedicated project for a neutrino telescope.

As a consequence, already in November of the same year a first 67-page IceCube proposal was submitted to NSF. It was signed essentially by the collaborators of the old AMANDA collaboration. Soon, a number of additional institutions became interested and a new collaboration was formed, the IceCube collaboration, which meanwhile has grown to more than 30 institutions. Paradoxically, the two collaborations co-existed until 2005, then joining to one collaboration, IceCube.

For IceCube construction, the thermal power of the hot-water drill factory was upgraded to 5 MW, compared to 2 MW for AMANDA. This reduced the average time to drill a 2 450 m deep hole to 35 hours. The commissioning of the drill during the first deployment season 2004/2005 turned out to be extremely challenging, but eventually a first, single string was deployed in January 2005: The first step was made! The following seasons resulted in 8, 13, 18, 19, 20 and 7 strings, respectively. The last of 86 strings was deployed at Dec. 18, 2010.

IceCube consists of 5 160 digital optical modules (DOMs) installed on 86 strings at depths of 1 450 to 2 450 m. A string carries 60 DOMs with 10-inch photomultipliers Hamamatsu R7081-02 housed in a 13-inch glass sphere. Signals are digitized in

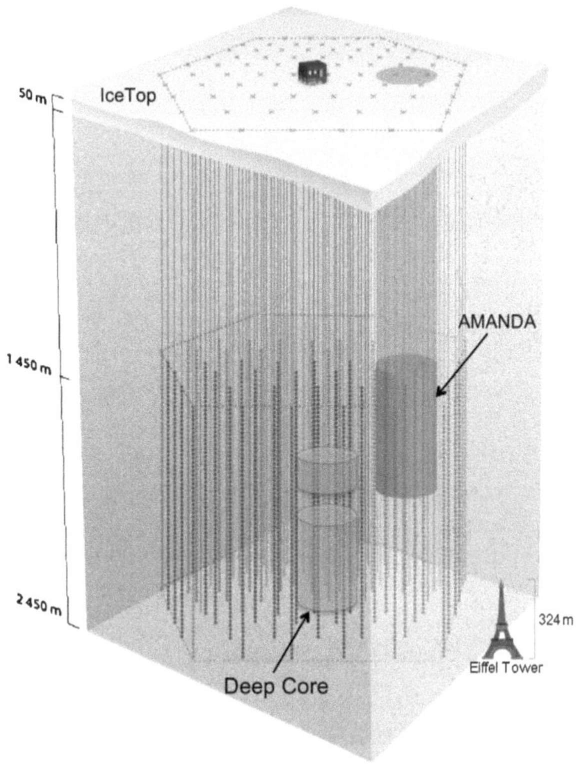


**Fig. 9.18** Schematic view of the IceCube neutrino observatory. AMANDA was replaced by DeepCore, a nested low-threshold array. At the surface, the air shower array IceTop and the IceCube counting house are indicated

the DOM and sent to shore via copper cables. 320 further DOMs are installed in IceTop, an array of detector stations on the ice surface directly above the strings (see Fig. 9.18). AMANDA, initially running as a low-energy sub-detector of IceCube, was decommissioned in 2009 and replaced by DeepCore, a high-density sub-array of six strings at large depths (i.e. in the best ice layer) at the center of IceCube. DeepCore collects photons with about six times the efficiency of full IceCube, due to its smaller spacing, the better ice quality and the higher quantum efficiency of new PMTs. Together with the veto provided by IceCube, this results in an expected threshold of about 10 GeV. This opens a new window for oscillation physics and indirect dark matter search.

The muon angular resolution achieved by present reconstruction algorithms is about 1° for 1 TeV muons and below 0.5° for energies above 10 TeV. Unlike underwater detectors with their environment of high optical noise, IceCube can be operated in a mode that is only possible in ice: The detection of burst neutrinos from supernovae. The low dark-count rate of the PMTs allows for detection of the feeble increase of the summed count rates of all PMTs during several seconds, which would be produced by millions of interactions of few-MeV neutrinos from a supernova burst (Abbasi et al., 2011b). IceCube records the counting rate of all PMTs in millisecond steps. A supernova in the center of the Galaxy would be detected with extremely high significance and the onset of the pulse could be measured in

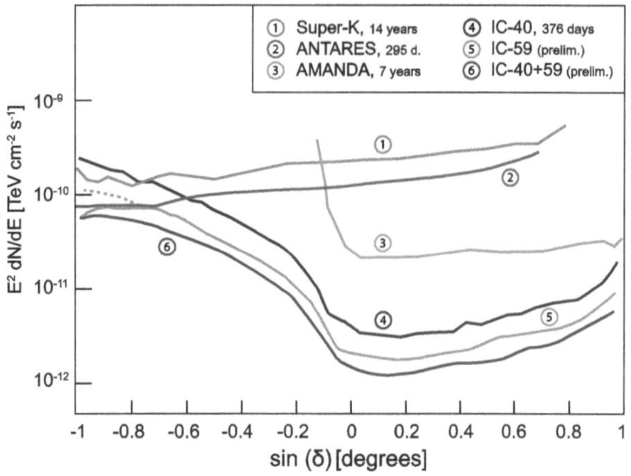


**Fig. 9.19** Point-source limits from various experiments. See Katz and Spiering (2012) for references.

unprecedented detail. Even a SN 1987A-type supernova in the Large Magellanic Cloud would provide a recognizable signal and be sufficient to provide a trigger to the SuperNova Early Warning System, SNEWS (Antonioli et al., 2004).

## 9.10 Where Do We Stand?

With IceCube, the sensitivity to point sources and to diffuse fluxes has been improved by nearly a factor of thousand when compared to the situation of the mid-1990s. But alas – no indication for extraterrestrial sources has found yet, but only ever tightening upper limits on fluxes have been established.

Figure 9.19 compiles the limits from previous experiments, from the different IceCube stages and from ANTARES. Note that the combined data of IceCube-40 and IceCube-59 surpass the mark of 1 km$^3$ $\times$ 1 year and thus exceed 1 year worth of data from the full IceCube detector. When this article is printed, a factor of 1 000 improvement of the sensitivity to point sources will have been reached when compared to the very first AMANDA point-source paper from the year 2000 (Andres et al., 2001).

Point-source searches use the directional and energy information to reduce the background from atmospheric neutrinos. Cosmic neutrinos from a given source would cluster around the source direction.

If the extraterrestrial signal is not concentrated to individual strong sources but distributed over all the sky, the signal has to be identified in diffuse fluxes. Searches for diffuse fluxes can only use the measured energy as criterion for separating cosmic and atmospheric neutrinos, searching for an excess at high energies. For these studies not only muon neutrino interactions with a muon in the final state are suitable. Since the directional information is not of prime importance, one can also study events without a long track but only a particle cascade in the final state. Such events emerge from electron- and tau-neutrino interactions and from all neutral current interactions. Due to neutrino oscillations, two third of the extraterrestrial neutrinos arrive as $\nu_e$ or $\nu_\tau$. No significant excesses over atmospheric neutrinos or other

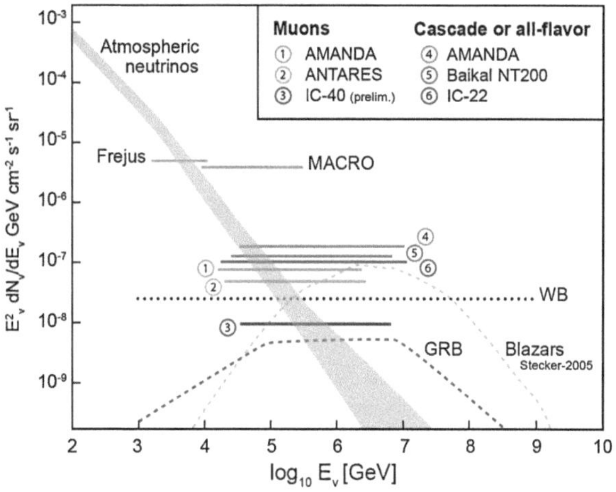


**Fig. 9.20** 90 % C.L. integral upper limits on the diffuse flux of extraterrestrial neutrinos. The *horizontal lines* extend over the energy range which would cover 90 % of the detected events from an $E^{-2}$ source (5 % would be below and 5 % above the range). All model predictions have been normalized to one flavor, i.e. all of the all-flavor limits have been divided by 3. The *colored band* indicates the measured flux of atmospheric neutrinos (see also Fig. 9.14), the broadening at higher energies reflects the uncertainties for prompt neutrinos. The limits on muon neutrinos are from 807 days AMANDA 334 days ANTARES and 375 days IceCube-40. Cascade/all flavor limits are from 807 days AMANDA 1038 days Baikal-NT200 257 days IceCube-22. See Katz and Spiering (2012) for references. Also indicated is the Waxman–Bahcall (WB) bound (Waxman and Bahcall, 1999)

kinds of background has been observed so far, resulting in upper limits on the diffuse flux of extraterrestrial high-energy neutrinos. Figure 9.20 summarizes the limits obtained in the TeV–PeV region. For each experiment and each method only the best limit is shown. Remarkably, from the first limit derived from the underground experiment Frejus to the 2010 IceCube-40 limit, a factor of 500 improvement has been achieved. Several models such as e.g. the blazar model of Stecker (2005) shown in the figure can be excluded. A further factor of 10 improvement is expected over the next 2–3 years, using the full IceCube detector and combining muon and cascade information. The expected sensitivity is more than an order of magnitude below a theoretical upper bound of Waxman and Bahcall (1999), and prompt atmospheric neutrinos will be detectable for all but the lowest predictions (Kowalski, 2005).

Five decades after the first conceptual ideas, and three decades after first practical attempts to build high-energy neutrino telescopes, we may be close to a turning point. IceCube, has started data taking in its full cubic-kilometer configuration but has not yet detected an extraterrestrial neutrino signal.

The strong case for high-energy neutrino astronomy has remained unchanged over time, but the requirements on the necessary sensitivity have tightened continuously. Whereas underground detectors on the kiloton mass scale (or on the $10^3$ m$^2$ fmuon area scale) seemed sufficient in the 1960s, predictions from the 1970s and

1980s already favored scales of $10^5$–$10^6$ $m^2$. Actually, DUMAND was conceived as a cubic kilometer configuration in 1978. On the other hand, underground detectors like MACRO or Super-Kamiokande were still given a certain potential for high-energy neutrino astronomy. Therefore, it is no surprise that, in spite of their declared goal of the kilometer scale, also the underwater/ice community did hope for early discoveries with NT200, AMANDA, NESTOR and ANTARES. This hope turned out to be illusory. Neither did the observations of GeV and TeV gamma rays in the last two decades support higher flux expectations nor has any of these detectors saw a signal indication with more than $3\sigma$ significance. Therefore the detection of first extraterrestrial high-energy neutrinos sources lies still ahead. With some optimism, we may expect it within the next few years. Galactic "Pevatrons" as those observed in gamma rays by the Milagro detector are within reach after a few years of IceCube data taking if the corresponding predictions are correct. Models assigning the most energetic cosmic rays to gamma-ray bursts are challenged by recent IceCube data (Abbasi et al., 2012) and will be more strongly scrutinized within a couple of years. However, clear detections are all but guaranteed.

## 9.11 What Next?

Whereas the identification of first extraterrestrial neutrinos IceCube has not yet been achieved, projects of similar or greater size on the Northern hemisphere are under preparation. In 2002, an expert committee installed by the International Union of Pure and Applied Physics (IUPAP) concluded (HENAP, 2002) that "a $km^3$-scale detector in the Northern hemisphere should be built to complement the IceCube detector being constructed at the South Pole". Following this recommendation, the Mediterranean neutrino telescope groups have formed the KM3NeT collaboration to prepare, construct and operate such a device. A design study from 2006 to 2009 resulted in a Conceptual Design Report (CDR) (Bagley et al., 2008) and a Technical Design Report (TDR) (Bagley et al., 2010). At present, the project is in a Preparatory Phase and envisages to install a detector with 6 $km^3$ volume from 2014 on. The total investment cost is estimated to be around 225 MEuro. A top view of a possible detector configuration consisting of two blocks, each 3 $km^3$, is sketched in Fig. 9.21. In Russia, the Baikal Collaboration plans the stepwise installation of a kilometer-scale array in Lake Baikal, the Gigaton Volume Detector, GVD (Aynutdinov et al., 2009). Realizing that the presently planned size of half a cubic kilometer is no longer enough, a three times larger array is presently being studied, as sketched at the bottom of Fig. 9.21. Note that due to the shallower depth of Lake Baikal, the hight of GVD will be shorter than for IceCube and KM3NeT.

The realization of these projects depends on several factors. First of all, IceCube results will play a strong role. Secondly, future gamma-ray data must provide stronger indications that the observed gamma rays are pion-decay counterparts of neutrinos and not only the result of inverse Compton scattering. Last but not least, the considerable funding must be found.

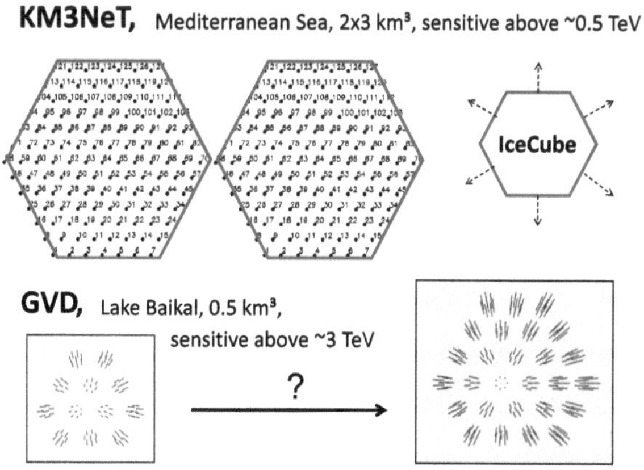


**Fig. 9.21** Top views of planned new detectors at the Northern hemisphere (KM3NeT and GVD). They are compared to the top view of IceCube. *Arrows* symbolize the possibility of an IceCube extension in case of discovery of extraterrestrial neutrinos

Missing or marginal evidence for sources from IceCube may have various consequences. If one is going to continue the venue of detectors which explore the energy range most characteristic for GRBs and AGNs, one has to envisage an order-of-magnitude step in sensitivity, i.e. beyond what is presently scheduled by KM3NeT and GVD.

The second option would be an even larger leap in size. It would address energies above 100 PeV with the help of new technologies like radio or acoustic detection and envisage 100–1 000 cubic kilometers of instrumented volume. This option might still have sensitivity to neutrinos from AGN jets but would also cover well the energy range of neutrinos from cosmic ray interactions with the 3-Kelvin microwave background. In contrast to optical detectors, new-technology detectors are still in the R&D phase and also have no natural calibration source like atmospheric neutrinos for optical detectors.

The third option would define, at least for the time being, an end to the search for neutrinos from cosmic accelerators. It would focus on optical detection with small spacing optimized to investigate oscillations with accelerator neutrinos (Mediterranean Sea) and atmospheric neutrinos, or, even more pretentious, to study supernova bursts beyond our own Galaxy or even proton decay.

Taken all together, we may be close to a turning point. We have made a factor-of-thousand step in sensitivity compared to a dozen years ago. This is far more than the traditional factor of ten which so often led to the discovery of new phenomena (Harwit, 1981). For instance, looking across our own field, the prospects for discovery had not been estimated too highly before launching the first X-ray rocket in 1962, or before detecting the Crab Nebula in TeV gamma rays in 1989. History has told another story, as we know today. The same may be the case for high-energy neutrino astronomy. The journey is not yet finished!

## References

Abbasi, R., et al. (IceCube Coll.): Phys. Rev. **D79**, 062001 (2009). arXiv:0809.1646

Abbasi, R., et al. (IceCube Coll.): Phys. Rev. **D83**, 012001 (2011a). arXiv:1010.3980
Abbasi, R., et al. (IceCube Coll.): Astron. Astrophys. **535**, A109 (2011b). arXiv.org/abs/1108.0171
Abbasi, A., et al. (IceCube Coll.): Neutrinos challenge gamma ray burst origin of cosmic rays. Nature **484**, 351 (2012)
Achar, C., et al.: Phys. Lett. **18**, 196 (1965)
Ackermann, M.: Searches for signals from cosmic point-like sources of high energy neutrinos in 5 years of AMANDA-II data. Ph.D. thesis, Humboldt University Berlin (2006)
Ackermann, M., et al. (AMANDA Coll.): J. Geophys. Res. **111**, D13203 (2006)
Ageron, M., et al. (ANTARES Coll.): Nucl. Instrum. Methods A **581**, 695 (2007)
Ageron, M., et al. (ANTARES Coll.): Nucl. Instrum. Methods A **656**, 11 (2011). arXiv:1104.1607
Aggouras, G., et al. (NESTOR Coll.): Nucl. Instrum. Methods A **552**, 420 (2005a)
Aggouras, G., et al. (NESTOR Coll.): Astropart. Phys. **23**, 377 (2005b)
Aguilar, J.A., et al. (ANTARES Coll.): Astropart. Phys. **34**, 652 (2011)
Ahrens, J., et al. (AMANDA Coll.): Nucl. Instrum. Methods A **524**, 169 (2004a). arXiv:astro-ph/0407044
Ahrens, J., et al. (IceCube Coll.): Astropart. Phys. **20**, 507 (2004b). arXiv:astro-ph/0305196
Amram, P., et al. (ANTARES Coll.): Astropart. Phys. **13**, 127 (2000)
Amram, P., et al. (ANTARES Coll.): Astropart. Phys. **19**, 253 (2003). arXiv:astro-ph/0206454
Andres, E., et al. (AMANDA Coll.): Nucl. Phys. Proc. Suppl. **91**, 423 (2001). arXiv:astro-ph/0009242
ANTARES homepage: http://antares.in2p3.fr
Antonioli, P., et al.: New J. Phys. **6**, 114 (2004). arXiv:astro-ph/0406214
Askebjer, P., et al. (AMANDA Coll.): Science **267**, 1147 (1995)
Aslanides, E., et al. (ANTARES Coll.): A deep sea telescope for high-energy neutrinos. arXiv:astro-ph/9907432 (1999)
Aynutdinov, V., et al. (Baikal Coll.): Astropart. Phys. **25**, 140 (2006a). arXiv:astro-ph/0508675
Aynutdinov, V., et al. (Baikal Coll.): Nucl. Instrum. Methods A **567**, 433 (2006b). arXiv:astro-ph/0609743
Aynutdinov, V., et al. (Baikal Coll.): Nucl. Instrum. Methods A **602**, 227 (2009). arXiv:0811.1110
Babson, E., et al. (DUMAND Coll.): Phys. Rev. D **42**, 3613 (1990)
Bagduev, R., et al. (Baikal Coll.): Nucl. Instrum. Methods A **420**, 138 (1999). arXiv:astro-ph/9903347
Bagley, P., et al. (KM3NeT Coll.): Conceptual design report (2008). isbn:978-90-6488-031-5. Available from: www.km3net.org
Bagley, P., et al. (KM3NeT Coll.): Technical design report (2010). isbn:978-90-6488-033-9. Available from: www.km3net.org
Balkanov, R.V., et al. (Baikal Coll.): Astropart. Phys. **12**, 75 (1999). arXiv:astro-ph/9705244
Belotti, E., Laveder, M.: In: Proc. 5th Int. Workshop on Neutrino Telescopes, Venice, p. 275 (1993)
Berezinsky, V.: In: Proc. Int. Workshop on Neutrino Telescopes, Venice, p. 125 (1990)
Bezrukov, L.B., et al. (Baikal Coll.): In: Proc. XI. Conf. on Neutrino Physics and Astrophysics, Nordkirchen, Germany, p. 550 (1984)
Bezrukov, L.B., et al.: In: Proc. 2nd Int. Symp. Underground Physics-87, Baksan Valley, USSR, p. 230 (1987)
Blondeau, F. (for the ANTARES Coll.): Prog. Part. Nucl. Phys. **40**, 413 (1998)
Bosetti, P., et al. (DUMAND Coll.): DUMAND II: Proposal to construct a deep-ocean laboratory for the study of high energy neutrino astrophysics and particle physics. Tech. Rep. HDC-2-88, Hawaii DUMAND Center. University of Hawaii (1988)
Capone, A., et al. (NEMO Coll.): Nucl. Instrum. Methods A **602**, 47 (2009)
Deneyko, A.O., et al.: In: Proc. 3rd Int. Workshop on Neutrino Telescopes, Venice, p. 407 (1991)
Domogatsky, G.V., et al.: In: Proc. XII. Conf. on Neutrino Physics and Astrophysics, Sendai, Japan, p. 737 (1986)
Eberl, T. (for the ANTARES Coll.): Prog. Part. Nucl. Phys. **66**, 457 (2011)
Feinstein, F. (for the ANTARES Coll.): Nucl. Phys. Proc. Suppl. **70**, 445 (1999)
Greisen, K.: Annu. Rev. Nucl. Part. Sci. **10**, 63 (1960)

Halzen, F.: Ice fishing for neutrinos (1995). Available from: http://icecube.berkeley.edu/amanda/ice-fishing.html
Halzen, F.: Antarctic dreams (1998). Available from: http://www.exploratorium.edu/origins/antarctica/tools/dreams1.html
Halzen, F., Hooper, D.: Astropart. Phys. **23**, 537 (2005). arXiv:astro-ph/0502449
Halzen, F., Learned, J.G.: In: Proc. 5th Int. Symp. on Very High-Energy Cosmic-Ray Interactions, Lodz, Poland (1988)
Harwit, M.: Cosmic Discovery. Basic Books, New York (1981)
IceCube homepage: http://www.icecube.wisc.edu/
Katz, U., Spiering, C.: High-energy neutrino astrophysics: status and perspectives. Prog. Part. Nucl. Phys. **67**, 651 (2012). arXiv:1111.0507
Kotzer, P. (ed.) DUMAND-75, Proc. 1975 Summer DUMAND Study, Western Washington State College, Bellingham (1976)
Kowalski, M.: J. Cosmol. Astropart. Phys. **0505**, 010 (2005). arXiv:astro-ph/0505506
Krishnaswamy, M., et al.: Proc. R. Soc. Lond. **323**, 489 (1971)
Learned, J. (ed.): DUMAND-1979. Proc. of Khabarovsk and Lake Baikal Summer Workshops (1979)
Lowder, D., et al.: Nature **353**, 331 (1991)
Markov, M.A.: In: Proc. 10th ICHEP, Rochester, p. 578 (1960)
Markov, M., Zheleznykh, I.: Nucl. Phys. **27**, 385 (1961)
NEMO homepage: http://nemoweb.lns.infn.it
NESTOR homepage: http://www.nestor.noa.gr/
Reimer, A., Böttcher, M., Postnikov, S.: Astrophys. J. **630**, 186 (2005). arXiv:astro-ph/0505233
Reines, F.: Annu. Rev. Nucl. Part. Sci. **10**, 1 (1960)
Reines, F.: In: Peterson, V. (ed.) Proc. 30th Int. Conf. on Neutrino Physics and Astrophysics, vol. 2, p. 496 (1981)
Reines, F., et al.: Phys. Rev. Lett. **15**, 9 (1965)
Resvanis, L.K., et al. (NESTOR Coll.): Nucl. Phys. Proc. Suppl. **35**, 294 (1994)
Roberts, A. (ed.): DUMAND-76, Proc. 1976 Summer DUMAND Workshop, University of Hawaii, Honolulu (1977)
Roberts, A.: Rev. Mod. Phys. **64**, 259 (1992)
Roberts, A., Wilkins, G. (eds.): DUMAND-78, Proc. 1978 Summer DUMAND Study (1978)
Sokalski, I., Spiering, C. (eds.) (Baikal Coll.): The Baikal Neutrino Telescope NT-200. Tech. Rep. Baikal-92-03, DESY/INR (1992)
Stecker, F.W.: Phys. Rev. D **72**, 107301 (2005). arXiv:astro-ph/0510537
Svoboda, R., et al.: Astrophys. J. **315**, 480 (1987)
Taiuti, M., et al. (NEMO Coll.): Nucl. Instrum. Methods A **626**, S25 (2011)
The High Energy Neutrino Astrophysics Panel: High energy neutrino observatories (2002). Available from: www.lngs.infn.it/lngs/infn/contents/docs/pdf/panagic/henap2002.pdf
van Aller, G., et al.: IEEE Trans. Nucl. Sci. **NS-30**, 1119 (1983)
Waxman, E., Bahcall, J.: Phys. Rev. D **59**, 023002 (1999). arXiv:hep-ph/9807282
Weekes, T.C., et al.: Astrophys. J. **342**, 379 (1989)
Wischnewski, R., et al.: In: Proc. 3rd Int. NESTOR Workshop, Pylos, Greece, p. 213 (1993)
Zheleznykh, I.: Int. J. Mod. Phys. A **21S1**, 1 (2006)

# Chapter 10
# From Waves to Particle Tracks and Quantum Probabilities

**Brigitte Falkenburg**

> Later I was asked to sit and count the tic-tac of the cosmic rays and only then did I accept for the first time the existence of cosmic rays. Until then I had been convinced that they were an invention of Bruno and other scientists
>
> Nora Lombroso (Rossi, 1990, 167)

## 10.1 Introduction

This chapter is on measurement theory. It investigates how to measure cosmic rays and how one came to know about their nature. In particular, it explains the laws involved in the methods of data analysis in a 'genetic' account that aims at making the growth of knowledge transparent. In the history and philosophy of physics it is well known that the experimental data are theory-laden. Of course, without a detailed theory of the measuring devices and the ways in which they measure the phenomena, no precise experimental results are available. This measurement theory should be well-confirmed and independent of the theory which is under test, in order to avoid circularity. But the way in which such an independent measurement theory is developed, empirically confirmed, and extended in the course of time is a neglected topic in the history and philosophy of physics.

The existing historical accounts of cosmic ray studies and particle physics omit this issue as a matter of tacit background knowledge. When the physicists describe the history of their discipline (Pais, 1986; Riordan, 1987), they more or less presuppose this background knowledge and focus on discoveries and new theories. Current historians of science, in turn, stand in the tradition of Thomas S. Kuhn (1962). They mainly focus on the schools, skills and pragmatic aspects of scientific practice, on "external", social factors (Pickering, 1984), or on the historical aspects of what scientists consider to be objective knowledge (Daston, 2000; Daston and Galison, 2007). In general these approaches miss the multiple ways in which theory and experiment are partially interwoven, partially independent, giving rise to a growing

B. Falkenburg (✉)
Fakultät 14, Institut für Philosophie und Politikwissenschaft, Technische Universität Dortmund, 44221 Dortmund, Germany
e-mail: brigitte.falkenburg@tu-dortmund.de

B. Falkenburg, W. Rhode (eds.), *From Ultra Rays to Astroparticles*,
DOI 10.1007/978-94-007-5422-5_10, © Springer Science+Business Media Dordrecht 2012

body of increasingly complex theoretical background knowledge, which, however, is kept apart from the theories under test as far as possible. Even though the neglect of experiment became an important topic in recent history of physics, the most prominent case studies on cosmic rays and particle physics (Franklin, 1986, 1990, 2001; Galison, 1987) deal with the relation between theory and experiment without giving attention to the details of the measurement theories involved in data analysis.

The measurement theory used in particle and astroparticle physics is far from having uniform theoretical foundations, if this very complex and heterogeneous aggregate of semi-empirical, phenomenological and theoretical laws may be called a *theory* at all. The laws used for the data analysis of particle tracks and scattering events in particle and astroparticle physics rest on semi-classical models and stem from the 1930s. Nevertheless, they are used up to the present day for the analysis of particle tracks, scattering events, and particle showers. The core of this measuring theory traces back to the early days of radioactivity and cosmic ray studies, quantum mechanics and quantum electrodynamics. Its laws were probed in the early cosmic ray studies of the 1930s–1950s, taken over from there in the 1960s to the high-energy scattering experiments of particle physics, and taken back from there to the investigation of cosmic rays in recent astroparticle physics. In all these periods, the measurement theory of cosmic rays and subatomic particles was enriched by new elements, whereas most of the old ones were kept.

The story to be told here is meandering between waves and particles, theory and experiment, classical and quantum laws, particle tracks and quantum probabilities. It reflects the way in which the investigation of cosmic rays moved forth and back between electromagnetic waves and charged particles, unexpected phenomena and unconfirmed theory, particle physics and astrophysics. It starts with the question of how to detect particles in contradistinction to waves, passes on to particle tracks and the question of how they relate to quantum mechanics, explains the semi-classical model of a particle track and its quantum limitations, crosses over to the discovery of the positron and sketches the puzzles of particle identification and quantum electrodynamics, which were unraveled due to precise mass measurements by means of nuclear emulsions. Finally the turn from particles back to waves is sketched, which in face of wave–particle duality, however, have to be interpreted in terms of probability amplitudes. (The chapter is partially based on Falkenburg, 2007.)

## 10.2 How to Detect Particles

The first particles ever detected in subatomic physics were the electron, the $\alpha$-particle, and the photon. It was not so easy, however, to establish evidence of the particle nature of the electron and the photon. The meandering between waves and particles began with cathode rays and the search for a measurement of isolated electron charges and ended up in wave–particle duality of the photon and all other particles.

### 10.2.1 Massive Charged Particles

The discovery of the electron is usually attributed to J.J. Thomson and dated to the year 1897, when the debate about the wave or particle nature of cathode rays was still ongoing. Thomson measured the ratio $e/m$ as a property of cathode rays, but he neither localized the mass nor the charge associated with it. It took a long way from his $e/m$ measurement of to Millikan's measurement of isolated electron charges. Whoever did not yet believe in 1897 that a massive fundamental charge unit existed, would not be convinced by Thomson's measurement result either. The measurement did not in any way test the hypothesis that cathode rays consist of single massive charged particles. It just confirmed that cathode rays are deflected by the Lorentz force and hence are carriers of mass and charge. Although this result was a strong indication for the existence of the electron, many physicists did not regard Thomson's conclusion as sufficient. This was in particular true of defenders of energetism like Ostwald, or the empiricist physicist and philosopher Mach.

Additional measurements were needed in order to identify the electrons as particles, i.e., as single carriers of charge and mass, rather than measuring only the rays that carry these properties. Still in 1897 and in Thomson's laboratory, Townsend made the first step towards such a measurement. He succeeded in using the principle of the cloud chamber (developed by Wilson in 1895) for localizing single charge carriers unsharply, and he determined their charges independently of mass. Townsend determined the charge of the single condensation droplets in the cloud chamber from the total charge of the steam and the number of droplets per volume. The total charge of the steam was measured electro-statically, and the number of droplets was calculated from the weight of the cloud and the mean weight of the single droplets (Townsend, 1897; Millikan, 1917, 43–47). As opposed to Thomson's $e/m$ measurement, the force law here is not only applied to invisible particles but to the macroscopic properties of the droplet cloud. After 1898, Thomson carried out measurements based on this principle, too (Thomson, 1899), even though in a more indirect way and adding considerably to the experimental uncertainties, as Millikan stressed (Millikan, 1917, 47–52).

It is not the $e/m$ determination from cathode rays but only such a charge measurement from condensation droplets that assigns a characteristic value of charge to entities identified one by one in a detector. However, the early measurements still allowed doubt as to whether every single condensation droplet carries an elementary charge unit, or at least an integer multiple of it. Only around 1910 were these doubts largely dispelled by Millikan's oil droplet experiments. In lengthy and difficult measurements, Millikan determined the force which an electric field exerts on single charged oil droplets (Millikan, 1911). When he published his value for $e$, he emphasized that it was due to the first method for measuring the charge of individual charge carriers (Millikan, 1911). Millikan himself, however, did not see the principal benefit of his results in the experimental validation of isolated elementary charges, but rather in the precision measurement for $e$ made possible by his method. Indeed he was so eager to have a high precision measurement that he even arbitrarily omitted some of the droplet events he had measured (as is known today from a careful analysis of his notebooks; see Franklin, 1986, 140–157). In this way, and probably in order to keep his measurement error small, he simply violated the rules of good scientific practice. However, at this time the existence of the electron and the atomistic constitution of matter were well-established due to an abundance of experimental evidence from other areas of physics. Indeed, Millikan's results were confirmed by many later experiments without any need of awkward data selection.

A few years after Thomson's $e/m$ measurement, the particle nature of $\alpha$-rays became evident. As Crookes and others discovered in 1903, a screen laminated with zinc sulfide starts to phosphoresce in total darkness when it is exposed to $\alpha$-rays. Observed with a magnifying glass, this glow could be resolved into a variety of single light flashes. Later, Rutherford, Chadwick and Ellis commented on this phenomenon as follows in their textbook on radioactivity:

> On viewing the surface of the screen with a magnifying glass, the light from the screen is seen not to be distributed uniformly but to consist of a number of scintillating points of light scattered over the surface and of short duration. Crookes devised a simple apparatus called a 'spinthariscope' to show the scintillations. A small point coated with a trace of radium is placed several millimeters away from a zinc sulfide screen which is fixed at one end of a short tube and viewed through a lens at the other end. In a dark room the surface of the screen is seen as a dark background dotted with brilliant points of light which come and go with great rapidity. This beautiful experiment brings vividly before the observer the idea that the radium is shooting out a stream of projectiles each of which causes a flash of light on striking the screen' (Rutherford et al., 1930, 54–55)

From 1906, Rutherford and his assistants carried out their famous scattering experiments with $\alpha$-rays. Rutherford measured the ratio $E/M$ of charge $E$ and mass $M$ of the $\alpha$-rays (Rutherford et al., 1930, 41–46), and then he proceeded to using the scintillation method to measure the scattering angle of single $\alpha$-particles. The discovery of backward scattering, the inference to the atomic nucleus, and the experimental check of Rutherford's famous scattering formula were later based on this method, too (Geiger and Marsden, 1913; Trigg, 1971). It turned out that the $\alpha$-particles are helium nuclei.

A decisive step towards establishing the atomistic structure of matter was made in 1908, when Perrin's measurements confirmed Einstein's 1905 theory of Brownian motion (Perrin, 1909). Now, the atomistic hypothesis was approved even by defenders of energetism like Ostwald (Blackmore, 1972, 217; Nye, 1972).

The story shows that in order to establish the particle nature of electrons and $\alpha$-rays, two kinds of evidence had to come together: the empirical observation of a local event, and the unambiguous attribution of particle properties such as mass and charge to this event. This requirement is in accordance with a twofold, causal and mereological particle concept (Falkenburg, 2007). According to the causal particle concept, particles in contradistinction to waves cause local events in a measurement device. This was first directly observed by means of the scintillation method. The mereological particle concept dates back to ancient atomism and its renaissance in Galileo's and Newton's beliefs. ("Mereology" is the logic of wholes and parts; see Simons, 1987.) It means that particles are the parts of matter, and that their dynamic properties add up to the properties of macroscopic matter according to well-defined sum rules for conserved quantities such as mass and charge.

So far, so good. When the cosmic rays were discovered by Victor Hess in 1912, no one thought that they might have mass and charge. Until the early 1930s, they were supposed to be gamma rays, i.e., electromagnetic waves rather than charged particles (see Chap. 2 and below).

### 10.2.2 The Photon

However, it turned out that electromagnetic waves have particle aspects, too. Due to quantum theory, the classical distinction of particles and waves did no longer hold. Einstein's light quantum hypothesis of 1905 seemed to step backwards to Newton's long-refuted atomistic theory of light, a fact that puzzled no one more than Einstein himself. Indeed, the physics community did not accept Einstein's light quantum hypothesis of 1905 for many years. Only after the observation made in 1919 that light is bent by gravity, as predicted by general relativity, Einstein's public reputation increased so enormously that in 1921 he was awarded the Nobel prize, not for relativity but for his light quantum hypothesis, because it made the bridge to Bohr's atomic model (Wheaton, 1983, 279–281). The Nobel prizes of 1921 and 1922 were given to Einstein and Bohr at the same ceremony in 1922.

Only afterwards was the photon hypothesis definitely confirmed by experimental proof of the Compton effect (Wheaton, 1983, 282–286). For the experimental confirmation the energy–momentum conservation of relativistic kinematics was decisive. It had similar significance as the Lorentz force for Thomson's $e/m$ measurement. It explained the measured decrease in frequency of scattered X rays in terms of the momentum transfer of the photon to an electron, i.e., in terms of a particle property.

From Einstein's light quantum hypothesis (Einstein, 1905) up to the experimental validation of the photon by Bothe and Geiger (1925) almost two decades passed. In Einstein's paper of 1905, the light quantum hypothesis had primarily a heuristic value, namely its unifying power. It explained in a uniform way the photo effect and several other experimental phenomena observed in the interaction of light and matter.

That light can cause a photocurrent was already known since the end of the 19th century. But the way in which the photocurrent depends on the frequency and intensity of the incoming light was not understood. Einstein explained the threshold of the photocurrent in terms of light quanta and the releasing energy of an electron. In 1914, Millikan tested this theoretical explanation experimentally and confirmed it with high precision (Wheaton, 1983, 238–241; Trigg, 1971). But the light quantum hypothesis was in conflict with the wave theory of light, and Millikan's results were not taken as sufficient experimental evidence. At that time, the light quantum hypothesis lacked both ingredients for indicating a particle. It was neither based on particles properties that could be measured nor did it correspond to any local events such as the observable scintillation flashes caused by $\alpha$-particles. In the years after 1905, it was only considered to be a well-confirmed phenomenological law which lacked theoretical understanding, like the other laws of early quantum theory such as

Planck's law of black-body radiation and the quantum postulates of Bohr's atomic model.

In 1916, Einstein extended his light quantum hypothesis of 1905 to a first, statistically well-founded theory of the absorption and emission of light (Einstein, 1917). This theory aggravated rather than cured Einstein's life-long discomfort about the probabilistic nature of quantum processes. However, it laid the grounds for the acceptance of the photon hypothesis by enlarging its empirical content in two decisive respects. On the one hand, it connected the light quantum hypothesis to Bohr's quantum postulates and made it possible to derive the frequency of the radiative transitions in the hydrogen atom. On the other hand, Einstein now attributed a momentum to the photon in addition to its energy $E = h\nu$ a momentum $p = \hbar k$, where $k$ is the wave vector of an electromagnetic wave with the wave number $k = 2\pi\nu/c$. In this way, the energy–momentum relation $E^2 = p^2c^2$ of relativistic kinematics became applicable to it. In this way, the photon hypothesis was transformed into the theory of a *relativistic particle of zero rest mass* with an *inertial mass* $m = h\nu/c^2$, which may give a *recoil* to a massive particle.

The breakthrough for the photon hypothesis came in 1922, when Compton applied relativistic kinematics to the scattering of gamma radiation and electrons. His quantitative prediction of the effect fitted surprisingly well with the existing data (Compton, 1923; Debye, 1923). As in the case of the discovery of the electron, the attribution of a typical particle property to the photon was a milestone toward accepting the photon hypothesis, but it was not yet regarded as sufficient. Classical electrodynamics was at stake, and Bohr, Kramers, and Slater invented the BKS theory in order to save it (Bohr et al., 1924). It predicted violations of energy–momentum conservation for individual subatomic processes and saved the classical picture only at the probabilistic level (thus paving the way for the probabilistic interpretation of quantum mechanics). To establish the photon hypothesis against the BKS theory, the localization of single photons of energy $h/\nu$ was needed. Unlike the measuring of single elementary electric charge units by Millikan, it was not long in coming. Bothe and Geiger proved in 1925 that relativistic energy–momentum conservation is not only valid in the time average for the scattering of light at electrons but also for single scattering processes. Their experiment tested energy–momentum conservation in the individual case, showing by means of a coincidence counter that in the Compton effect every single photon is actually correlated with a recoil electron (Bothe and Geiger, 1925). The coincidences were accepted empirical evidence for the effects of individual photons.

But this is not the whole story. There are some parallels with Millikan's $e$ measurement about 15 years after the electron hypothesis found support by Thomson's measurement; and with the confirmation of the neutrino hypothesis by indirect $\beta$-decay, about two and a half decades after establishing this hypothesis. Since the early days of quantum theory, several semi-classical theories of the interaction of light and matter have been developed which quantize the atom but keep classical electromagnetic radiation. They do indeed explain the experimental phenomena that were taken as evidence for the photon, including the photoelectric effect and the Compton effect (Greenstein and Zajonc, 1997, 23–26). The most prominent defender of a semi-classical explanation of the photoelectric effect was Planck, and a semi-classical explanation of the Compton effect was already given by Schrödinger (1927). The decisive proof of single photons only came by the coincidence experiments of modern quantum optics, about six decades after Einstein got the Nobel Prize! The morals of this history is that it is not easy to establish empirical evidence for single particles.

## 10.3 From Waves to Particle Tracks

As noted, cosmic rays were first considered to be waves. The electrometers used until the late 1920s did not differentiate between $\gamma$- and $\beta$-rays (see Chap. 2). Only when Geiger–Müller tubes were employed for coincidence measurements and when the Wilson chamber was operated in a magnetic field, it became possible to identify charged particles. Even after the discovery of the positron, however, doubts remained concerning the question of whether the charged particles belonged to primary or secondary cosmic rays.

To speak of waves in contradistinction to particles may be slightly misleading here. The classical distinction of waves and particles had already become blurred. Early cosmic ray studies were made at the time of the rise of the light quantum hypothesis (Einstein, 1905, 1917), old quantum theory (Bohr, 1913, 1922), and wave–particle duality (de Broglie, 1923). But old quantum theory before quantum mechanics was far from being a well-defined theory. So the physicists continued to use the traditional terms of wave and particle, justified by lack of better concepts as well as by Bohr's correspondence principle. Indeed the classical distinction did not break down at once but step-by-step. The traditional terms were maintained as far as possible in certain contexts and associated with specific new meanings (such as probability waves, field quanta, etc.) in others (Falkenburg, 2007). In particular, in the domain of non-relativistic quantum mechanics a far-reaching correspondence to the classical terms could be stated, giving rise to Bohr's "complementarity" interpretation of quantum mechanics (Bohr, 1920, 1927) as well as to a generalized version of Bohr's correspondence principle and the well-known procedure of quantizing the classical theories (Heisenberg, 1930).

At first sight, the above mentioned shift from the wave picture to a particle picture of cosmic rays had nothing to do with wave–particle duality. It just referred to the distinction of massive charged particles such as electrons, protons, and $\alpha$-particles on the one hand, and $\gamma$-rays or photons, on the other hand. The belief that cosmic rays consist of charged particles was established by the use of coincidence detectors that were able to measure repeated particle detections, i.e., rudimentary particle tracks. In contradistinction to the electrometer, Geiger–Müller coincidence counters and the Wilson chamber discriminate massive charged particles (which give rise to repeated position measurements) and photons (which do not). Hence, to take in consideration that cosmic rays are charged particles was independent of the rise of quantum mechanics. It just happened around 1930, a few years after Schrödinger's wave equation and Born's probabilistic interpretation of it, and in the early days of quantum electrodynamics.

Defenders of the wave picture of cosmic rays like Millikan, however, still resisted against applying quantum theory to them, when defenders of the particle picture interpreted the charged particles in terms of quantum electrodynamics. In the late 1920s, Millikan still maintained a classical picture of the interactions of atoms with matter and believed that cosmic rays are $\gamma$-rays, i.e., electromagnetic waves. He defended an atom-building theory according to which these $\gamma$-rays stem from the synthesis of atoms. To this "birth cry of atoms" theory, he adhered for religious reasons, and he explained it in intuitive classical terms rather than within the framework of quantum mechanics (Galison, 1987, 80–89). In opposition to Millikan's views, Bothe and Kolhörster (1929) invented the coincidence measurement method in order to prove that cosmic rays consist of charged particles, and in 1930 Rossi substantially improved their method (see Rossi, 1990). In 1932, Anderson (who was a member of Millikan's group) discovered the positron and turned from Millikan's views over to the interpretation of cosmic rays in terms of the Dirac equation. In 1933, the *New York Times* even published a sharp controversy between Millikan and Compton (see Galison, 1987, 93): The classical physicist and Nobel prize winner Millikan insisted that cosmic rays are electromagnetic waves. The quantum physicist and Nobel prize winner Compton on the contrary argued that cosmic rays are charged particles.

At that time, however, the particle picture used in cosmic ray studies remained to be associated with the model of a classical trajectory, as if there was no quantum theory. According to quantum mechanics, there is no continuous track. The appearance of a particle track stems from a sequence of discrete position measurements. It is just due to the repeated localization of a conserved quantity of mass and charge. Indeed, non-relativistic quantum mechanics predicts the appearance of a quasi-classical track with extremely high precision in terms of ionization events that occur with extremely high probability along the corresponding classical trajectory (Mott, 1929; Heisenberg, 1930; see below). In view of these quasi-classical tracks recorded by the Wilson chamber, the traditional term "particle" was kept and adopted for the discipline of particle physics that emerged from the cosmic ray studies of the 1930s and 1940s. But after almost a century of philosophical debates about the foundations of quantum mechanics and its probabilistic interpretation, it has to be emphasized that *there is no (relativistic) quantum theory of individual particles* that describes an individual, quasi-classical track. The discovery of the positron, which was a first and crucial confirmation of quantum electrodynamics, gave rise to a less naive picture of the quantum processes that may happen along a particle track.

It is worth looking into the details of the phenomenological analysis and the quantum mechanics of particle tracks, on which the data analysis of particle and astroparticle physics rests up to the present day. Today, after 100 years of quantum theory, cosmic ray studies and particle physics, the meaning of the term "particle" is predominantly operational. It is based on the detection of repeated "clicks", particle tracks, and local scattering events measured by particle detectors.

### *10.3.1 The Phenomenology of Particle Tracks*

The Geiger–Müller coincidence counters and the Wilson chamber paved the way to modern particle physics (see Chap. 2). The efficiency of the cloud chamber was improved by means of triggering it with coincidence counters, and starting with the discovery of the positron in 1932, many new kinds of particle were found in the tracks of cosmic rays. In order to determine their mass and charge from the measured tracks, it became crucial to refine the measurement methods

The most important marks for identifying the particles were the track curvature in the magnetic field, the range of the particles in the vapor of the cloud chamber, and the ionization density or density of the measurement points. The latter is a measure of the frequency of the interactions of the particle with the hydrogen atoms of the Wilson chamber, and hence a measure for the ionization degree which in turn indicates the mass of an unidentified particle as compared to the mass of well-known particles. For $\alpha$-particles or protons, the ionization degree is substantially larger than for electrons, as was known from the first photos of particle tracks since 1912. The tracks of $\alpha$-particles, protons, and electrons in the Wilson chamber look significantly different. For $\alpha$-rays only the tracks and no condensation droplets are seen on the cloud chamber photographs, while for $\beta$-rays the individual measurement points of the tracks can be clearly distinguished:

> Owing to the density of the ionization, the path of the $\alpha$-particle shows as a continuous line of water drops. A swift $\beta$-particle, on the other hand, gives so much smaller ionization that the individual ions formed along its track can be counted." (Rutherford et al. 1930, 57)

The ionization density depends on the velocity of a particle of given mass and charge and gives no hint to the absolute mass value (Skobeltzin, 1985, 114). The same is true of the Lorentz force and the curvature of a charged particle due to a magnetic field. The range of a particle in matter gives some more information. The range of a particle (or the length of its track) is related to its energy loss during its passage through the detector until being stopped or absorbed. A half-empirical law, the so-called energy–range relation, was already formulated in the early days of particle physics. It was based on the scattering experiments with particles from radioactive radiation sources, as performed in Rutherford's laboratory. The relation connects the kinetic energy of a massive charged particle to its range (or track length) in different materials (Rutherford et al., 1930, 294).

These laws and relations (which were half classical, half empirical) made up the measurement theory of particle physics and cosmic ray studies in the 1930s and 1940s. They made it possible to give a rough estimation of the mass of a charged particle from the track signature. In this way, it was tricky though not too difficult to discriminate the electron and the proton mass. By doing so, Anderson discovered the positron in 1932.

But as soon as particles of medium mass such as the muon or the pion came into play, a rough mass estimation did no longer help to identify the particles. Only a detailed theory of what happens to charged particles during their passage through matter would have helped to make the mass measurement more precise. Such a theory was *in principle* available since Bethe's seminal work on the scattering processes of charged particles in matter (Bethe, 1930) and its extension to quantum electrodynamic processes such as *bremsstrahlung* and pair creation. After the discovery of the positron, quantum electrodynamics got much credit, and many calculations were performed. But quantum electrodynamics did give rise to divergences beyond first-order perturbation theory, and its first-order predictions for electron scattering drastically disagreed with the cosmic ray measurements.

Hence, for two decades the theoretical predictions were elaborated and confirmed in a zig-zag between cosmic ray measurements and phenomenological calculations based on quantum electrodynamics. This zig-zag had to deal with the puzzles of quantum electrodynamics and particle identification mentioned above, which are explained in the next section. In addition, it had to bridge a certain mismatch between the quantum theory of scattering and the semi-classical model of an individual particle track. The physicists knew that particle tracks are due to ionization processes described by the quantum mechanics of scattering, on the one hand, but they had to use more or less classical methods in order to interpret the tracks, on the other hand. With regard to the energy loss due to ionization or "collision loss", this problem was clearly stated in Rossi's textbook of 1952 (Rossi, 1952, 29):

> The energy loss of a charged particle in matter is a statistical phenomenon because the collisions that are responsible for this loss are independent of each other. Thus particles of a given kind and a given energy do not all lose exactly the same amount of energy traversing a given thickness of material. The quantity $k_{\text{col}}(E)$ defined as 'collision loss' [...] represents only an average value.

The same is obviously true of the range of charged particles in matter, which in the semi-classical model of an individual track derives from the energy loss along the track. For non-relativistic particles, however, the statistical effects are low, as Rossi also stated (Rossi, 1952, ibid.):

> The statistical fluctuations in the energy loss by collision are comparatively small because the average transfer of energy in each individual collision process is small and the number of collisions necessary to cause any appreciable energy change is correspondingly large.

For relativistic particles, however, the nice correspondence of the quantum mechanics of scattering to the classical picture of a track breaks down. Quantum electrodynamic processes such as *bremsstrahlung* or pair creation give rise to an abrupt drastic energy loss of a charged particle. The corresponding track signatures are observable kinks, the emergence of a pair of tracks of opposite curvature, or the appearance of a particle shower. For these processes, the nice correspondence of the quantum mechanical scattering to a classical trajectory breaks down, and with it the validity of the semi-classical explanation of an individual particle track.

### *10.3.2 The Semi-classical Model of a Particle Track*

In order to sketch the physicists' understanding of the relation between quantum mechanics and particle tracks around 1930 (and its limitations), let us have a closer look at Mott's and Bethe's semi-classical model of particle tracks. This model is still in use for data analysis up to the present day. It has first been employed in cosmic ray studies, then in the high-energy scattering experiments of particle physics, and finally in recent astroparticle physics. Given its limitations in the relativistic domain, of course in current high-energy physics and astroparticle physics it is no longer applied to individual particle tracks but used for *statistical* data analysis.

The formulas for the energy loss along a track are genuine, probabilistic quantum laws. They describe the dissipation of energy in a sequence of irreversible quantum processes. As far as these processes are observable, i.e., give rise to a particle detection or position measurement, each of them results in an irreversible change of the momentum state of the particle. Hence, in terms of quantum mechanics the particle states after the measurement points of one-and-the-same track do not belong to one-and-the-same quantum ensemble. Nevertheless it became experimental practice in particle physics to apply them to individual particle tracks, *as if there was no quantum measurement problem*. This is no wonder. When Mott and Bethe developed their semi-classical model of a track, there was no quantum theory of measurement. When von Neumann laid the foundations for it (von Neumann, 1932) and the Bohr–Einstein debate on the foundations of quantum mechanics went on (Bohr, 1927, 1949; Einstein et al., 1935), the physicists working on cosmic rays or on quantum electrodynamics neglected these foundational problems. They considered them not to be relevant for their practice.

And they did so for good pragmatic reasons. The semi-classical model of a particle track is based on trust in Bohr's correspondence principle, or Heisenberg's generalized version of it. As mentioned above, Mott and Heisenberg showed in 1930 that Born's quantum mechanics of scattering predicts particle tracks with a classical shape. Thus, the energy loss of charged particles in matter was calculated on the basis of a naive, quasi-classical, realistic picture of subatomic reality. As Heisenberg stressed in his 1930 book on quantum mechanics, the probability of $\alpha$-particle deflection due to repeated ionization of molecules in the vapor is non-zero

> only if the connecting line of the two molecules runs parallel to the velocity direction of the $\alpha$-particles (Heisenberg, 1930, 53; my translation).

The corresponding calculation was carried out by Mott in 1929. According to Born's quantum mechanical scattering theory, the scattering is not due to an impact but due to diffraction, i.e., it is described in a wave model. In particular, the quantum mechanical description of scattering lacks the classical trajectory of a deflected particle and the corresponding classical impact parameter. The squared wave function predicts only the probability (and hence the relative frequency) of particle detections at a certain scattering angle (Born, 1926a,b).

Mott calculated the probability for two subsequent collisions of an $\alpha$-particle and a hydrogen atom with the effect of the ionization of both atoms. The ionized atoms give rise to observable measurement points. They are the core of droplets which condense in the vapor of the Wilson chamber. The observation of a droplet is a position measurement, whereas the observation of the particle deflection given by straight lines drawn between the adjacent droplets is a momentum measurement.

Heisenberg showed in his 1930 book by a heuristic consideration that the uncertainty relation for position and momentum holds for any ionization process along the track (Heisenberg, 1930, 18; Engl. transl., 24). Hence, the quantum mechanical explanation of the single measurement points of a particle track is in perfect correspondence to the classical particle picture, as long as the (unobservable) path between the position measurements is neglected. Thus, for all practical purposes it predicts a classical particle track and supports the application of quantum mechanical scattering theory to individual particle tracks.

Mott's calculation is probabilistic and it is performed in a quantum mechanics without measurement. According to Born's quantum mechanics of scattering, the wave function is diffracted at two atoms at a given distance $R$. Mott's main result is that the first two orders of perturbation theory predict a classical track. To first order (incoherent scattering), the outgoing wave is concentrated in a cone of very small angle behind the ionized atom, in the direction of the incoming wave. To second order (coherent scattering at two atoms), the contribution is non-zero if and only if both atoms lie inside that angle in the same direction (Mott, 1929; Heisenberg, 1930, 56 and Engl. transl., 75–76). Generalized to $N$ atoms, Mott's results predict the following results: To first order, the scattering probability is $N$ times the probability for incoherent scattering at a single atom. Coherent scattering at more than one atom contributes only to second, third, ..., $N$th order. However, to *any* order the scattered wave propagates along a classical path.

As noted above, Mott's 1929 calculation is based on the unrealistic idealization that the energy loss associated with ionization is not taken into account. The particle is described as if it did not *really* transfer a definite amount of energy to the hydrogen atom when ionizing it. Although the calculation deals with the amplitudes of inelastic collisions (and hence with the dissipative process of energy loss), it is performed as if the momentum state of the charged particle remained unaffected by the energy transfer to the hydrogen atom which gives rise to ionization. That is, the charged particle which gives rise to subsequent ionization processes and observable droplets in the vapor of the Wilson chamber is treated as if its collisions with the hydrogen atoms were elastic. This unrealistic idealization looks reasonable once we notice that e.g. the energy loss of an $\alpha$-particle due to ionization of hydrogen atoms can be neglected. The ionization energy of hydrogen is *very* small compared to the kinetic energy of the $\alpha$-particle. Therefore the momentum of the $\alpha$-particle remains practically unchanged along its track in the Wilson chamber.

### *10.3.3 Energy Loss by Ionization*

In the case of a substantial amount of energy loss along a particle track, however, the agreement of the classical and the quantum descriptions vanishes. Nevertheless, Mott's semi-classical model of the scattering processes along an observable track was maintained in all later calculations of the energy loss of charged particles in matter. This semi-classical model comes together with the following intuitive classical picture of what happens when a charged particle loses its energy along a particle track. The particle is slowed down repeatedly by inelastic collisions with detector atoms, the track curvature in an external magnetic field increases, and the track ends when the particle has transferred its total kinetic energy and momentum to the detector atoms. Indeed the discovery of the positron was based on this picture. In order to identify the sign of the charge of the particles from cosmic rays, Anderson measured the flight direction by putting a lead plate into the Wilson chamber. The lead plate caused a substantial energy loss and gave rise to an observable increase of the curvature of particle tracks in the magnetic field.

Until the early 1930s, there was no satisfying quantum theory of ionization. When the exploration of cosmic rays with the cloud chamber started in the late 1920s, there was only Bohr's classical calculation of ionization (Bohr, 1913, 1915), which was known to give wrong results for fast particles. Its predictions differed from the half-empirical knowledge about particle tracks accumulated until the late 1920s, they were particularly in disagreement with the semi-empirical energy–range relation. Around 1930, the classical theory of the energy loss of charged particles in matter was known to give some correct results on average, but to be in need of quantum theoretical corrections and to have no validity for individual particle tracks (Rutherford et al., 1930, 439). In turn, Mott's and Heisenberg's quantum mechanical explanation of the appearance of a quasi-classical track neglected the energy loss due to ionization into account, not to speak of *bremsstrahlung* and pair creation.

The quantum theory for describing the interactions of charged particles with matter, however, already existed. Born's seminal papers on the probabilistic interpretation of quantum mechanics laid the grounds for the quantum mechanics of scattering (Born, 1926a,b). Using it in the first order of perturbation theory (today known as Born approximation), Bethe developed the quantum theory of the passage of charged particles through matter in 1930 (Bethe, 1930). His paper made the first reliable calculation of a non-negligible energy loss. The results only agree with Bohr's classical result for *vanishing* particle velocity $v$ and energy loss (for a detailed discussion, see Falkenburg, 2007, 178–183). This limit of zero velocity and no energy loss is exactly the idealized case of Mott's calculation discussed above. Indeed, for the energy loss along a particle track, Bohr's correspondence principle in general fails, as becomes evident for relativistic processes such as *bremsstrahlung* and pair creation.

Bethe calculated the quantum mechanical expectation value $\langle E \rangle$ for the energy loss at a given atom and implemented it into Mott's semi-classical model by applying it to the scattering processes along a particle track. From a quantum mechanical point of view, however, $\langle E \rangle$ is the mean energy loss per atom and per incoming particle in the limit of infinitely many incoming particles ($N_{\text{in}} \rightarrow \infty$). In the classical part of his semi-classical model, Bethe applied his formula for $\langle E \rangle$ to the subsequent scattering processes along an individual particle track, be it with or without an observable effect. Then he calculated the number of observable and unobservable inelastic collisions along a track for $N$ atoms per volume $\Delta x^3$ in a given material

(Bethe, 1930, 358). Finally, he gave the following simple expression for the mean energy loss $\Delta E$ per length $\Delta x$ of matter (Bethe, 1930, 360):

$$\frac{\Delta E}{\Delta x} = N\langle E\rangle.$$

Here, Bethe interprets the expectation value $\langle E\rangle$ as the average energy loss of a charged particle by successive scattering from many detector atoms along an individual track, normalized to the number of atoms per path $\Delta x$. Of course, Bethe discussed neither the physical interpretation of his results nor their philosophical justification. He simply suggested that $\langle E\rangle$ applies to the subsequent individual quantum transitions along a track, whether their results be observable or not. This was completely in the spirit of Mott's quasi-classical results.

One of Bethe's quantitative results was the ratio of observable to unobservable collisions. For hydrogen, 28.5 % of all inelastic collisions give rise to ionization (Bethe, 1930, 360), causing observable droplets in the vapor of the Wilson chamber. The calculation shows that a substantial amount of the energy lost along a particle track in the Wilson chamber is indeed measured, in striking contrast to Mott's idealized model in which the energy loss along a track was neglected. Hence, the situation is no longer comparable with Mott's idealized model in which the preparation of the particle and the expectation value of the scattering results do not change along the track.

From a strict quantum mechanical point of view, Bethe's semi-classical model for the energy loss of a charged particle in matter is incoherent. Within a quantum mechanics without measurement, Bethe calculates a formula which holds for the energy dissipation due to the position measurements along a track. It can be shown, however, that for the fast particles which were the subject of his calculation this may be neglected for all practical purposes, at least as long as the particles are not *too* fast, i.e., have relativistic velocity (Falkenburg, 2007, 182). In this case, the momentum dependence of $\langle E\rangle$ is very weak, making practically no difference for the particle states after the measurement points of an individual particle track. Therefore, Rossi's observation quoted above, according to which for ionization loss the statistical effects along a track are of negligible (Rossi, 1952, 29), is not only supported by Mott's highly idealized case but also by Bethe's results. In this way, the application of Bethe's formula to the energy loss along an individual particle track is justified *for all practical purposes*. (What looks queer from a philosophical point of view may be a good approximation in physical practice.)

### 10.3.4 Bremsstrahlung and Pair Creation

Like Born, Bethe used the non-relativistic quantum mechanics of scattering. His theory did therefore not apply to the tracks of the high-energy particles from cosmic rays. In order to describe the interactions of relativistic charged particles, quantum electrodynamics was needed. The basic equations were given by Dirac in 1927.

For relativistic particles, however, the naive quasi-classical picture of a track breaks down. Quantum electrodynamics predicts that a particle does not lose its energy smoothly. Due to *bremsstrahlung* and pair creation, the energy loss along a particle track may become completely irregular and extreme deviations from the classical path may occur. Therefore, with increasing particle energy the correspondence between the shape of individual particle tracks and the classical case breaks down stepwise, and the semi-classical model of a particle track does, too.

Møller had already shown in 1931 how Dirac's theory of the electron can be combined with Born's quantum mechanics of scattering. After the discovery of the positron, Bethe, Bloch, and Heitler extended Bethe's 1930 approach to calculations of *bremsstrahlung* and pair creation (Bethe, 1932; Bloch, 1933; Heitler and Sauler, 1933). The expectation value $\langle E \rangle$ of these processes depends on the energy of the charged particles. Only at non-relativistic particle energies, ionization is predominant and the relative frequency of *bremsstrahlung* or pair creation is negligible. In the relativistic domain, the relative frequency of the latter processes increases rapidly with increasing particle energy (Rossi, 1952, 29–30 and 60). Thus, in the transition from the non-relativistic to the relativistic domain the smooth quasi-classical shape of the particle tracks predicted by Mott's and Bethe's 1929/1930 calculations gets lost for an increasing number of particle tracks. Today, in the data analysis of high-energy scattering experiments or astroparticle physics, this is corrected at the probabilistic level (see next section). There is no other way to take the quantum electrodynamic fluctuations of the energy loss along a particle track into account.

Therefore, the way in which these quantum electrodynamic formulas were used did not change. All the same, they were simply inserted into Bethe's expression for the *mean energy loss* $\Delta E$ *per length* $\Delta x$ *of matter* along a particle track. Again, this procedure is justified by the incoherent first-order contributions to the quantum mechanics of scattering. The fact that they dominate makes it unproblematic to apply classical stochastic methods to the analysis of particle tracks. Bethe's and Bloch's 1932–1933 calculations gave rise to the Bethe–Bloch formula for the mean energy loss of fast charged particles per path length in matter which consists of heavy atoms (Bloch, 1933; for a semi-classical calculation see Rossi, 1952, 17). The Bethe–Bloch formula made it possible to calculate a theoretical value for the average range of charged particles in a given kind of matter or detector material (see Rossi, 1952, 22–27).

## 10.4 Quantum Electrodynamics and Particle Identification

Confidence in such calculations stood or fell with confidence in the Dirac equation. Before the discovery of the positron, the Dirac equation did not have much credit with physicists. Its solutions corresponding to negative energy values had no empirical interpretation and were considered to be "unphysical". After Anderson's discovery, this equation had much more credit and many quantum electrodynamic formulas were calculated, giving rise to a phenomenology of particle reactions in between the experimental results and the field theoretic approach of Heisenberg, Pauli, and others.

However, even though the degree of acceptance of this theory increased considerably after the discovery of the positron, its empirical content was far from being satisfactory and severe puzzles about quantum electrodynamics and particle identification showed up. The trust in quantum electrodynamics depended on resolving the divergences by means of renormalization, on the one hand, and on independent empirical evidence, on the other hand. Such evidence, in turn, depended on identifying the particles that generated the tracks from cosmic rays. So the theory of quantum electrodynamics and the experimental methods were trapped in a vicious circle. Indeed, the mass measurement from particle tracks and the calculations of quantum electrodynamics were not only stuck in circularity. As long as the calculations were only made for electrons and protons, the theoretical predictions and the shape of the tracks drastically disagreed. In this way, neither particle identification was possible, nor was quantum electrodynamics considered to be reliable. The puzzles could only be resolved when nuclear emulsions made more precise mass measurements possible, i.e., when a new independent measurement method became available that made no recourse to the formulas obtained from quantum electrodynamics.

Hence, even though the detailed theory of ionization and other collision processes used today stems from the 1930s, it was not available for data analysis before the consolidation of quantum electrodynamics in the late 1940s. Two decades after the discovery of the positron, Rossi noted at the beginning of his 1952 textbook *High-Energy Particles*, a landmark in the methods of data analysis (Rossi, 1952, 10):

> Theoretical physicists have not yet succeeded in their attempts to formulate the principles of quantum electrodynamics in a completely general manner, free from internal contradictions. They have, however, established a formalism that answers unambiguously most problems arising in the study of electromagnetic interactions between radiation and matter. Whenever the theoretical predictions have been submitted to experimental tests, they have found to be accurate, within the limits of experimental errors and the mathematical approximations made in the development of the theory. Confidence in the theory of electromagnetic interactions has grown to the point where one may grant its validity beyond the limits of experimental accuracy and perhaps even apply it to fields where experimental tests are still lacking. In the past, study of high-energy phenomena, cosmic rays in particular, was mainly a means for testing the theory of electromagnetic interactions. Today, however, one may justifiably use the results of this theory as a basis for the interpretation of the observed phenomena.

### 10.4.1 The Positron Track

The positron was the first new particle found in cosmic radiation. Its discovery in 1932 was completely based on the phenomenological features of cloud chamber tracks. Anderson identified it from a track with a density of ionization typical of an electron, but wrong sign of the curvature. Supervised by Millikan, he had been working with a cloud chamber since 1931 to examine cosmic rays (Anderson and

Anderson, 1983; Pais, 1986, 351–352). In order to identify the charge of the particles, he used a strong magnetic field. On his photographs he found quite a number of tracks which indicated positive particles. At first he attributed them to protons, as the proton was the only positively charged particle known at that time. But evidence spoke against protons. The low ionization degree of the tracks indicated a mass of the order of the electron rather than the proton mass, which is almost 2 000 times larger. Under the assumption that it were electrons from cosmic rays, however, the observed flight direction was not compatible with the track curvature in the magnetic field. It seemed absurd that they should be electrons from cosmic rays traveling upwards. To determine the flight direction of the particles unambiguously, Anderson put a lead plate of width 6 mm into the center of the cloud chamber. When passing the lead plate, the particles lost energy and momentum. This energy loss gave rise to an increase in the track curvature.

Anderson discovered a track particularly suitable for particle identification in August 1932 (Anderson, 1932, 1933; see Fig. 2.11, Sect. 2.6.1). In his analysis of the track, he discussed all degrees of freedom for the interpretation: the mass, the amount of charge, the sign of the charge and the number of particles to which the track may be imputed. From the ionization degree of the track he inferred that the charge could not differ in magnitude more than around a factor two from the electron. The magnitude of the mass was estimated indirectly using the track length, the track curvature and the known values of the electron and the proton mass.

His interpretation of the track was based on the following reasoning. According to the Lorentz force, for a particle of the proton charge and mass the track curvature indicated an energy of 300 MeV. According to the semi-empirical energy–range relation for protons, however, a proton of 300 MeV could only have a range of around 5 mm (Rutherford et al., 1930, 294). And due to the ionization density of the track, the mass had to be substantially smaller than the proton mass. But the assumption that it was an electron would have implied a drastic violation of energy conservation: due to the curvatures of the partial tracks, an electron causing the complete track would not have lost energy in the 6 mm thick lead plate, but rather have been accelerated by 40 MeV, as Anderson emphasized. Finally, it was very implausible to assume that it consisted of two independent electron tracks which met by chance at the lead disk. The probability of such a coincidence was extremely low. Therefore, only two possibilities remained. Either the track was due to a single particle of positive charge which had lost energy at the lead plate and which had a mass and charge comparable to the electron. Or it was due to a pair of particles with opposite charges and equal mass which had arisen from one and the same reaction in the lead plate, and of which one was an electron. The existence of a positive electron, the positron, resulted from both possibilities.

Anderson's way of proceeding teaches an important lesson about the relation between quantum theory and experimental cosmic ray studies. In the first decades of particle physics, many decisive discoveries happened independently of theory formation. The carrying out and evaluation of many crucial experiments was largely autonomous with regard to the simultaneous development of new theoretical approaches. This autonomy of the experiment was also emphasized by Galison (1987).

It concerned the measurement theories. In the early cosmic ray studies as well as in the current experiments of particle and astroparticle physics, the physicists only use reliable, well-confirmed theoretical background knowledge for the analysis of their experimental data. This is above all true when they explore new empirical grounds, like in the investigation of cosmic rays. (And it is no less true when they search for data that confirm a theoretical hypothesis, as in the case of the neutrino. See Chap. 7.) Sometimes it happens that they disregard an existing theory, as in the case of the positron. Anderson knew the Dirac equation, but he did not use it (Anderson and Anderson, 1983, 140):

> It has often been stated in the literature that the discovery of the positron was a consequence of its theoretical prediction by Paul A.M. Dirac, but this is not true. The discovery of the positron was wholly accidental. Despite the fact that Dirac's relativistic theory of the electron was an excellent theory of the positron, and despite the fact that the existence of this theory was well known to nearly all physicists, including myself, it played no part whatsoever in the discovery of the positron.

His careful analysis of the positron tracks available in his data since 1931 did not consider the Dirac equation, as his published papers show as well (Anderson, 1932; 1933). In the 1983 review of his work he did not explain why he neglected the Dirac equation. Probably he simply did so because Millikan was his supervisor. And probably, like Bothe and Kolhörster or Rossi, he was more interested in the particle content of cosmic rays than in the search for Dirac particles. In contradistinction to him, Blackett and Occhialini were (Blackett and Occhialini, 1933; see Chap. 2). Anderson tried as long as possible to interpret the atypical tracks in terms of the proton charge and mass, as did Millikan. Indeed, Dirac himself had also tried to give such a conservative interpretation, namely to assign the negative energy solutions of his equation to the proton (Pais, 1986, 346–348). Blackett and Occhialini confirmed Anderson's discovery of the positron. In addition they identified processes of pair creation, to which Anderson could track the positrons of the cosmic rays back, too, still in 1933 (Anderson and Anderson, 1983). The process of pair creation was also explained by the Dirac equation.

### *10.4.2 The Phenomenology of Quantum Electrodynamics*

This unexpected success of the supposed "unphysical" negative energy solutions of the Dirac equation encouraged Bethe and other theoreticians to apply quantum electrodynamics to the interactions of charged particles of high energy with matter. In turn, it encouraged Anderson and other experimenters to use the theoretical results for the analysis of their cosmic ray data. All this happened in a step-by-step manner and several puzzles had to be resolved. The development of quantum electrodynamics and its empirical successes took several detours in the 1930s (Cassidy, 1981; Galison, 1987; Schweber, 1994). Abstract quantum field theory was further developed more or less independently of cosmic ray studies and vice versa. Between both fields of research, however, a new intermediate field of research was established, a phenomenology of calculations of the quantum electrodynamic processes relevant for understanding the tracks from cosmic rays. As soon as the Dirac equation was confirmed by the discovery of the positron, Bethe, Bloch, Heitler, and others started to apply it to the reactions of charged particles with matter.

Essential steps towards the calculation of the interactions of relativistic charged particles with matter were taken long before the consolidation of quantum electrodynamics. In 1931, Møller managed to derive the scattering of two electrons from the Dirac equation in the Born approximation of quantum mechanics in a relativistically invariant description (Møller, 1931). In 1932, Bethe used Møller's scattering formula to derive a relativistic formula for the energy loss of electrons due to ionization processes in the matter of a particle detector. However, this formula could be exactly calculated only for the hydrogen atom, that is, for the energy loss of electrons in the cloud chamber (Bethe, 1932). In 1933, Bloch completed Bethe's formula for more complex atoms in a Fermi gas model of the electrons of an atom (Bloch, 1933). In this way the Bethe–Bloch formula was derived for high-energy charged particles, as a law describing the stopping power of atoms. In 1934, the formulas for *bremsstrahlung* and pair production were introduced (Bethe and Heitler, 1934; Pais, 1986, 375–376; Rossi, 1952, 151; Galison, 1987, 103–110). Pair production was discovered in 1933 by Blackett and Occhialini and explained in terms of Dirac's hole theory. With the Bethe–Bloch formula and the formulas for *bremsstrahlung* and pair creation, the energy loss of charged particles due to passing through matter could be completely calculated.

In 1932, by further improving his coincidence method Rossi had already measured the first particle showers (Rossi, 1990, 21–25). Electromagnetic showers were also observed by Blackett and Occhialini and further investigated by Auger (see also Chap. 6). In 1936, the showers were calculated in terms of cascades of *bremsstrahlung* and pair creation by Oppenheimer, and independently by Bhabha and Heitler (see Cassidy, 1981 and the literature given there). In this way, by the middle of the 1930s the foundations of the measurement theory of modern particle physics were completed.

But at that time quantum electrodynamics was still far from consolidation. When quantum field theory was developed and refined parallel to relativistic quantum mechanics, severe divergence problems remained, which were resolved only in the 1950s by the renormalization approach. In addition, there was great confusion around the interpretation of the tracks from cosmic rays. There were tracks due to particles of medium mass (then called "mesotrons", now called mesons), which were only identified in 1936. Prior to this, they gave rise to substantial confusion as regards the origin and the particle content of particle showers from cosmic radiation (Rossi, 1983, 110–124; Cassidy, 1981; and Chap. 6). In 1935, many field theorists including Dirac himself thought that quantum electrodynamics should be given up (Schweber, 1994, 84). Even the 'pragmatic' theorists working on the phenomenology of quantum electrodynamics despaired. They thought that quantum electrodynamics fails at high energies. As Cassidy put it (Cassidy, 1981, 14), they

> were using experiments to unravel experimental data at low energies, while using experiments to justify theoretical failure at high energies. Such dualism typified their expectations

> and goals, their strong belief that applications of QED failed at high energies, and their need to find a rough limit below which they could safely apply the theory.

The calculations seemed to be valid only for the "soft", non-penetrating electron-photon component of showers, but not for the "hard", high-energetic shower components. Hence, there was a phenomenology of quantum electrodynamics that turned out to be correct for low energies but seemed to go astray for high energies.

### *10.4.3 Refining the Mass Measurement*

Only in 1936, Anderson concluded that there was a new charged particle with a mass between the electron and proton masses. The lack of confidence in quantum electrodynamics and in the calculation of the energy loss by scattering processes at high particle energies delayed its identification for two years. The delay was due to the vicious circle described above. With the insufficiently selective experimentation methods of 1933, Anderson found some particle tracks which he regarded as tracks of electrons in his photographs from the cosmic rays. However, these were the still unidentified muon tracks. Their signature confirmed the suspicion that quantum electrodynamics fails at high particle energies (Cassidy, 1981, 2 and 12–15; Anderson and Anderson, 1983, 143–146).

Anderson used the same cloud chamber as for the discovery of the positron, but worked with a better developed experimenting technique (Galison, 1987, 137). Later, he further refined his measuring methods by the installing a platinum absorber of width 1 cm into the cloud chamber. This absorber stopped the electrons and let the muons pass through. The obvious conclusion was that the platinum absorber was passed by a charged particle which was heavier than the electron. In their 1936 paper, Anderson and Neddermeyer finally concluded that there is another option than the failure of quantum electrodynamics at high energies, namely (after Cassidy, 1981, 22):

> that either the theory of absorption breaks down for energies greater than about 1000 MeV, or else that these high energy particles are not electrons.

Quantum electrodynamics was much better validated for the energy loss by ionization processes. It predicted that the new particle had to be lighter than the proton, if its predictions for particles of low to medium energy were correct at all (Pais, 1986, 432; Cassidy, 1981, 14).

The process which led step by step to the identification of the muon has been investigated in detail by the historian of science Galison (1987, Chap. 3; in particular 126–133). He emphasizes that the exact instant of the muon discovery cannot be defined because the discovery was due to a collective learning process amongst particle physicists (as for the neutrino). In this collective learning process, more and more explanatory options were carefully eliminated:

> The move towards acceptance of the muon was not the revelation of a moment. But by tracing an extended chain of experimental reasoning like this one, we have seen a dynamic

process that, while sometimes compressed in time, has occurred over and over in particle physics. With the discovery of the neutrino, for instance, one sees such a gradual elimination of alternatives (Galison, 1987, 133).

Due to the discovery of the muon, the confidence in quantum electrodynamics increased. But further puzzles remained. Until 1947, it was not possible to distinguish the muon discovered in 1936 and the pion which had already been predicted in 1935 in Yukawa's theory of the strong interaction. It had been mistaken for the muon from 1936 to 1947, since there was no precise mass measurement available. The muon and the pion have masses of 106 $\text{MeV}/c^2$ and 140 $\text{MeV}/c^2$, respectively (Rossi, 1952, 162–163; Lattes, 1983). They could only be discriminated when more precise independent methods of mass measurement became available. Anderson could not have dreamed of resolving such a small mass difference with his methods of 1936.

The puzzles of particle identification of the mid-1930s–1940s could only be resolved when nuclear emulsions became available. They were developed by the physicist Marietta Blau (Halpern, 1993; Galison, 1997; Strohmaier and Rosner, 2006) who worked at the Vienna Institute of Radium Research from 1927 until 1937, without any salary, however. Since 1932, she was working with her former PhD student Hertha Wambacher on the improvement of photographic plates. In 1937, she had the opportunity to expose her plates in Hess' laboratory at the top of the Hafelekar (Innsbruck) for five months, in 2 300 m of height. Blau and Wambacher discovered not only very long tracks of protons of extremely high energy but also star-shaped tracks which stemmed from nuclear disintegration, i.e., a scattering event in which a high particle from cosmic rays of extremely high energy made an atomic nucleus burst, giving rise to several tracks of protons or $\alpha$-particles. The discovery gave rise to a publication in Nature (Blau and Wambacher, 1937). In 1938, she emigrated first to Oslo and later to Mexico and the USA, where she arrived in 1944, without any possibility to return to her scientific work there, however. Blau was finally awarded for her merits by the Schrödinger prize of the Austrian Academy of Science in 1962, a few years before she died almost forgotten in 1970.

It was Powell who improved the new photographic method further and used it in order to investigate the particle content of cosmic rays, in the 1940s. His nuclear emulsions made it possible to record the tracks of charged particles and to develop their photographs with a very high spatial resolution (of 1 μm) (Rossi, 1952, 127–142; Powell et al., 1959, 26–32). Now, a semi-empirical method of mass measurement was available that was completely independent of quantum electrodynamics. They allowed one to determine the mass of a particle with high precision, independently of its range, using the density of the measurement points (Rossi, 1952, 138–142). So Powell discovered the pion, in 1947, and received the Nobel prize in 1950 for the development of the photographic method and the discovery of the pion.

Further improvements in the recording and analyzing of particle tracks from cosmic rays came with the bubble chamber invented by Glaser in 1952. When particle physics shifted from cosmic ray studies to scattering experiments at particle accelerators, around 1960, the bubble chamber made it possible to measure the semi-empirical energy–range relation with high precision for many particle types and over a large energy range.

## 10.5 From Particle Tracks to Quantum Probabilities

While the observation of particle tracks on a photo plate is also possible for the layman, the data analysis of these tracks makes use of a very complex measurement theory. This measurement theory has historically grown, and it has a layer structure. Indeed it is based on Mott's and Bethe's semi-classical model of a particle track up to the present day. Its foundations were laid in 1930 by Bethe's seminal paper, based on the idea of correspondence to the classical case and the quantum mechanics of scattering. The formulas for the energy loss of charged particles in matter were elaborated by Bethe, Bloch, Heitler, and others in the early days of quantum electrodynamics. However, before the consolidation of quantum electrodynamics these formulas could not be used for the mass measurement of cosmic rays. In order to resolve the puzzles of particle identification, independent semi-empirical formulas such as the energy–range relation were employed.

The resulting collection of laws for the analysis of particle tracks was probed and refined in cosmic ray studies during the 1930s–1950s. After the consolidation of quantum electrodynamics, more and more new particles were detected, and conservation laws for quantum properties such as spin, parity, isospin, etc. were added in order to understand their properties and their reactions. The conservation laws brought theoretical structure (i.e., dynamic symmetries) in the increasing particle zoo.

The measurement theory was taken over to particle physics in the 1960s. It was complemented by the Breit–Wigner formula for the energy and width of a resonance (which is due to the decay of an unstable particle), and further refined by radiative corrections based on higher-order terms of perturbation theory. In the era of particle accelerators, with increasing data sets, it was completed by statistical methods. Finally, advanced computer methods such as Monte Carlo simulations entered. The resulting collection of theoretical and semi-empirical laws has been used for the analysis of high-energy scattering experiments up to the present day. With the rise of modern astroparticle physics in the late 1980s, it was finally taken back to the cosmic ray studies with neutrino telescopes and other most advanced particle detectors.

### *10.5.1 A Multi-layered Measurement Theory*

It is worth to have a closer look at this measurement theory. It has a multi-layer structure in two regards.

*First*, it has historically grown by adding one layer of well-established laws after the other. The body of safe background knowledge incorporated in the current measurement theory of particle physics and astroparticle physics has grown stepwise. During this growth, each new law was established by only using well-confirmed laws of previous layers as background knowledge for empirical confirmation. Once quantum electrodynamics was well-confirmed by the use of nuclear emulsions and semi-empirical laws, the formulas for *bremsstrahlung* and pair creation were used for the analysis of particle tracks, and higher-order contributions to the scattering amplitudes for the statistical data analysis of high-energy scattering experiments, but before they were not.

*Second*, it has different theoretical layers. It is made up of classical and semi-classical laws that apply to particle tracks, conservation laws that apply to scattering events, quantum laws applied at the probabilistic level, and statistical laws for the analysis of large data samples.

Let us now have a look at these theoretical layers and the way in which they are connected.

On the one hand, it is a complex aggregation of semi-empirical and theoretical background knowledge which has historically grown, is theoretically and empirically well-justified in all parts, but is far from being a well-defined theory. On the other hand, however, it demonstrates that the shift from waves to particles in the cosmic ray studies of the late 1920s was not the last word about the nature of cosmic rays, given that their *quantum* nature has to be taken into account.

Due to this quantum nature, the analysis of individual particle tracks has to be corrected by quantum laws. But due to the irreducibly probabilistic character of quantum laws, the corrections can only be applied at the probabilistic level, taking into account sophisticated statistical methods such as unfolding procedures in order to correct the data samples. Taking all non-quantum measurement errors (such as mere statistical effects) aside, these methods indicate a shift from particles back to waves, or to be more precise, to quantum waves with probability amplitudes. This shift is due to the wave–particle duality and the probabilistic interpretation of quantum theory, according to which the relative frequency of particles corresponds to the (squared) scattering amplitudes of probability waves. So let us have a closer look at the layers of this measurement theory and the shift from particles to quantum waves, which it implies.

### *10.5.2 Particle Tracks*

The quantities attributed to individual particle tracks are mass and charge, which are measured based on Mott's and Bethe's semi-classical model explained above. The analysis of particle tracks employs classical and semi-classical laws. For charged particles, the Lorentz force (Lorentz, 1895) connects the ratio of mass and charge of a particle to its acceleration in electric and/or magnetic fields. It was applied to charged particles from the very beginnings of subatomic physics (Thomson, 1897). The mass of charged particles is measured from a large number of semi-empirical laws such as the empirical energy–range relation (Rutherford et al., 1930; Rossi, 1952), which were measured and used since the early radioactivity and cosmic ray studies. In addition, classical considerations suggest that the track length corresponds to the absorption length of a stable particle or to the decay time of an unstable particle. Even though both are genuinely quantum, i.e., probabilistic quantities, there is no problem to attribute them to an individual particle track once the track is generated (and all quantum measurements are done).

For the measurement of the dynamic properties of uncharged particles, there are few laws at the level of individual particle tracks. Even though the energy-range relation may be used for them, too, in order to determine the mass of uncharged particles more precise, scattering events or resonances have to be measured.

### *10.5.3 Scattering Events*

The analysis of scattering events is based on conservation laws. The conserved quantities attributed to scattering events are genuine quantum properties. The corresponding conservation laws, however, hold for any individual particle reaction. Hence, they apply to the individual particle tracks observed in cloud chamber or bubble chamber photographs, on nuclear emulsions, or in the computer reconstruction of scattering events recorded by electronic devices.

For the analysis of scattering events, the laws of relativistic kinematics (Einstein, 1905, 1917) are basic. They describe the propagation and scattering processes of relativistic particles. They have been used in cosmic ray studies and particle physics since the validation of the Compton effect. Later, the conservation laws for momentum–energy and quantum properties such as spin, parity, isospin, etc. were added to them. They were connected to the symmetries of subatomic particles according to the Noether theorem (Noether, 1918), first used for the concept of isospin (Heisenberg, 1932), taken up in Yukawa's theory of mesons (Yukawa, 1935; see Chap. 2) and generalized to any relativistic field theory in Wigner's mathematical approach (Wigner, 1939). The conservation laws and symmetries brought theoretical structure into the increasing particle zoo of the 1940s and 1950s. In a history that took many detours, they paved the way to the gauge invariant quantum field theories beyond quantum electrodynamics and to the current standard model of particle physics.

### *10.5.4 Probabilistic Scattering Cross Sections*

All quantities which derive from the quantum theory of scattering are irreducibly probabilistic. This is in particular true of the cross section of a given kind of particle reaction, the quantity in which a quantum field theory comes down to earth. Quantum field theory and the experiment meet in the scattering cross section of a particle reaction. As an empirical quantity, the cross section is obtained by counting the relative frequency of scattering events of a given dynamic type. As a theoretical quantity, it is calculated within the quantum mechanics of scattering from the S-matrix element of a quantum field theory for the corresponding particle reaction.

In the experiments of particle physics and astroparticle physics, some probabilistic measurement laws can be embedded in the semi-classical measurement theory of particle tracks. The quantum laws for the dissipation of energy along a particle track and the deflection of charged particles in matter are based on the quantum electrodynamic formulas of ionization, *bremsstrahlung*, pair creation, and multiple scattering calculated in the 1930s. However, average values calculated from these formulas may be used in the semi-classical model of an individual particle track. The experimental practice of analyzing particle tracks relies on the belief that the validity of the semi-classical model of particle tracks and particle reactions does not break down at once but only step by step. This assumption is empirically supported by the energy–range relation, which can be measured at a particle accelerator by detecting the particles of a beam of well-defined energy, serving as an empirical test of the quantum electrodynamic predictions for energy loss.

Even though in the relativistic domain the classical picture of energy loss along a particle track no longer holds, in high-energy physics it became experimental practice to apply them to individual particle tracks in the spirit of Bethe's semi-classical calculation. There was simply no other way to handle them. The resulting semi-classical formulas for $\Delta E/\Delta x$ have been used for decades. They belong to the familiar background knowledge of particle physics, which has been taken over to astroparticle physics in the late 1980s. At the individual level, this procedure may give wrong results because the energy loss $\Delta E$ per length $\Delta x$ is a probabilistic mean value whereas the single particle tracks are due to stochastic scattering effects. At high energies, the distribution of the actual energy loss per detector length around the mean value is highly non-Gaussian due to the processes of *bremsstrahlung* and pair creation which give rise to large fluctuations. The systematic errors produced in this way are corrected by means of statistical unfolding methods.

In the remaining probabilistic quantum formulas used for data analysis, the concept of a particle which causes an individual particle track is abandoned in favor of the probabilistic account of the propagation of quantum waves. The Breit–Wigner formula for the mean energy and width of a resonance relates the energy and width of a resonance to the mass and lifetime of an unstable quantum particle (Breit and Wigner, 1936). Here, 'particle' means an unstable quantum state with a certain probabilistic decay time.

Another typically probabilistic quantity used for the measurement of subatomic structures is the scattering cross section of pointlike or structureless particles. The calculation of this quantity is needed for the measurement of form factors and structure functions of the proton and neutron. With increasing energy of the probe particles, its calculation is based on the formulas for Rutherford scattering (the one and only case with exact classical correspondence), Mott scattering, Dirac scattering on the proton, etc. Like the formulas for energy loss, they stem from the quantum mechanics of scattering and the phenomenology of quantum electrodynamics of the early 1930s. But it makes no sense to apply them to individual particle tracks. According to them, the scattering is due to the diffraction of a quantum wave at a pointlike scattering center. In the case of Rutherford scattering, the correspondence to the classical case only holds on the probabilistic level of the probability of the scattering results. The particle aspect only shows up in the relative frequency of particle detections in a given direction. However, there is no classical path. In particular, the classical impact parameter of an individual scattering process has no quantum correlate.

The probabilistic nature of quantum waves is most obviously employed in the measurement of CP violations or of neutrino oscillations. Here, quantum superpositions of different particles (or quantum states of given dynamic properties) are measured. CP violations are measured in the scattering experiments of particle physics, whereas neutrino oscillations have been detected by measuring the solar neutrino flux. It has been argued that CP violations are a good example of quantum superpositions that cannot be interpreted in terms of an ignorance interpretation of quantum probabilities (Müller, 1993). (The ignorance interpretation of quantum theory claims that quantum probabilities are just due to missing knowledge about the quantum properties, but not to something like objectively undetermined properties.) I suppose that neutrino oscillations are, too.

### *10.5.5 Statistical Methods*

In addition to the measurement laws based on classical physics and quantum physics, statistical methods also used in other fields of empirical science are employed in the data analysis of current particle and astroparticle physics. The methods for calculating the relative frequencies of scattering events from the cross sections of the respective particle reaction include Monte Carlo simulations and the above mentioned unfolding methods. They are employed in the high-energy scattering experiments of particle physics as well as in the measurement of the flux of cosmic rays in astroparticle physics, in order to get results of high precision from the analysis of very large data samples. Today, also the most advanced methods of multivariate data analysis from statistics are employed.

## 10.6 Changing the Focus

The measurement theory of cosmic ray studies has been taken over to particle physics, elaborated further in the era of the big accelerators, and finally taken back to the measurement of cosmic rays in recent astroparticle physics. We have seen that the foundations of this measurement theory date back to the first half of the 20th century. Its core is a collection of classical laws, like the Lorentz force, the laws of relativistic kinematics, the quantum mechanics of scattering, and the formulas of quantum electrodynamics calculated in the early 1930s.

As sketched above, for almost two decades there were no independent measurement methods for the mass attributed to a particle track. The analysis of particle tracks was trapped in a vicious circle which was not easy to resolve, as the puzzles of quantum electrodynamics and particle identification of the 1930s–1940s show. Only due to the use of nuclear emulsions it became possible to escape the trap. Hence, the notorious theory-ladennes of the measurement results of particle physics and astroparticle physics may be a question of principle for philosophers or historians of science who are not familiar with the details of data analysis. For physicists, it is not. For them, it is just pragmatic problem. As long as there is no vicious circle in the methods of data analysis, there is no case for any objection. As long as there would be one, the search for independent measurement methods goes on. And when they are found, the physicists attempt to confirm the laws which are not well-established in order to employ them in further data analysis.

After the consolidation of quantum electrodynamics, the measurement theory discussed here was extended by the conservation laws for dynamic quantum properties such as spin, isospin, parity, etc., and the cross sections of different particle reactions. The further development of the measurement theory is a case of cumulative scientific progress in Popper's or even Carnap's sense. The background knowledge having grown in this way is much more stable than the theoretical speculations that paved the way to the quantum field theories of the standard model of particle physics. The measurement theory of particle and astroparticle physics indicates the autonomy of experiment stated by Galison (1987). It consists of well-established phenomenological laws (Cartwright, 1983) of various theoretical levels. For such phenomenological laws, incommensurability in Kuhn's sense (Kuhn, 1962) plays almost no role. This is even true of the relation between the individual particle tracks and the probabilistic quantum laws of energy loss in the relativistic domain, where the correspondence to the classical case completely breaks down. This hardest known case of incommensurability is simply overcome by means of statistical unfolding methods.

However, there is another important philosophical issue. The background knowledge did not substantially change in the transition from cosmic ray studies to particle physics and from there to astroparticle physics. But the focus of interest changed, and with it the use of the different parts of the measurement theory and the inclusion of new measurement laws from astrophysics. Cosmic ray studies focused on the analysis of individual particle tracks and scattering events detected by means of the cloud chamber and nuclear emulsions. The main task was to identify the dynamic properties of the particles that made the tracks. The scattering experiments of particle physics aim at hadron spectroscopy, precise measurement of structure functions, etc., testing the current quantum field theories and at detecting new particles predicted by theories beyond the standard model of particle physics. Here, the main task became the generation of a beam of well-defined particle type and energy, the investigation of the particle reactions to which it gives rise, and the precise measurement of many kinds of differential and total scattering cross section. In contradistinction to these goals, the experiments of astroparticle physics aim at identifying cosmic rays of given type and their sources. In particular, the experiments of astroparticle physics attempt to distinguish primary cosmic rays from the secondary particles produced by scattering in the atmosphere. They measure the flux of cosmic rays of given type and energy in order to identify their cosmic sources and to explain the mechanisms of their generation.

This shift of interest has several pragmatic consequences, which open a new and interesting field of investigation in the philosophy of physics. Let me just mention two of them and sketch some of their philosophical implications.

*First*, astroparticle physics is mainly interested in uncharged particles which point to their cosmic origin. Charged particles are deflected by cosmic magnetic fields. Hence, they carry no information about their origin. Therefore, the classical Lorentz force plays no longer a crucial role in the data analysis of the particle tracks measured in astroparticle physics. The Lorentz law still serves to sort out the charged particle background of the photon or neutrino flux which is measured. But a precise measurement of the mass and momentum from a particle track is no longer needed. Hence, it is no longer necessary to reconstruct the individual particle tracks. The particle tracks are simply fitted as straight lines; their $\chi^2$ is optimized by means of multivariate statistical analysis, and the probability that the track belongs to a specific kind of particle of given energy is calculated. The sample of the individual particle tracks reconstructed in the high precision measurements of particle physics is now replaced by a probability cloud.

*Second*, there is an important kinematic distinction between the collider experiments of recent high-energy physics and the measurements of astroparticle physics. The dominating kinematic domain of collider experiments are scattering events with a high transverse momentum. For a collider experiment, scattering events in forward direction are neither experimentally accessible, nor can they be calculated with sufficient precision. In the experiments of astroparticle physics, however, the accessible kinematics is vice versa. This matter of fact indicates that the collider experiments of particle physics and the cosmic particle flux measurements of astroparticle physics complement each other.

The first consequence raises an interesting question concerning the interpretation of quantum theory. To replace the reconstruction of individual particle tracks by a probability cloud means in a certain sense that the flux measurement of cosmic rays directly captures the reference of quantum states to probability waves. Instead of attributing dynamic properties to individual particle tracks in a semi-classical model and correcting for the errors thus produced, the data analysis takes only place at the probabilistic level. Here, an obvious philosophical issue is: In which terms may this probabilistic data analysis be understood? In terms of an ensemble interpretation of quantum theory, in terms of decoherence, or in other terms?

The second consequence indicates that the scattering experiments of particle physics and the flux measurements of cosmic rays are complementary regarding their respective domains. Indeed they complement each other in a more general sense, by exploring the grounds of 'new physics'. The physicists of both fields claim that their respective experiments are relevant for the 'big questions' of unified physics at a small scale and at a large scale. The experiments of particle physics attempt to make the bridge from the current standard model of particle physics to physics beyond: to the reign of supersymmetry, superstring theory, loop quantum gravity, or some other unifying theory. The experiments of astroparticle physics make the bridge from subatomic particles to the cosmic sources which emit them. Both approaches are complementary. Both rely on the same measurement theory used with a different focus. But astroparticle physics employs measurement methods and concepts of astrophysics in addition to those of particle physics.

The complementarity of the respective experimental methods and concepts should be investigated in more detail. Here, the philosophical issue in view of two competing fields of investigation is: What is the specific significance of astrophysical methods and concepts in astroparticle physics? For example, which role plays the concept of messenger particles for unifying the physics at a small scale and at a large scale? (For a first approach see Falkenburg and Rhode, 2007; Falkenburg, 2012.) These philosophical topics are beyond the scope of this historical introduction to astroparticle physics. However, the field for their investigation now is open.

## References

Anderson, C.D.: The apparent existence of easily deflectible positives. Science **7**, 238 (1932)

Anderson, C.D.: The positive electron. Phys. Rev. **43**, 491–494 (1933)

Anderson, C.D., Anderson, H.L.: Unraveling the particle content of cosmic rays. In: Brown, L.M., Hoddeson, L. (eds.) The Birth of Particle Physics, pp. 131–154. Cambridge University Press, Cambridge (1983)

Bethe, H.: Zur Theorie des Durchgangs schneller Korpuskularstrahlen durch Materie. Ann. Phys. **5**, 325–400 (1930)

Bethe, H.: Bremsformel für Elektronen relativistischer Geschwindigkeit. Z. Phys. **76**, 293–299 (1932)

Bethe, H., Heitler, W.: On the stopping of fast particles and on the creation of positive electrons. Proc. R. Soc. A **146**, 83–112 (1934)

Blackett, P.M.S., Occhialini, G.: Some photographs of the tracks of penetrating radiation. Proc. R. Soc. Lond. Ser. A **139**, 699–726 (1933)

Blackmore, J.T.: Ernst Mach; His Work, Life, and Influence. University of California Press, Berkeley (1972)

Blau, M., Wambacher, H.: Disintegration processes by cosmic rays with the simultaneous emission of several heavy particles. Nature **140**, 585 (1937)

Bloch, F.: Bremsvermögen von Atomen mit mehreren Elektronen. Z. Phys. **81**, 363–376 (1933)

Bohr, N.: On the theory of the decrease of velocity of moving electrified particles on passing through matter. Philos. Mag. **25**, 10–31 (1913). Repr. in: Bohr Collected Works (BCW), vol. 2, pp. 18–39. North-Holland (Elsevier), Amsterdam

Bohr, N.: On the decrease of velocity of swiftly moving electrified particles on passing through matter. Philos. Mag., **30**, 581–612 (1915). Repr. in: BCW, vol. 8, pp. 127–160

Bohr, N.: Über die Serienspektren der Elemente. Z. Phys. **2**, 423–469 (1920). Engl. transl. in: Bohr (1922)

Bohr, N.: The Theory of Spectra and Atomic Constitution. Cambridge University Press, Cambridge (1922)

Bohr, N.: Como Lecture (1927). Modified version: The quantum postulate and the recent development of atomic theory. Nature **121**, 580–590 (1927). Repr. in: Wheeler and Zurek, 1983, 87–126. Both versions in: BCW, vol. 6, pp. 109–158

Bohr, N.: Discussion with Einstein on epistemological problems of atomic physics. In: Schilpp, P.A. (ed.) Albert Einstein: Philosopher–Scientist. Library of Living Philosophers, pp. 115–150. Evanston, Illinois (1949). Repr. in: Wheeler and Zurek, 1983, pp. 9–49

Bohr, N., Kramers, H.A., Slater, J.C.: The quantum theory of radiation. Philos. Mag. **47**, 785–802 (1924). Repr. in: van der Waerden, B.L. (ed.): Sources of Quantum Mechanics, pp. 159–176. Dover, New York (1967)

Born, M.: Zur Quantenmechanik der Stoßvorgänge. Z. Phys. **37**, 863–867 (1926a). Engl. transl. in: Wheeler and Zurek (1983), 52–55

Born, M.: Quantenmechanik der Stoßvorgänge. Z. Phys. **38**, 803–827 (1926b)
Bothe, W., Geiger, H.: Über das Wesen des Comptoneffekts. Z. Phys. **32**, 639–663 (1925)
Bothe, W., Kolhörster, W.: Das Wesen der Höhenstrahlung. Z. Phys. **56**, 751–777 (1929)
Breit, G., Wigner, E.: Capture of slow neutrons. Phys. Rev. **49**, 519–531 (1936)
Cartwright, N.: How the Laws of Physics Lie. Clarendon Press, Oxford (1983)
Cassidy, D.: Cosmic ray showers, high energy physics, and quantum field theories: programmatic interactions in the thirties. Hist. Stud. Phys. Sci. **12**, 1–39 (1981)
Compton, A.H.: A quantum theory of the scattering of X-rays by light elements. Phys. Rev. **21**, 483–502 (1923)
Daston, L. (ed.): Biographies of Scientific Objects. University of Chicago Press, Chicago (2000)
Daston, L., Galison, P.: Objectivity. MIT Press, Cambridge (2007)
de Broglie, L.: Ondes et quanta. Comptes Rendus **177**, 507–510 (1923)
Debye, P.J.W.: Zerstreuung von Röntgenstrahlen und Quantentheorie. Phys. Z. **24**, 161–166 (1923). Engl. transl. in: The Collected Papers of P.J.W. Debye, pp. 80–88. Interscience, New York (1954)
Einstein, A.: Über einen die Erzeugung und Verwandlung des Lichts betreffenden heuristischen Gesichtspunkt. Ann. Phys. **17**, 132–148 (1905)
Einstein, A.: Zur Quantentheorie der Strahlung. Phys. Z. **18**, pp. 121–128 (1917)
Einstein, A., Podolsky, B., Rosen, N.: Quantum mechanical description of reality be considered complete? Phys. Rev. **47**, 777–780 (1935)
Falkenburg, B.: Particle Metaphysics. A Critical Account of Subatomic Reality. Springer, Heidelberg (2007)
Falkenburg, B.: Pragmatic unification, observation and realism in astroparticle physics. J. Gen. Philos. Sci. (Z. Allg. Wiss.theor.) (2012, to appear)
Falkenburg, B., Rhode, W.: Astroteilchenphysik: Die Brücke zwischen Mikro- und Makrokosmos. In: Falkenburg, B. (ed.) Natur – Technik – Kultur, pp. 57–90. Mentis, Paderborn (2007)
Franklin, A.: The Neglect of Experiment. Cambridge University Press, Cambridge (1986)
Franklin, A.: Experiment, Right or Wrong. Cambridge University Press, Cambridge (1990)
Franklin, A.: Are There Really Neutrinos? An Evidential History. Perseus Books, Cambridge (2001)
Galison, P.: How Experiments End. University of Chicago Press, Chicago (1987)
Galison, P.: Marietta Blau: between nazis and nuclei. Phys. Today **50**, 42 (1997)
Geiger, H., Marsden, E.: Die Zerstreuungsgesetze der $\alpha$-Strahlen bei großem Ablenkungswinkel. Philos. Mag. **25**, 604–623 (1913)
Greenstein, G., Zajonc, A.G.: The Quantum Challenge. Modern Research on the Foundations of Quantum Mechanics. Jones & Bartlett, Boston (1997)
Halpern, L.: Marietta Blau (1894–1970). In: Grinstein, L.S., Rafailovich, M.H. (eds.) Women in Chemistry and Physics. A Biobibliographic Sourcebook, p. 58. Greenwood Press, London (1993)
Heisenberg, W.: Physikalische Prinzipien der Quantentheorie. Hirzel, Leipzig (1930). Repr. in: Bibliographisches Institut, Manheim (1958). Engl. transl. by C. Eckart and F.C. Hoyt: Physical Principles of Quantum Theory. University of Chicago Press, Chicago
Heisenberg, W.: Über den Bau der Atomkerne. Z. Phys. **77**, 1–11 (1932)
Heitler, W., Sauter, F.: Stopping of fast particles with emission of radiation and the birth of positive electrons. Nature **132**, 892 (1933)
Kuhn, Th.S.: The Structure of Scientific Revolutions. University of Chicago Press, Chicago (1962). 2nd edn. (1970)
Lattes, C.M.G.: My work in meson physics with nuclear emulsions. In: Brown, L.M., Hoddeson, L. (eds.) The Birth of Particle Physics, pp. 307–310 Cambridge University Press, Cambridge (1983)
Lorentz, H.A.: Versuch einer Theorie der electrischen und optischen Erscheinungen in bewegten Körpern, Sect. 12. Brill, Leiden (1895). Repr. in: Collected Papers, vol. 5. The Hague: Martinus Nijhoff (1936)

Millikan, R.A.: The isolation of an ion, a precision measurement of its charge, and the correction of Stokes's law. Phys. Rev. **32**, 349–397 (1911)
Millikan, R.A.: The Electron. University of Chicago Press, Chicago (1917)
Møller, Ch.: Über den Stoß zweier Teilchen unter Berücksichtigung der Retardation der Kräfte. Z. Phys. **70**, 786–795 (1931)
Mott, N.F.: The wave mechanics of $\alpha$-rays. Proc. R. Soc. A **126**, 79–84 (1929). Repr. in: Wheeler and Zurek (1983), 129–134
Müller, A.: Complementarity in the neutral kaon system. Philos. Nat. **30**, 247–253 (1993)
Nye, M.: Molecular Reality: A Perspective on the Scientific Work of Jean Perrin. MacDonald, London (1972)
Noether, E.: Invariante Variationsprobleme. Nachr. Ges. Wiss. Gött., Math.-Phys. Kl., 235–257 (1918)
Pais, A.: Inward Bound. Clarendon Press, Oxford (1986)
Perrin, J.: Ann. Chim. Phys. **18**, 1–14 (1909) Engl. transl. by F. Soddy: Brownian Motion and Molecular Reality. Dover Phoenix Editions, Book on Demand, (1909)
Pickering, A.: Constructing Quarks. Edinburgh University Press, Edinburgh (1984)
Powell, C.F., Fowler, P.H., Perkins, D.H.: The Study of Elementary Particles by the Photographic Method. An Account of the Principal Techniques and Discoveries. Illustrated by an Atlas of Photomicrographs. Pergamon Press, London (1959)
Riordan, M.: The Hunting of the Quark. Simon & Schuster, New York (1987)
Rossi, B.: High-Energy Particles. Prentice-Hall, New York (1952)
Rossi, B.: The decay of "Mesotrons" (1939–1943). Experimental particle physics in the age of innocence. In: Brown, L.M., Hoddeson, L. (eds.) The Birth of Particle Physics, pp. 183–205 Cambridge University Press, Cambridge (1983)
Rossi, B.: Moments in the Life of a Scientist. Cambridge University Press, Cambridge (1990)
Rutherford, E., Chadwick, J., Ellis, C.D.: Radiations from Radioactive Substances. Cambridge University Press, Cambridge (1930). Repr.: (1951)
Schrödinger, E.: Über den Comptoneffekt. Ann. Phys. **82**, 257–264 (1927)
Schweber, S.S.: QED and the Men Who Made It. Dyson, Feynman, Schwinger, and Tomonaga. Princeton University Press, Princeton (1994)
Simons, P.: Parts. A Study in Ontology. Clarendon Press, Oxford (1987)
Skobeltzyn, D.V.: The early stage of cosmic particle research. In: Sekido, Y., Elliot, H. (eds.) Early History of Cosmic Ray Studies – Personal Reminiscences with Old Photographs, pp. 47–51. Reidel, Dordrecht (1985)
Strohmaier, B., Rosner, R. (eds.): Marietta Blau. Stars of Disintegration. Biography of a Pioneer of Particle Physics. Ariadne, Riverside (2006)
Thomson, J.J.: Cathode rays. Philos. Mag. **44**, 293–316 (1897)
Thomson, J.J.: On the masses of the ions in gases at low pressures. Philos. Mag. **48**, 547–567 (1899)
Townsend, J.S.: On electricity in gases and the formation of clouds in charged gases. Proc. Camb. Philos. Soc. **9**, 244–258 (1897)
Trigg, G.L.: Crucial Experiments in Modern Physics. Crane Russah & Co, New York (1971)
von Neumann, J.: Mathematische Grundlagen der Quantenmechanik. Springer, Berlin (1932). Engl. transl. by R. Beyer: Mathematical Foundations of Quantum Mechanics. Princeton University Press, Princeton (1955)
Yukawa, H.: On the interaction of elementary particles. Proc. Phys.-Math. Soc. Jpn. **17**, 48–57 (1935)
Wheaton, B.R.: The Tiger and the Shark. Cambridge University Press, Cambridge (1983)
Wheeler, J.A., Zurek, W.H. (eds.): Quantum Theory and Measurement. Princeton University Press, Princeton (1983)
Wigner, E.P.: On unitary representations of the inhomogeneous Lorentz group. Ann. Math. **40**, 149–204 (1939)

# Appendix A
# Timetable

1054 Chinese astronomers observe a supernova explosion.
1521 Observation of the Magellanic Clouds.
1572 Brahe discovers the supernova SN 1572.
1604 Kepler discovers the supernova SN 1604.
1609 Galilei is the first to use optical instruments for astronomical observations.
Kepler publishes the fundamental laws of planetary movement.
1693 Newton notes the instability of a stable, infinitely extended universe.
1733 D'Ortous de Mairan suggests the solar origin of the polar lights.
1750 Wright discusses galaxies and the shape of the Milky Way.
1755 Kant publishes his *Allgemeine Naturgeschichte und Theorie des Himmels* (*Universal Natural History and Theory of the Heavens*).
1781 The final version of the *Messier Catalogue of Nebulae and Star Clusters* is published.
1785 Coulomb recognizes the loss of charge by isolated charged conductors.
1845 Discovery of a spiral nebula by Rosse.
1857 Geißler develops the electrical discharge tube.
1888 Dreyer publishes *The General Catalogue of Nebulae and Clusters of Stars*.
1895 Wilson develops a first version of the cloud chamber.
Röntgen discovers X-rays.
1896 Becquerel discovers radioactivity.
1897 Townsend localizes single charge carriers unsharply.
J.J. Thomson measures $e/m$ for cathode rays and concludes that they consist of electrons.
1898 Presenting the results of their research Marie and Pierre Curie introduce the term *Radioactivity*.
1899 Villard discovers $\gamma$ rays.
Hartmann observes calcium lines in the interstellar medium spectroscopically.
1900 As explanation for the ionisation of charged conductors, Wilson suggests small amounts of radioactive substances.
Julius Elster and Hans Geitel arrive at the same conclusion.
Planck suggests the quantization of the interaction between radiation and matter.

B. Falkenburg, W. Rhode (eds.), *From Ultra Rays to Astroparticles*,
DOI 10.1007/978-94-007-5422-5, 

1901 Wilson suggests to look for the source of the ionisation of charged conductors outside the atmosphere.
1902 Linke performs twelve balloon-flights in altitudes up to 5 500 meters with the result of increasing ionisation in height over 3 000 meters.
Kaufmann convincingly shows that $\beta$-rays are electrons.
Kennely and Heaviside suggest an electrically conducting layer in the upper atmosphere.
1903 Rutherford suggests terrestrial, radioactive substances to be the reason of the ionisation of charged conductors.
1905 Einstein publishes his light quantum hypothesis, the theory of Brownian motion and Special Relativity.
1906 Rutherford and his assistants carry out their scattering experiments.
1908 Perrin's measurements confirm Einstein's theory of Brownian motion.
Flemming and Bergwitz start a series of balloon-flights but due to problems with their measurement apparatus, they yield no convincing results.
Gockel uses the term *cosmic radiation*.
1909 In three balloon flights Gockel confirms that ionisation decreases only slowly with the altitude.
Under the direction of Rutherford, Geiger and Marsden measure unexpected backward scattering and suggest that atoms have a nucleus.
1910 Millikan performs his oil droplet experiment.
After designing and building the electrometer, Wulf conducts measurements on the Eiffel Tower. The results contradict the assumption that the radiation leading to the ionisation of charged conductors is terrestrial. The same conclusion is reached by Pacini, Simpson and Wright.
1911 Wilson uses the cloud chamber to visualize $\alpha$- and $\beta$-rays.
Hertzsprung reveals the main sequence.
1912 Victor Hess recognizes an increase of radiation in altitudes of over 3 000 meters. As a reason he suggests the existence of *Höhenstrahlung*.
Leavitt discovers the correlation of brightness and period of Cepheid variable stars.
1913 Bohr develops his first quantum theory of atoms.
1914 Walter Kolörster confirms the results of Hess with further balloon flights.
Russell is able to obtain a diagram similar to the one of Hertzsprung.
1915 Einstein publishes the final version of his field equations of General Relativity.
Schweidler excludes several terrestrial and solar possibilities as origin of the cosmic rays.
Gockel and Hess give further confirmation of *Höhenstrahlung* with long term measurements.
1916 Einstein extends his light quantum hypothesis.
Schwarzschild shows the possibility of the existence of black holes.
1917 Einstein extends his field equations by introducing the cosmological constant.
Slipher finds first evidence for the expansion of the universe.
1919 Eddington gives, influenced by the work of de Sitter, evidence for the theory of General Relativity by observing the eclipse on May 29th.

Kolhörster proposes the possibility of a latitude dependence of the cosmic rays.
Rutherford finds first evidence for the proton.

1920 The discussion about the nature of nebulae culminates in the Great Debate between Shapley and Curtis.
Eddington proposes that the energy of the sun might be produced by the fusion of hydrogen to helium.
Michelson and Pease perform the first measurements of stellar diameters.

1921 Nernst speculates about cosmic rays as by-products of the explosion of stars.
Chadwick and Bieler predict the strong force.

1922 Compton applies relativistic kinematics to the scattering of gamma radiation and electrons.
Friedman publishes his solutions to Einstein's field equations.
Wirtz formulates the idea of an expanding universe.

1923 Compton confirms Photons as particles.

1924 Bohr, Kramers and Slater develop the BKS-theory.
Bothe invents the coincidence method.
De Broglie proposes the wave–particle duality of matter.

1925 Bothe and Geiger validate the Compton-effect for single scattering processes.
Hubble measures the distance to the Andromeda Nebula and in doing so settles the Shapley–Curtis–Debate.
Wooster and Ellis try to solve the issue of monoenergetic electrons in $\beta$-decay calorimetrically.
Uhlenbeck and Goudsmit discover that the electron has spin $1/2$.
Pauli formulates the exclusion principle.
Heisenberg develops matrix mechanics.
Adams observes the redshift.

1926 Millikan finally reaches the same conclusion as Hess and Kolhörster and confirms the existence of the *Millikan Rays* – as he calls the cosmic rays.
Schrödinger develops his wave mechanics.
Born gives the probability interpretation of quantum mechanics.
The term *photon* is proposed by Lewis.

1927 Lemaître obtains the same solutions to Einstein's field equations as Friedman and in doing so gives evidence for the expansion of the universe. His work can be seen as the starting point for modern cosmology.
Dirac gives the basic equations of quantum electrodynamics.
First detection of cosmic rays in a cloud chamber by Skobeltzyn.
Heisenberg states his uncertainty principle.

1928 The Geiger–Müller counter is developed.
Gamow discovers the $\alpha$-decay by quantum tunneling and applies.

1929 Hubble and Humason discover the proportionality between radial velocities and distances of galaxies; they introduce the Hubble constant.

1930 Rossi improves the coincidence method given by Bothe.
Dirac predicts the negative energy solutions of the Dirac equation.
Mott and Heisenberg show that Born's quantum mechanics of scattering predicts particle tracks with a classical shape.

Bethe develops the quantum theory of the passage of a charged particle through matter.
Bothe and collaborators detect the neutron, but fail to recognize it.
Jansky discovers extraterrestrial radio-noise.

1932 Regener confirms the results of Hess and Kolhörster with high accuracy measurements.
Chadwick discovers the neutron.
Using a cloud chamber, Anderson detects the positron in cosmic rays.
Blackett and Occhialini develop the method of triggering the cloud chamber by means of a coincidence counter.

1933 The Bothe–Kolhörster experiment confirms together with the results of Skobeltzyn, Clay and Compton that cosmic rays are charged particles.
Blackett and Occhialini detect $e^+e^-$ pair-production as well as the first particle showers.
Fermi gives the correct theory for $\beta$-decay.
Rossi observes that the coincidence rate increases if an absorber is placed above the counter and thus measures the first particle showers.
Alvarez shows that cosmic radiation consists mainly of positively charged particles.

1934 Zwicky and Baade use the term *supernova* and are able to demonstrate that supernovae are sources of cosmic rays.
Bethe and Heitler lay down the key processes of cascade multiplication in extended air showers.
While examining galaxies in the Coma cluster, Zwicky points out the necessity of the existence of dark matter to hold together the clusters of stars.
The Cherenkov effect is discovered.
Pauli introduces the term *neutrino*.
Bethe and Peierls calculate the cross section for neutrinos interacting with a nucleus.

1935 Yukawa predicts the pion.
The Compton–Getting effect is predicted.
Goeppert-Mayer estimates the lifetime of double beta decay.

1936 The Breit–Wigner formula is stated.
Neddermeyer and Anderson detect a *meson*.
Hubble introduces the classification into spiral, elliptical, barrel spiral, and irregular galaxies.

1937 Blau develops nuclear emulsions.
Majorana proposes a two-component theory of the neutrino.

1938 Auger et al. can show the existence of air showers with coincidence measurement of two counters separated by some distance.
The Bhabha–Heitler theory gives further insights into extended air showers; Carlson and Oppenheimer complete the theory and prove that it qualitatively agrees with the experimental results of Regener and Pfotzer.
Bethe and Weizsäcker propose the CNO-cycle as a source of energy generation in the sun.

1940 Williams and Roberts present the first photograph of a decaying muon.
Bethe and Critchfield propose the pp-chain.
Reber discovers the radio source Cygnus A.

1941 Rasetti is the first to measure the lifetime of a meson.
The term *nucleon* is introduced.

1944 Walter Baade distinguishes between two different stellar populations and in doing so can correct the Hubble constant by a factor 2.7. Further corrections follow.
Discovery of the Seyfert galaxies.

1947 Blackett predicts that relativistic particles passing the atmosphere produce Cherenkov light.
The Feynman diagrams are introduced.

1948 Alpher and Gamow formulate their theory of the origin of the elements.
Hoyle, Bondi and Gold support the steady-state model of the universe.
The first artificial pions are produced in the synchotron cyclotron.

1949 Hoyle is the first to use the term *Big Bang*.
Gamow and collaborators predict the cosmic microwave background.
Fermi is the first to describe a power law distribution of cosmic rays.
Discovery of the extragalactic radio sources NGC 4486 (M87) and NGC 5128 (Centaurus A).
The $K^+$ is detected.

1950 The $\Pi^0$ is detected at the cyclotron at the university of California.

1951 The detector *El Monstro* ist designed and built.
Biermann predicts the solar winds.
The $\lambda^\circ$ and the $K^\circ$ are detected in cosmic rays.

1952 Glaser invents the bubble chamber.
Discovery of the $\Delta$.

1953 Bassi, Clark and Rossi show that the disk of air showers is quite thin.
Galbraith and Jelley prove that air showers generate Cherenkov light.
First measurements of *Project Poltergeist*.

1954 Baade and Minkowsky identify the optical counterpart of Cygnus A.
Yang and Mills develop the gauge theories, providing the theoretical foundations for the later standard model.

1956 Kulikov and Khristiansen discover the *knee* of the cosmic ray spectrum.
First detection of neutrinos by Cowan and Reines near a nuclear reactor.

1957 Hoyle works on the synthesis of the elements in stars with impressive results.
Sputnik is launched.
Schwinger proposes the unification of the weak and electromagnetic interactions.

1958 Porter is the first to succeed in preventing bacterial growth in unfiltered water long enough to realise a stable detector.
Morrison puts forward strong arguments for observing very high energy gamma rays from the Crab nebula.
The Bolivian Air Shower Joint Experiment (BASJE) at Mount Chacaltaya is established.

Discovery of the Van Allen Belt.
1959 The work to build the Volcano Ranch Array begins.
Like Morrison earlier, Guiseppe Cocconi presents strong arguments for having observed very high energy gamma rays from the Crab nebula.
1960 Sandage discovers an unusual blue quasi stellar object by optically observing 3C48.
1961 Ryle and Clarke publish the results of their survey of radio galaxies which contradict the steady-state model.
Linsley discovers the first ultra-high energy cosmic ray event.
1962 The Schrödinger Prize of the Australian Academy of Science is awarded to Blau.
Giacconi discovers the first X-ray sources outside the Solar System.
The fluorescence technique is first discussed at an international forum in La Paz.
Maki, Nakagawa and Sakata introduce neutrino flavour mixing and flavour oscillations.
Muon neutrinos are discovered and distinguished from electron neutrinos.
Hazard, Mackey and Shimmins determine the precise position for 3C273 and state that it is a double source.
1963 First detection of quasars by Schmidt and identifying its Balmer lines.
1964 Zweig and Gell-Mann predict quarks.
1965 Penzias and Wilson discover the cosmic micro wave background and in doing so finally disprove the steady-state model.
Dicke gives a proper explanation of the microwave background.
Reines and colleagues set first astrophysical limits for the energy of neutrinos.
1966 Gould and Schréder make the first prediction of the opacity of the universe.
The Greisen–Zatsepin–Kuzmin cutoff is stated.
1967 Sachs and Wolfe calculate the temperature fluctuations in the cosmic microwave background.
First detection of a gamma ray burst by the Vela satellite.
First detection of solar neutrinos by the Davis experiment.
Wheeler uses the term *black hole*.
1968 The concept of Silk damping is formulated.
Hewish is the first to detect a pulsar.
First measurements at the Sydney University Giant Air Shower Recorder (SUGAR) and Haverah Park.
Confirmation of the quark-parton model at SLAC.
First measurement of solar neutrinos and statement of the *solar neutrino problem*.
1970 The Sunyaev–Zeldovich effect is stated.
The Yakutsk Array begins taking data.
The Uhuru satellite is launched.
1972 Bolton confirms the existence of black holes by detection of Cygnus X-1.
1973 Ostriker and Peebles discover that the amount of visible matter in the disks of typical spiral galaxies is not sufficient for keeping it from flying apart.

1974 Perl discovers the $\tau$-lepton.
Fanaroff and Riley distinguish between FR I and FR II radio sources.
1976 The Faber–Jackson relation is discovered.
1977 The Tully–Fisher relation is discovered.
1978 Gregory and Thompson describe the Coma supercluster.
1981 The Fly's Eye array starts taking data.
1983 Samorski and Stamm report the observation of ultra-high-energy gamma-ray emission from Cygnus X-3.
The Kamiokande Nucleon Decay Experiment (KamiokaNDE) starts taking data.
CERN discovers the $W^{\pm}$- and the $Z^{\circ}$-bosons.
1985 IMB and Kamiokande discover the *atmospheric neutrino anomaly*.
1986 The Kamiokande-II experiment starts taking data and confirms the deficit of solar neutrinos.
1987 Discovery of supernova 1987A.
Detection of solar neutrinos by the Kamiokande detector in real time.
The First International School of Astroparticle Physics takes place.
1989 Cosmic Background Explorer (COBE) is launched and can confirm the cosmic microwave background as the thermal afterglow of the hot early universe as well as giving support of the theory of existence of cold dark matter.
The CASE-MIA array starts its work as does the EAS-TOP-array.
The Whipple collaboration detects TeV gamma rays from the Crab nebula.
Kiel physicists start to build an improved scintillation counter array, the High Energy Gamma Ray Astronomy (HEGRA) experiment.
The LEP accelerator experiments and the SLAC state the existence of only three light neutrino species.
1990 First data are taken at the Akeno Giant Air Shower Array (AGASA).
1991 The *ankle* is detected at Haverah Park, Akeno and Fly's Eye.
The Gallium Experiment (GALLEX) starts taking data.
SAGE and GALLEX confirm the solar neutrino deficit.
1992 Fleury and Vacanti invite the community to a conference at Palaiseau with the aim of forming a project of a major imaging Cherenkov telescope – no consensus can be achieved.
First detection of pp-neutrinos.
The Large Volume Detector (LVD) starts taking data.
Perl discovers the $\tau$.
1996 KArlsruhe Shower Core and Array DEtector (KASCADE) starts taking data.
Super-Kamiokande begins searching for neutrinos.
1997 The High-Resolution Fly's Eye (HiRes) becomes the successor of Fly's Eye.
The HEGRA-collaboration confirms the very high energy $\gamma$-emission of Mkn 501.
1998 The Balloon Observation Of Millimetric Extragalactic Radiation ANd Geophysics (BOOMERanG) starts and finds further support for the existence of dark matter.
Riess and others publish measurements based on several supernovae.

Turner introduces the term *Dark Energy* to explain the acceleration of the expansion of the universe.
The Gallium Neutrino Observatory (GNO) succeeds GALLEX.
The Sudbury Neutrino Observatory (SNO) starts taking data.
The Super-Kamiokande collaboration confirms the existence of neutrino-oscillations and thus the existence of non-zero neutrino mass.

1999 Perlmutter and others publish the cosmological results from the investigations of several supernovae.
The Collaboration of Australia and Nippon for Gamma-Ray Observation in the Outback (Cangoroo III) starts taking data.

2000 The DONUT experiment detects the first tau neutrino; it is stated as oscillating partner to the muon neutrino.

2001 AMANDA II is completed.
SNO announces the observation of neutral currents from solar neutrinos and finally solves the problem of missing solar neutrinos.

2002 The High Energy Stereoscopic System (HESS) starts taking data as well as the first really successful tail-catcher detector Milargo.
Kamioka Liquid-scintillator Anti-Neutrino Detector (KamLAND) begins data taking.
Using neutrinos from an accelerator, KamLAND confirms the results of the observations of the results of the observations of solar neutrinos.

2003 The KASKADE array is expanded to KASKADE-Grande.

2004 MAGIC starts taking data.
Physics data collecting begins at the Pierre Auger Observatory.

2006 The Very Energetic Radiation Imaging Telescope System (VERITAS) Starts taking data.

2007 Data taking starts at the BOREXINO experiment.
A first real-time detection of monoenergetic $^{7}$Be neutrinos is announced.

2008 The Telescope Array (TA) near Utah starts taking data.
HiRes and Auger discover the suppression at the GZK threshold.

2010 The neutrino detector IceCube is completed.

# Appendix B
# Nobel Prizes

All quotes taken from: http://www.nobelprize.org/nobel_prizes/physics/laureates.

1901 Wilhelm Conrad Röntgen "in recognition of the extraordinary services he has rendered by the discovery of the remarkable rays subsequently named after him".

1903 Antoine Henri Becquerel, Pierre Curie and Marie Curie, née Sklodowska. The Nobel Prize in Physics 1903 was divided, one half awarded to Antoine Henri Becquerel "in recognition of the extraordinary services he has rendered by his discovery of spontaneous radioactivity", the other half jointly to Pierre Curie and Marie Curie, née Sklodowska "in recognition of the extraordinary services they have rendered by their joint researches on the radiation phenomena discovered by Professor Henri Becquerel".

1905 Philipp Eduard Anton von Lenard "for his work on cathode rays".

1906 Joseph John Thomson "in recognition of the great merits of his theoretical and experimental investigations on the conduction of electricity by gases".

1907 Albert Abraham Michelson "for his optical precision instruments and the spectroscopic and metrological investigations carried out with their aid".

1918 Max Karl Ernst Ludwig Planck "in recognition of the services he rendered to the advancement of Physics by his discovery of energy quanta".

1921 Albert Einstein "for his services to Theoretical Physics, and especially for his discovery of the law of the photoelectric effect".

1922 Niels Henrik David Bohr "for his services in the investigation of the structure of atoms and of the radiation emanating from them".

1923 Robert Andrews Millikan "for his work on the elementary charge of electricity and on the photoelectric effect".

1925 James Franck and Gustav Ludwig Hertz "for their discovery of the laws governing the impact of an electron upon an atom".

1927 Arthur Holly Compton "for his discovery of the effect named after him "and Charles Thomson Rees Wilson "for his method of making the paths of electrically charged particles visible by condensation of vapour".

1929 Prince Louis-Victor Pierre Raymond de Broglie "for his discovery of the wave nature of electrons".

B. Falkenburg, W. Rhode (eds.), *From Ultra Rays to Astroparticles*,
DOI 10.1007/978-94-007-5422-5, © Springer Science+Business Media Dordrecht 2012

1932 Werner Karl Heisenberg "for the creation of quantum mechanics, the application of which has, inter alia, led to the discovery of the allotropic forms of hydrogen".
1933 Erwin Schrödinger and Paul Adrien Maurice Dirac "for the discovery of new productive forms of atomic theory".
1935 James Chadwick "for the discovery of the neutron".
1936 Victor Franz Hess "for his discovery of cosmic radiation" and Carl David Anderson "for his discovery of the positron".
1938 Enrico Fermi "for his demonstrations of the existence of new radioactive elements produced by neutron irradiation, and for his related discovery of nuclear reactions brought about by slow neutrons".
1939 Ernest Orlando Lawrence "for the invention and development of the cyclotron and for results obtained with it, especially with regard to artificial radioactive elements".
1945 Wolfgang Pauli "for the discovery of the Exclusion Principle, also called the Pauli Principle".
1948 Patrick Maynard Stuart Blackett "for his development of the Wilson cloud chamber method, and his discoveries therewith in the fields of nuclear physics and cosmic radiation".
1949 Hideki Yukawa "for his prediction of the existence of mesons on the basis of theoretical work on nuclear forces".
1950 Cecil Frank Powell "for his development of the photographic method of studying nuclear processes and his discoveries regarding mesons made with this method".
1951 Sir John Douglas Cockcroft and Ernest Thomas Sinton Walton "for their pioneer work on the transmutation of atomic nuclei by artificially accelerated atomic particles".
1954 Max Born "for his fundamental research in quantum mechanics, especially for his statistical interpretation of the wavefunction" and Walther Bothe "for the coincidence method and his discoveries made therewith".
1955 Divided equally between Willis Eugene Lamb "for his discoveries concerning the fine structure of the hydrogen spectrum" and Polykarp Kusch "for his precision determination of the magnetic moment of the electron".
1957 Chen Ning Yang and Tsung-Dao (T.D.) Lee "for their penetrating investigation of the so-called parity laws which has led to important discoveries regarding the elementary particles".
1958 Pavel Alekseyevich Cherenkov, Il'ja Mikhailovich Frank and Igor Yevgenyevich Tamm "for the discovery and the interpretation of the Cherenkov effect".
1959 Emilio Gino Segré and Owen Chamberlain "for their discovery of the antiproton".
1960 Donald A. Glaser "for the invention of the bubble chamber".
1961 Divided equally between Robert Hofstadter "for his pioneering studies of electron scattering in atomic nuclei and for his thereby achieved discoveries concerning the structure of the nucleons" and Rudolf Ludwig Mössbauer "for his researches concerning the resonance absorption of gamma radiation and his discovery in this connection of the effect which bears his name".

1963 Divided, one half awarded to Eugene Paul Wigner "for his contributions to the theory of the atomic nucleus and the elementary particles, particularly through the discovery and application of fundamental symmetry principles", the other half jointly to Maria Goeppert Mayer and J. Hans D. Jensen "for their discoveries concerning nuclear shell structure".

1965 Jointly to Sin-Itiro Tomonaga, Julian Schwinger and Richard P. Feynman "for their fundamental work in quantum electrodynamics, with deep-ploughing consequences for the physics of elementary particles".

1967 Hans Bethe "for his contributions to the theory of nuclear reactions, especially his discoveries concerning the energy production in stars".

1969 Murray Gell-Mann "for his contributions and discoveries concerning the classification of elementary particles and their interactions".

1970 Divided equally between Hannes Olof Gösta Alfvén "for fundamental work and discoveries in magnetohydro-dynamics with fruitful applications in different parts of plasma physics" and Louis Eugéne Félix Néel "for fundamental work and discoveries concerning antiferromagnetism and ferrimagnetism which have led to important applications in solid state physics".

1974 Martin Ryle and Antony Hewish "for their pioneering research in radio astrophysics: Ryle for his observations and inventions, in particular of the aperture synthesis technique, and Hewish for his decisive role in the discovery of pulsars".

1977 Jointly to Philip Warren Anderson, Sir Nevill Francis Mott and John Hasbrouck van Vleck "for their fundamental theoretical investigations of the electronic structure of magnetic and disordered systems".

1978 Divided, one half awarded to Pyotr Leonidovich Kapitsa "for his basic inventions and discoveries in the area of low-temperature physics", the other half jointly to Arno Allan Penzias and Robert Woodrow Wilson "for their discovery of cosmic microwave background radiation".

1979 Jointly to Sheldon Lee Glashow, Abdus Salam and Steven Weinberg "for their contributions to the theory of the unified weak and electromagnetic interaction between elementary particles, including, inter alia, the prediction of the weak neutral current".

1980 Jointly to James Watson Cronin and Val Logsdon Fitch "for the discovery of violations of fundamental symmetry principles in the decay of neutral K-mesons".

1981 Divided, one half jointly to Nicolaas Bloembergen and Arthur Leonard Schawlow "for their contribution to the development of laser spectroscopy" and the other half to Kai M. Siegbahn "for his contribution to the development of high-resolution electron spectroscopy".

1983 Divided equally between Subramanyan Chandrasekhar "for his theoretical studies of the physical processes of importance to the structure and evolution of the stars" and William Alfred Fowler "for his theoretical and experimental studies of the nuclear reactions of importance in the formation of the chemical elements in the universe".

1984 Jointly to Carlo Rubbia and Simon van der Meer "for their decisive contributions to the large project, which led to the discovery of the field particles W and Z, communicators of weak interaction".

1986 Divided, one half awarded to Ernst Ruska "for his fundamental work in electron optics, and for the design of the first electron microscope", the other half jointly to Gerd Binnig and Heinrich Rohrer "for their design of the scanning tunneling microscope".

1988 Jointly to Leon M. Lederman, Melvin Schwartz and Jack Steinberger "for the neutrino beam method and the demonstration of the doublet structure of the leptons through the discovery of the muon neutrino".

1990 Jointly to Jerome I. Friedman, Henry W. Kendall and Richard E. Taylor "for their pioneering investigations concerning deep inelastic scattering of electrons on protons and bound neutrons, which have been of essential importance for the development of the quark model in particle physics".

1992 Georges Charpak "for his invention and development of particle detectors, in particular the multiwire proportional chamber".

1993 Jointly to Russell A. Hulse and Joseph H. Taylor Jr. "for the discovery of a new type of pulsar, a discovery that has opened up new possibilities for the study of gravitation".

1995 Awarded "for pioneering experimental contributions to lepton physics" jointly with one half to Martin L. Perl "for the discovery of the tau lepton" and with one half to Frederick Reines "for the detection of the neutrino".

1996 Jointly to David M. Lee, Douglas D. Osheroff and Robert C. Richardson "for their discovery of superfluidity in helium-3".

2002 One half jointly to Raymond Davis Jr. and Masatoshi Koshiba "for pioneering contributions to astrophysics, in particular for the detection of cosmic neutrinos" and the other half to Riccardo Giacconi "for pioneering contributions to astrophysics, which have led to the discovery of cosmic X-ray sources".

2006 Jointly to John C. Mather and George F. Smoot "for their discovery of the blackbody form and anisotropy of the cosmic microwave background radiation".

2011 Divided, one half awarded to Saul Perlmutter, the other half jointly to Brian P. Schmidt and Adam G. Riess "for the discovery of the accelerating expansion of the Universe through observations of distant supernovae".

# Appendix C
# Textbooks

## C.1 Textbooks 1987–2012

AHARONIAN, F. A., 2004, *Very High Energy Cosmic Gamma Radiation – A Crucial Window on the Extreme Universe*. Singapur: World Scientific.

AITCHISON, I. J., HEY, A. J. G., AND AITCHISON, I. J. R., 2004, *Gauge Theories in Particle Physics*. Bristol: Taylor & Francis.

ALLDAY, J., 2001, *Quarks, Leptons and the Big Bang*. Bristol: Institute of Physics Pub.

BEREZINSKII, V. S., BULANOV, S., DOGIEL, V., GINZBURG, V., AND PTUSKIN, V., 1990, *Astrophysics of Cosmic Rays*. Amsterdam: North-Holland.

BERGSTRÖM, L., AND GOOBAR, A., 2004, *Cosmology and Astroparticle Physics*. Berlin, Heidelberg, New York: Springer, and Praxis Publishing Ltd, Chichester, UK.

BESKIN, V., HENRI, G., MENARD, F., PELLETIER, G., AND DALIBARD, J., 2003, *Accretion Discs, Jets, and High Energy Phenomena in Astrophysics*. Berlin, Heidelberg, New York: Springer.

BIEBER, J., EROSHENKO, E., EVENSON, P., FLÜCKIGER, E., AND KALLENBACH, R., 2000, *Cosmic Rays and Earth*. Berlin, Heidelberg, New York: Springer.

BOTHUN, G., AND BOTHUN, B., 1998, *Modern Cosmological Observations and Problems*. Gießen: Taylor & Francis.

CALDWELL, D. O., 2001, *Current Aspects of Neutrino Physics*. Berlin, Heidelberg, New York: Springer.

CARROLL, B. W., AND OSTLIE, D. A., 2007, *An Introduction to Modern Astrophysics*. San Francisco: Pearson Addison-Wesley.

CHABRIER, G., 2011, *Structure Formation in Astrophysics*. Cambridge: Cambridge University Press.

CHARAP, J. M., 2004, *Explaining the Universe*. Hyderabad: Princeton University Press.

B. Falkenburg, W. Rhode (eds.), *From Ultra Rays to Astroparticles*,
DOI 10.1007/978-94-007-5422-5,  

COMMITTEE ON THE PHYSICS OF THE UNIVERSE, N. R. C., 2003, *Connecting Quarks with the Cosmos – Eleven Science Questions for the New Century*. Washington D.C.: National Academies Press.

DERMER, C. D., AND MENON, G., 2009, *High Energy Radiation from Black Holes – Gamma Rays, Cosmic Rays, and Neutrinos*. Princeton: Princeton University Press.

DIEHL, R., HARTMANN, D. H., AND PRANTZOS, N., 2010, *Astronomy with Radioactivities*. Berlin, Heidelberg, New York: Springer.

DIRAC, P. A. M., 1981, *The Principles of Quantum Mechanics*. London: Oxford University Press.

DODELSON, S., 2003, *Modern Cosmology*. Amsterdam, Boston: Academic Press.

DORMAN, L. I., 2004, *Cosmic Rays in the Earth's Atmosphere and Underground*. Berlin, Heidelberg, New York: Springer.

DURIC, N., 2004, *Advanced Astrophysics*. Cambridge: Cambridge University Press.

FERGUSON, K., 1998, *Prisons of Light – Black Holes*. Cambridge: Cambridge University Press.

FONTAINE, G., AND VAN, J. T. T., 1993, *Particle Astrophysics*. Paris: Atlantica Séguier Frontiére.

FRIEDLANDER, M. W., 2002, *A Thin Cosmic Rain – Particles from Outer Space*, reprint ed. Cambridge: Harvard University Press.

GAISSER, T. K., 1991, *Cosmic Rays and Particle Physics*. Cambridge: Cambridge University Press.

GIACOMELLI, G., SPURIO, M., AND DERKAOUI, J. E., 2001, *Cosmic Radiations – From Astronomy to Particle Physics*. Berlin, Heidelberg, New York: Springer.

GRATTON, L., 1970, *Non-Solar X- and Gamma-Ray Astronomy*. Berlin, Heidelberg, New York: Springer.

GRIEDER, P. K. F., 2001, *Cosmic Rays at Earth – Researcher's Reference Manual and Data Book*. Orlando: Gulf Professional Publishing.

GRUPEN, C., 2001, *Astroteilchenphysik – Das Universum im Licht der kosmischen Strahlung*. Wiesbaden: Gabler Wissenschaftsverlage.

GRUPEN, C., COWAN, G., EIDELMAN, S., AND STROH, T., 2005, *Astroparticle Physics*. Berlin, Heidelberg, New York: Springer.

HELLER, M., 2009, *Ultimate Explanations of the Universe*. Berlin, Heidelberg, New York: Springer.

JONES, J., 1990, *Particle Astrophysics – The NASA Cosmic Ray Program for the 1990s and Beyond, Greenbelt, MD 1989*. College Park: American Institute of Physics.

KENNEDY, D., 2007, *Particle Astrophysics – Special Section*. Freuburg.

KIRSHNER, R. P., 2002, *The Extravagant Universe – Exploding Stars, Dark Energy and the Accelerating Cosmos*. Kassel: Princeton University Press.

KOLB, E. W., AND TURNER, M. S., 1994, *The Early Universe*. New York: Westview Press.

LAVRO, A. S., 2002, *Neutrinos – A Bibliography With Indexes*. Nova Publishers.
LEMOINE, M., AND SIGL, G., 2001, *Physics and Astrophysics of Ultra-High-Energy Cosmic Rays*. Berlin, Heidelberg, New York: Springer.
LIDDLE, A., 2003, *An introduction to Modern Cosmology*. New York: Wiley.
LIDDLE, A., AND OTTERSTEIN, S., 2008, *Einführung in die Moderne Kosmologie*. Weinheim: Wiley.
LIDDLE, A. R., AND LYTH, D. H., 2000, *Cosmological Inflation and Large-Scale Structure*. Cambridge: Cambridge University Press.
LIEBSCHER, D.-E., 2005, *Cosmology*. Berlin, Heidelberg, New York: Springer.
LONGAIR, M. S., 2011, *High Energy Astrophysics*. Cambridge: Cambridge University Press.
MARTIENSSEN, W., AND RÖSS, D., 2011, *Physik im 21. Jahrhundert – Essays zum Stand der Physik*. Berlin, Heidelberg, New York: Springer.
MATTHEWS, J. M., 1994, *High Energy Astrophysics – Models and Observations from MeV to EeV*. Singapur: World Scientific.
MELIA, F., 2009, *High-Energy Astrophysics*. Princeton: Princeton University Press.
MUKHANOV, V., 2005, *Physical Foundations of Cosmology*. Cambridge: Cambridge University Press.
MURTHY, P. V. R., AND WOLFENDALE, A. W., 1993, *Gamma-Ray Astronomy*. Cambridge: Cambridge University Press.
PEACOCK, J. A., 1999, *Cosmological Physics*. Cambridge: Cambridge University Press.
PERKINS, D. H., 2009, *Particle Astrophysics*, 2nd ed. New York: Oxford University Press.
RAYCHAUDHURI, A., BANERJI, S., AND BANERJEE, A., 2003, *General Relativity, Astrophysics, and Cosmology*. Berlin, Heidelberg, New York: Springer.
RUBIN, V., 1997, *Bright Galaxies, Dark Matters*. Berlin, Heidelberg, New York: Springer.
SARKAR, U., 2008, *Particle and Astroparticle Physics*, illustrated ed. Gießen: Taylor & Francis.
SCHLICKEISER, R., 2002, *Cosmic Ray Astrophysics*. Berlin, Heidelberg, New York: Springer.
SCHNEIDER, P., 2008, *Einführung in Die Extragalaktische Astronomie und Kosmologie*. Berlin, Heidelberg, New York: Springer.
SCHRAMM, D. N., AND SCHRAMM, S., 1996, *The Big Bang and Other Explosions in Nuclear and Particle Astrophysics*. Singapur: World Scientific.
SHAPIRO, M., STANEV, T., AND WEFEL, J. P., 2001, *Astrophysical Sources of High Energy Particles and Radiation*. Berlin, Heidelberg, New York: Springer.
SHAPIRO, S. L., AND TEUKOLSKY, S. A., 2008, *Black Holes, White Dwarfs and Neutron Stars – The Physics of Compact Objects*. New York: Wiley.
SPARKE, L. S., AND GALLAGHER, J. S. III, 2007, *Galaxies in the Universe – An Introduction*. Cambridge: Cambridge University Press.
SPIER, F., 1996, *The Structure of Big History from the Big Bang Until Today*. Amsterdam: Amsterdam University Press.

TICER, B., 2004, *Gravity, Quasi Black Holes, and Cosmic Relativity*. Bloomington: AuthorHouse.
UNSÖLD, A., AND BASCHEK, B., 2002, *Der neue Kosmos – Einführung in die Astronomie und Astrophysik*. Wiesbaden: Gabler Wissenschaftsverlage.
WEIGERT, A., WENDKER, H. J., AND WISOTZKI, L., 2009, *Astronomie und Astrophysik – Ein Grundkurs*. Weinheim: Wiley-VCH.
WEINBERG, S., 1993, *The First Three Minutes – A Modern View of the Origin of the Universe*, updated ed. United States: Basic Books.
WEINBERG, S., 2000, *The Quantum Theory of Fields – Supersymmetry*. Cambridge: Cambridge University Press.
WEINBERG, S., 2008, *Cosmology*. New York: Oxford University Press.
ZEIDLER, E., 2006, *Quantum Field Theory*, 1st ed. 2006. Corr. 2nd printing. Berlin, Heidelberg, New York: Springer.
ZUBER, K., AND KLAPDOR-KLEINGROTHAUS, H. V., 1997, *Teilchenastrophysik*. Wiesbaden: Teubner.

## C.2 Textbooks 1962–1986

ADAMS, D. J., 1980, *Cosmic X-Ray Astronomy*. Bristol: Hilger.
BAGGE, E., 1966, *Die Entstehung der kosmischen Ultrastrahlung und das Expansionsphänomen der Welt*. München: K. Thiemig.
CRANSHAW, T. E., 1963, *Cosmic Rays*. Oxford: Oxford University Press.
ALFVEN, H., 1966, *Worlds – Antiworlds: Antimatter in Cosmology*. San Francisco: Freeman.
ALLKOFER, O. C., 1971, *Teilchen-Detektoren – Particle Detectors*. München: Thiemig.
ALLKOFER, O. C., 1975, *Introduction to Cosmic Radiation*. München: Thiemig.
BRADT, H., AND GIACCONI, R., 1973, *X- and Gamma-Ray Astronomy*. Dordrecht: D. Reidel.
GRATTON, L., 1970, *Non-Solar X- and Gamma-Ray Astronomy*. Berlin, Heidelberg, New York: Springer.
HAYAKAWA, S., 1969, *Cosmic Ray Physics – Nuclear and Astrophysical Aspects*. New York: Wiley.
HILLAS, A. M., 1972, *Cosmic Rays*. Oxford: Pergamon Press.
HILLIER, R., 1984, *Gamma Ray Astronomy*. Oxford: Clarendon Press.
ROSSI, B. B., 1965, *High-Energy Particles*. New York: Prentice-Hall.
SCHAEFER, H., 1985, *Elektromagnetische Strahlung – Informationen aus dem Weltall*, 1. aufl. Wiesbaden: Vieweg.
SHAPIRO, M., 1983, *Composition and Origin of Cosmic Rays*, 1. aufl. Berlin, Heidelberg, New York: Springer.
SHAPIRO, M., 1986, *Cosmic Radiation in Contemporary Astrophysics*, 1. aufl. Berlin, Heidelberg, New York: Springer.
SILK, J., 1980, *The Big Bang – The Creation and Evolution of the Universe*. New Jersey: Freeman.

VAN, J. T. T., 1981, *Cosmology and Particles*. Paris: Atlantica Séguier Frontiéres.
WEINBERG, S., 1972, *Gravitation and Cosmology: Principles and Applications of the General Theory of Relativity*. New York: Wiley.

## C.3 Textbooks 1937–1961

GEIGER, H., 1940, *Die kosmische Ultrastrahlung als Forschungsproblem*. Berlin: de Gruyter.
HEISENBERG, W., 1953, *Kosmische Strahlung*. Berlin.
HILLAS, A. M., 1972, *Cosmic Rays*. Oxford: Pergamon Press.
MIEHLNICKEL, E., 1938, *Höhenstrahlung* (*Ultrastrahlung*). Berlin: Steinkopff.
POWELL, C. F., FOWLER, P. H. AND PERKINS, D. H., 1959, *The Study of Elementary Particles by the Photographic Method. An Account of the Principal Techniques and Discoveries. Illustrated by an Atlas of Photomicrographs*. London: Pergamon Press.
SHAPLEY, H., 1960, *Source Book in Astronomy, 1900–1950*. Harvard University Press.
SITTE, K., 1961, *Kosmische Strahlung*. New York, Berlin, Heidelberg: Springer.
WEISS, H., 1940, *Harte und Haufigkeit der von der kosmischen Ultrastrahlung ausgeloster Schauer*. Tübingen: Eberhard-Karls-Universitat Tübingen.

## C.4 Textbooks 1912–1936

MESSERSCHMIDT, W., 1936, *Untersuchungen zur Ionisation durch Ultrastrahlung – 1. Untersuchungen über den Reststrom von Ionisationsdruckkammern und über die Druckabhängigkeit der Ionisation durch Ultrastrahlung; 2. Untersuchungen über Ultrastrahlungsstöße*. Berlin, Heidelberg, New York: Springer.
MILLIKAN, R. A., 1921, *The Electron – Its Isolation and Measurement and the Determination of Some of Its Properties*. Chicago: The University of Chicago Press.
NERNST, W., 1921, *Das Weltgebäude im Lichte der neueren Forschung*. Berlin, Heidelberg, New York: Springer.
RUTHERFORD, E., CHADWICK, J., AND ELLIS, C. D., 1930, *Radiations from Radioactive Substances*. Cambridge: Cambridge University Press.
WEISS, H., 1940, *Harte und Haufigkeit der von der kosmischen Ultrastrahlung ausgeloster Schauer*. Tübingen: Eberhard-Karls-Universitat Tübingen.
WÖLCKEN, K., 1931, *Ueber Messungen und Registrierungen der Ultrastrahlung*. (*Durchdringende Höhenstrahlung*). Göttingen: Göttingen.

# Appendix D
# Books in History of Physics


ALPHER, R., AND HERMAN, R., 2001, *Genesis of the Big Bang*. New York: Oxford University Press.

ASHTEKAR, A., 2005, *One Hundred Years of Relativity*. Singapore: World Scientific.

BERENDZEN, R., HART, R., AND SEELEY, D., 1984, *Man Discovers the Galaxies*. New York: Columbia University Press.

BERNSTEIN, J., Ed., 1979, *Hans Bethe: Prophet of Energy*. Basic Books.

BERTOTTI, B., BALBINOT, R., BERGIA, S., AND MESSINA, A., 1990, *Modern Cosmology in Retrospect*, illustrated ed. Cambridge: Cambridge University Press.

BLEEKER, J., GEISS, J., AND HUBER, M., Eds., 2001, *The Century of Space Science*. Berlin, Heidelberg, New York: Springer.

BRECHT, P., 1990, *History of Cosmology and Its Significance*. Charlottesville: University of Virginia.

BROWN, L. M., DRESDEN, M., AND HODDESON, L., Eds., 2009, *Pions to Quarks – Particle Physics in the 1950s*. Cambridge: Cambridge University Press.

BROWN, L. M., AND HODDESON, L., Eds., 1986, *The Birth of Particle Physics*. Cambridge: Cambridge University Press.

CAHN, R. N., AND GOLDHABER, G., 2009, *The Experimental Foundations of Particle Physics*. Cambridge: Cambridge University Press.

CHARON, J. E., AND WALDHEIM, H. V., 1970, *Geschichte der Kosmologie*. München: Kindler.

CRANSHAW, T. E., 1963, *Cosmic Rays*. Oxford: Oxford University Press.

CROPPER, W. H., 2004, *Great Physicists – The Life and Times of Leading Physicists from Galileo to Hawking*. Oxford: Oxford University Press.

DA COSTA ANDRADE, E. N., 1973, *Rutherford and the Nature of the Atom*. New York: Peter Smith Pub.

DAHL, P. F., 1997, *Flash of the Cathode Rays – A History of J.J. Thomson's Electron*, illustrated ed. Bristol: Institute of Physics Pub.

DICK, W. R., HAMEL, J., AND DUERBECK, H. W., Eds., 2008, *Beiträge zur Astronomiegeschichte*. Frankfurt am Main: Harri Deutsch Verlag.



B. Falkenburg, W. Rhode (eds.), *From Ultra Rays to Astroparticles*,
DOI 10.1007/978-94-007-5422-5, © Springer Science+Business Media Dordrecht 2012

DUERBECK, H. W., AND DICK, W. R., 2005, *Einsteins Kosmos – Untersuchungen zur Geschichte der Kosmologie, Relativitätstheorie und zu Einsteins Wirken und Nachwirken*. Frankfurt am Main: Harri Deutsch Verlag, Thun.

EVANS, J., AND THORNDIKE, A. S., 2007, *Quantum Mechanics at the Crossroads – New Perspectives From History, Philosophy and Physics*. Berlin, Heidelberg, New York: Springer.

FALKENBURG, B., 2007, *Particle Metaphysics – A Critical Account of Subatomic Reality*. Berlin, Heidelberg, New York: Springer.

FRANKLIN, A., 1989, *The Neglect of Experiment*. Cambridge: Cambridge University Press.

FRANKLIN, A., 1990, *Experiment, Right or Wrong*. Cambridge: Cambridge University Press.

FRANKLIN, A., 2003, *Are There Really Neutrinos? – An Evidential History*. New York: Westview Press.

GALISON, P., 1987, *How Experiments End*. Chicago: University of Chicago Press.

GREGORY, J., 2005, *Fred Hoyle's Universe*. New York: Oxford University Press.

HAMBLIN, J. D., BURNS, W. E., AND SERVICES, A.-C. I., 2005, *Science in the Early Twentieth Century – An Encyclopedia*. United States of America: ABC-CLIO.

HEILBRON, J. L., 2005, *The Oxford Guide To The History of Physics and Astronomy*. New York: Oxford University Press.

HENTSCHEL, K., 1998, *Zum Zusammenspiel von Instrument, Experiment und Theorie: Rotverschiebung im Sonnenspektrum und verwandte spektrale Verschiebungseffekte von 1880 bis 1960, Schriftenreihe naturwissenschaftliche Forschungsergebnisse*, vol. 58. Hamburg: Kovac.

HENTSCHEL, K., 2007, *Unsichtbares Licht? Dunkle Wärme? Chemische Strahlen?: Eine wissenschaftshistorische und -theoretische Analyse von Argumenten für das Klassifizieren von Strahlungssorten 1650–1925 mit Schwerpunkt auf den Jahren 1770–1900*. Diepholz: GNT.

HETHERINGTON, H., AND HETHERINGTON, N. S., 1993, *Cosmology – Historical, Literary, Philosophical, Religious, and Scientific Perspectives*. New York: Garland.

JONES, B. Z., 1965, *Lighthouse of the skies – the Smithsonian Astrophysical Observatory: background and history, 1846–1955*. Washington: Smithsonian Institute.

JONES, B. Z., 1966, *The Golden Age of Science – Thirty Portraits of the Giants of 19th-Century Science, by Their Scientific Contemporaries*. Washington: Simon and Schuster.

JONES, B. Z., MENZEL, D. H., AND BOYD, L. G., 1971, *The Harvard College Observatory – The First Four Directorships, 1839–1919*. Cambridge: Harvard University Press.

KANITSCHEIDER, B., 1984, *Kosmologie – Geschichte und Systematik in philosophischer Perspektive*. Stuttgart: Reclam.

KRAGH, H., 1999, *Cosmology and Controversy – The Historical Development of Two Theories of the Universe*, new ed. Princeton: Princeton University Press.

KRAGH, H., 2004, *Matter and Spirit in the Universe – Scientific and Religious Preludes to Modern Cosmology*. Paris: OECD Publishing.

KRAGH, H. S., 2007, *Conceptions of Cosmos – From Myths to the Accelerating Universe: A History of Cosmology*. New York: Oxford University Press.

KRÖGER, B., 1981, *Der Nachlass von Julius Elster und Hans Geitel*. Frankfurt am Main: Klostermann.

KROPP, W., SOBEL, H., AND SCHULTZ, J., 1990, *Neutrinos and Other Matters – Selected Works of Frederick Reines*. Singapur: World Scientific.

KUHN, TH. S., 1962, *The Structure of Scientific Revolutions*. University of Chicago Press. 2nd ed.: 1970.

LANG, K. R., AND GINGERICH, O., 1979, *A Source Book in Astronomy and Astrophysics, 1900–1975*. Cambridge: Harvard University Press.

LAVRO, A. S., 2002, *Neutrinos – A Bibliography with Indexes*. Nova Publishers.

LONGAIR, M. S., 2006, *The Cosmic Century – A History of Astrophysics and Cosmology*. Cambridge: Cambridge University Pres.

LOSEE, J., 2001, *A Historical Introduction to the Philosophy of Science*. New York: Oxford University Press.

MUNITZ, M. K., 1957, *Theories of the Universe – From Babylonian Myth to Modern Science*. New York: Simon and Schuster.

NORTH, J., AND SENGERLING, R., 2001, *Viewegs Geschichte der Astronomie und Kosmologie*. Berlin, Heidelberg, New York: Springer.

NORTH, J. D., 2008, *Cosmos – An Illustrated History of Astronomy and Cosmology*, revised ed. Chicago: University of Chicago Press.

NYE, M. J., 2003, *The Cambridge History of Science: The Modern Physical and Mathematical Sciences*. Cambridge: Cambridge University Press.

PAIS, A., 1988, *Inward Bound – Of Matter and Forces in the Physical World*. Oxford: Clarendon Press.

PICKERING, A., 1984, *Constructing Quarks*. Edinburgh University Press.

REEVES, R., 2008, *A Force of Nature – The Frontier Genius of Ernest Rutherford*. New York: Norton.

RIORDAN, M., 1987, *The Hunting of the Quark – A True Story of Modern Physics*. New York: Simon & Schuster.

ROSNER, R., AND STROHMAIER, B., 2003, *Marietta Blau – Sterne der Zertrümmerung: Biographie einer Wegbereiterin der modernen Teilchenphysik*. Wien: Böhlau Verlag.

ROSNER, R., AND STROHMAIER, B., Eds., 2003, *Marietta Blau. Stars of Disintegration. Biography of a Pioneer of Particle Physics*. Ariadne, Riverside, California 2006. Engl. transl. of Rosner and Strohmaier.

ROSSI, B. B., 1965, *High-Energy Particles*. New York: Prentice-Hall.

ROSSI, B., 1990, *Moments in the Life of a Scientist*. Cambridge: Cambridge University Press.

SCHMEIDLER, F., 1962, *Alte und moderne Kosmologie*. Berlin: Duncker & Humblot.

SCHWEBER, S. S., 1994, *QED and the Men Who Made It – Dyson, Feynman, Schwinger, and Tomonaga*. Princeton: Princeton University Press.

SCHWEBER, S. S., 2009, *Einstein and Oppenheimer – The Meaning of Genius*. Harvard University Press.
SHAPLEY, H., 1960, *Source Book in Astronomy, 1900–1950*. Harvard University Press.
SEKIDO, Y., AND ELLIOT, H., Eds., 1985, *Early History of Cosmic Ray Studies – Personal Reminiscences with Old Photographs*, pp. 47–51. Dordrecht: Reidel.
ULRICH, R., JR, DAVIES, R., AND BAHCALL, J. N., 2002, *Solar Neutrinos – The First Thirty Years*, reprint ed. New York: Westview Press.
WHEATON, B. R., 1983, *The Tiger and the Shark*. Cambridge University Press.
ZIMMERMAN, R., 2010, *The Universe in a Mirror – The Saga of the Hubble Space Telescope and the Visionaries who Built It*, new ed. Kassel: Princeton University Press.

# Citation Index



B. Falkenburg, W. Rhode (eds.), *From Ultra Rays to Astroparticles*,
DOI 10.1007/978-94-007-5422-5, © Springer Science+Business Media Dordrecht 2012

## C

**Y**



**Z**

# Name Index



B. Falkenburg, W. Rhode (eds.), *From Ultra Rays to Astroparticles*,
DOI 10.1007/978-94-007-5422-5, © Springer Science+Business Media Dordrecht 2012

# Subject Index




B. Falkenburg, W. Rhode (eds.), *From Ultra Rays to Astroparticles*,
DOI 10.1007/978-94-007-5422-5, © Springer Science+Business Media Dordrecht 2012

MIX
Papier aus verantwortungsvollen Quellen
Paper from responsible sources
FSC® C105338

If you have any concerns about our products,
you can contact us on
**ProductSafety@springernature.com**

In case Publisher is established outside the EU,
the EU authorized representative is:
**Springer Nature Customer Service Center GmbH**
**Europaplatz 3, 69115 Heidelberg, Germany**

Printed by Libri Plureos GmbH
in Hamburg, Germany